T0250986

Introduction to the Micromechanics of Composite Materials

Huiming Yin
Columbia University, New York
New York, USA

Yingtao Zhao
Beijing Institute of Technology
PR of China

CRC Press
Taylor & Francis Group
Boca Raton London New York

CRC Press is an imprint of the
Taylor & Francis Group, an **informa** business
A SPON PRESS BOOK

CRC Press
Taylor & Francis Group
6000 Broken Sound Parkway NW, Suite 300
Boca Raton, FL 33487-2742

First issued in paperback 2017

ISBN 13: 978-1-138-49049-9 (pbk)
ISBN 13: 978-1-4987-0728-2 (hbk)

Library of Congress Cataloging-in-Publication Data

Yin, Huiming.
Introduction to the micromechanics of composite materials / Huiming Yin and Yingtao Zhao.
pages cm
"A CRC title."
Includes bibliographical references and index.
ISBN 978-1-4987-0728-2 (alk. paper)
1. Composite materials. 2. Micromechanics. 3. Composite construction--Fatigue. I. Zhao, Yingtao. II. Title.

TA418.9.C6Y54 2016
620.1'1892--dc23 2015030297

Visit the Taylor & Francis Web site at
http://www.taylorandfrancis.com

and the CRC Press Web site at
http://www.crcpress.com

Contents

Preface

IN 2004, WHEN author Huiming Yin worked at the University of Illinois at Urbana-Champaign as a postdoctoral research associate, his supervisor, Professor Glaucio H. Paulino, asked him to give a series of short courses to introduce micromechanics to a group of students including PhD, master's, and senior undergraduate students, so that they could learn general concepts and conduct original research. He found that it was difficult to recommend an appropriate textbook for the course. Although the masterpiece by Toshio Mura, titled *Micromechanics of Defects in Solids*, covered the most important topics in micromechanics, students were often puzzled because they did not know where to begin and how to select from the broad topics in this textbook that presented micromechanics as a profound subject.

As various emerging composite materials have been increasingly used in civil, mechanical, biomedical, and materials engineering, engineers and practitioners are seeking tools and methods for the design and selection of appropriate composite materials for different applications. Micromechanics-based formulae have attracted significant interest for their concise form and accuracy. It is important for senior undergraduates and master's students to learn this subject so they can use it in the engineering practice and establish a foundation to advance the state-of-the-art for modeling and characterization of composite materials. The course Micromechanics of Composite Materials, which is described below, provides students with the fundamental understanding of the mechanical behavior of composite materials and prepares them for future research and development of new composite materials.

After he moved to Columbia University in 2008, Dr. Yin found that an introductory course about micromechanics was absent at the university, but the need existed in several departments, i.e., civil engineering, mechanical engineering, biomedical engineering, and materials engineering, etc., for composite materials information. He took time to provide references and answer questions from students from different backgrounds, which motivated him to develop an introductory course to micromechanics. After several years of preparation, the course Micromechanics of Composite Materials was first taught in 2012. Although Mura's book and several others have been recommended as references in this course, the classes are mainly organized by the learning sequence and experience of students through Dr. Yin's design and interaction with students.

Author Yingtao Zhao visited Columbia University as a visiting scholar in 2014. He finished his PhD thesis at Peking University on the topic of Eshelby's problem, which is the basic problem in micromechanics. After he taught courses for more than ten years in mechanics of materials and elasticity at Beijing Institute of Technology, many notes on understanding the basic concepts of micromechanics and composite materials were provided to help the students learn the topic. While a TA in Dr. Yin's course on micromechanics, Dr. Zhao discussed details of the basic concept and derivation, provided more examples and exercises to help the students understand the topic, and enriched the scope of this course with additional research problems.

To provide an introductory reference and textbook to the next generation of engineers and students, after some preparation, we decided to assemble the teaching notes and some relevant research results of our ongoing projects and include them in this book. This book has been designed to facilitate the teaching and learning of a 3-credit course for 12–14 weeks.

This course is placed in a sequence of the solid mechanics courses in the Department of Civil Engineering and Engineering Mechanics at Columbia University: After juniors finish Mechanics of Solids, they are required to take Experimental Mechanics of Materials to test the mechanical behavior of a variety of engineering materials and conduct stress and strain analysis of the testing results. Then, they can take Advanced Mechanics of Solids in the first semester of their senior year with some graduate students. With some preparation in mathematics and solid mechanics, senior students can take this course with graduate students in the last semester. However, the majority of the class is from our graduate study programs.

As minimum requirements to use this book or to take this course, the readers should have a basic understanding of stress and strain, ordinary and partial differential equations, and constitutive behavior of materials. This book introduces the approach of microstructure-based solid mechanics and the corresponding effective mechanical properties of heterogeneous materials through homogenization, which can be used in design, processing, testing, and control of composite materials. The overall objectives of this course are that, upon completion of it, the student will be able to: (1) understand the basic concepts of micromechanics such as RVE, eigenstrain, inclusions, and inhomogeneities; (2) master the constitutive law of general composite materials; (3) use the tensorial indicial notation to formulate the Eshelby's problem; (4) learn the common homogenization methods; and (5) introduce some research areas in micromechanics and composite materials, which may continue flourishing in the coming decades.

To these ends, we wrote this introductory textbook following the learning curves of our students. We tried to keep this book as concise as possible and thus only cited some necessary references. For students who would like to explore some advanced topics, the authors suggest that they read Mura's book or delve into other references and the current literature for details. Micromechanics has been a dynamic subject for decades, and the literature is very rich and diverse. Obviously, this book can only cover a very small part of the literature. We apologize not to those authors who have done excellent work but to those whom we did not cite or give deserved credit.

Acknowledgments

During the preparation of this book, we have received a great deal of help and support from our colleagues and students. First, we thank Dr. Po-Hua Lee, who provided significant help to transfer the teaching notes to the electronic version. My students, Richard L. Li, Siyu Zhu, Yingjie Liu, Zheyu Shou, and Qiliang Lin among others, have carefully read the book chapters and pointed out a number of typos. In addition, we thank the editor, Tony Moore, for his encouragement and diligent work on this book. We would also like to give our special thanks to our families for their patience and support. Finally, we acknowledge the National Science Foundation (NSF), Air Force Office of Scientific Research (AFOSR), and National Natural Science Foundation of China (NSFC) for their support of our research in the area of micromechanics of composite materials.

CHAPTER 1

Introduction

Micromechanics is a branch of continuum mechanics. It studies the effective mechanical behavior of heterogeneous materials based on their microstructures of material phases. Continuum mechanics deals with materials under the assumption that the material distribution is uniform, so stress and strain satisfy the constitutive law uniformly. Obviously, at a lower scale the material distribution is not uniform, so that local stress and strain are not uniform, even if we apply a uniform loading at the boundary. However, when the material phases are distributed homogeneously in a statistical sense, at a larger length scale, the composite can be treated as a statistically uniform material, and the stress and strain at a material point can be evaluated by the averages of stress and strain on a representative volume element (RVE) [1–4]. Here an RVE in a continuum body is a material volume that statistically represents the neighborhood of a material point. From the relation between averaged stress and strain, we can derive an effective mechanical constitutive law of the composite. Inevitably, the overall mechanical properties of the composite are affected by factors such as the volume fraction of the particles, the elastic moduli of the matrix and particles, and the microstructure of the composites. Micromechanics has become a powerful tool in design and characterization of composite materials.

1.1 COMPOSITE MATERIALS

Composite materials are engineered or natural materials made of two or more phases with significant different physical or chemical properties, which remain separate and distinct at a lower scale within the finished structure. Early examples of composite materials can be traced back to the bible story that the ancient Egyptian used straw and mud to make bricks about 3500 years ago. Nowadays, concrete and asphalt mix are the two most used man-made composites in the world, which form buildings, roadways, dams, and bridges among other infrastructure. However, for a long history, the design and manufacture of those composites are based on empirical formulations and trial-error tests. Micromechanics provides a quantitative way to optimize the material design and to predict the material behavior. It can be imagined that 1% increase of the lifetime or strength of material caused by micromechanics-based optimization may save billions of dollars of material cost annually. With the development of nanotechnologies, new composites are invented with an exponentially increasing trend. Some features, which generally do not exist in natural materials, can be achieved through micromechanics-based design of composite materials, such as materials with negative Poisson's ratio and zero thermal expansion coefficient.

A composite material is typically composed of the matrix and dispersed phases. The matrix phase represents a continuous material that binds the material together and thus plays a dominant role in the effective material behavior, whereas the dispersed phase may significantly modify the effective mechanical and physical properties of the composites. In different applications, the matrix is also called binder or cement, whereas the dispersed phase has aliases of filler, modifier, or reinforcement. Composite materials have been widely used in civil engineering, mechanical engineering, aerospace engineering, and biomedical engineering among others. They can be classified by matrices, dispersed phases, and microstructures.

With respect to matrices, there are commonly three types of composites: metal matrix composite (MMC), polymer matrix composite (PMC), and ceramic matrix composite (CMC). Generally, an MMC exhibits effective elasto-plastic behavior characterized by stiffness, yield strength, and flow rule after yielding; a PMC shows effective visco-elastic or visco-elasto-plastic behavior, which can be studied in frequency or time domain; a CMC is generally brittle exhibiting an elastic behavior in a large temperature range with certain fracture properties. Different matrices lead to different effective mechanical behavior with different sets of material constants we need to model or characterize through micromechanics.

Classified by dispersed phases, fiber reinforced composite (FRC) and particle reinforced composite (PRC) are two common examples. An FRC typically uses a small amount of fibers that may significantly increase the strength and ductility of the composite. Short fibers or long fibers may be used for different applications. For example, Figure 1.1 illustrates a

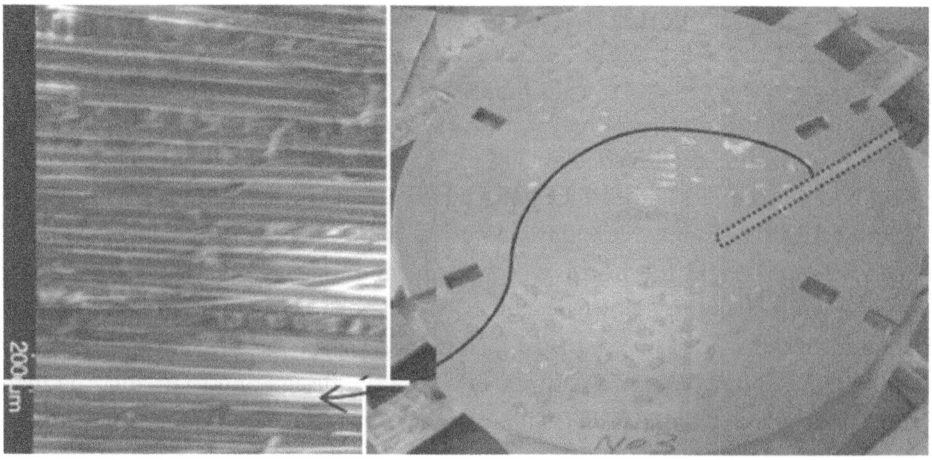

FIGURE 1.1 Glass fiber reinforced epoxy for a composite utility cover.

composite utility cover made of glass fiber reinforced epoxy, which has been used by Con Edison in New York City to replace the traditional cast iron covers for its excellent electric and thermal insulation performance. Typical fibers include glass fibers, carbon fibers, aramid fibers, and metal fibers. In a PRC, the concentration of particles may vary in a large range. For example, asphalt mix may contain a volume fraction of mineral aggregate as high as 90%. The particle shape and size may be nonuniform. The particles can be metal, ceramic, and oxide particles. In addition, although air voids are not a physical material, it can be considered a material phase with zero stiffness. In this sense, foamed materials can be considered as particulate composites, which are used to produce lightweight materials or increase the volume of materials. The recent development of nanotubes and nanoparticles has created a new era of nanocomposites, which may exhibit the similar microstructures to FRCs and PRCs, respectively, but with lower dimensions.

According to microstructures, the most straightforward way to make composites is to mix material phases together uniformly for a randomly dispersed composite. Statistically the particle distribution is homogeneous so that the effective material behavior is isotropic. To improve the material efficiency or to produce some unique features in a certain orientation, the dispersed phase can be aligned in a plane or orientation. For example, aligning fibers in a certain direction may significantly increase the strength of the material along that direction. In our recent work, we have aligned ferromagnetic particles into chains in a silicone matrix to make strain gauge. Figure 1.2 illustrates the microstructure change after we applied a magnetic field, in which the randomly dispersed ferromagnetic particles in Figure 1.2a is transformed into chains parallel to the direction of the applied field in Figure 1.2b due to the magnetic interaction force. Upon curing the silicone matrix, the chain structure is locked into place so that the chain-like microstructures are obtained in the composites. Periodic microstructure has also been studied in the literature, which mainly follows the crystal lattice structures using the unit cell. However, no ideal periodic distribution can be made in real

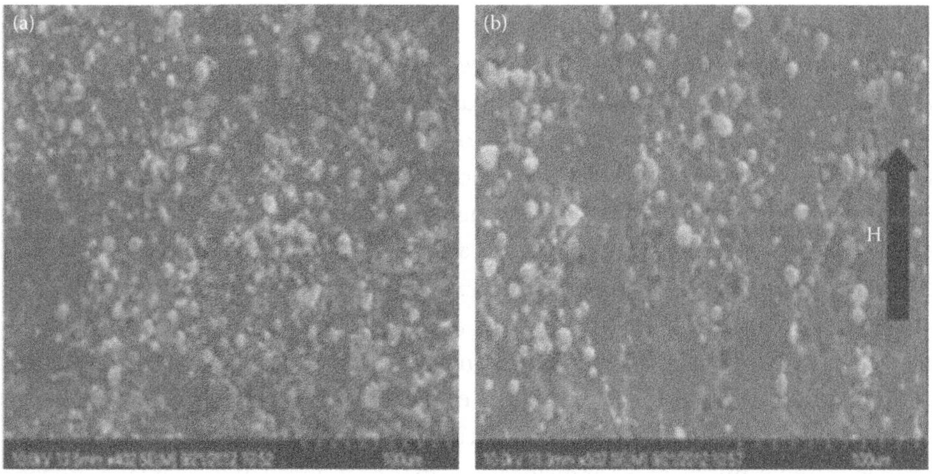

FIGURE 1.2 Randomly dispersed ferromagnetic particles in (a) are aligned into chains under a magnetic field in (b).

composites. This microstructure is used for the mathematical convenience in the modeling, which commonly leads to anisotropic material properties.

In addition, functionally graded materials belong to another class of composites, which use the volume fraction change in a certain direction to achieve some special functions or properties. For example, metal–ceramic functionally graded materials have been used in thermal barrier coatings since 1990s. The ceramic component provides acceptable strength at high temperature at the hot side; and protects the metal from melting; while the metal component provides structural support to the whole unit.

Before micromechanics was established, people commonly estimated the overall material properties of composites by the volume fraction of material phases. For a composite containing n types of dispersed phases, the volume fraction of each phase is written as ϕ_i, for $i = 1, 2, \ldots, n$, which makes $\sum_{i=0}^{n} \phi_i = 1$. Here, the term for $i = 0$ represents the phase of the matrix. Obviously, the effective density can be obtained by $\bar{d} = \sum_{i=0}^{n} d_i \phi_i$ with the density of each phase written as d_i. However, as for the effective mechanical properties, a simple volume average of the elastic constants may produce contradictory predictions. For example, a uniaxial tension test of a two-phase composite may provide the effective Young's modulus $\overline{E}$ or the effective compliance, say $1/\overline{E}$. Following the rule of volume average of Young's modulus $\overline{E}$ or $1/\overline{E}$, we can obtain

$$\overline{E} = E_0\phi_0 + E_1\phi_1 \tag{1.1}$$

or

$$\frac{1}{\overline{E}} = \frac{\phi_0}{E_0} + \frac{\phi_1}{E_1} \quad \text{or} \quad \overline{E} = \frac{E_0 E_1}{\phi_0 E_1 + \phi_1 E_0} \tag{1.2}$$

respectively, which provide distinct predictions of Young's modulus for the same composite. Later on, we can see they are two bounds of the effective Young's modulus, that is, the Voigt and Reuss approximations. Therefore, the simple volume fraction of material constants may not provide physical predictions of its effective material constants.

When the concentration of a dispersed phase is low, typically there exists a linear range that the effective material properties proportionally change with the volume fraction of the phase. However, when the concentration reaches a certain value, the so-called threshold, the material properties may exhibit a nonlinear or disruptive change. In general, experimental, analytical, and numerical approaches are used in the studies of emerging composite materials. Analytical models validated with experiments attracted much attention for the convenience to form some design parameters and formulation.

To model and predict the effective mechanical behavior of different composite materials, the initial microstructure is used for the elastic analysis. However, the microstructure may change with the loading path and history, and thus change the effective material performance. For example, microcracks or dislocations may be induced in the damage and yielding process. Micromechanics provides a viable way to address these features and thus predict the overall material behavior.

1.2 HISTORY OF MICROMECHANICS

The beginning of micromechanics has been commonly traced back to Eshelby's pioneer work [5,6] to solve the elastic field for one ellipsoid embedded in an infinite medium. Mura has done tremendous work to normalize the concepts, integrate the research results at his era into a uniform theoretical framework, and thus establish this subject with his masterpiece book [3]. Although micromechanics theory has been well established, it is still a very active research area as new materials and phenomena are emerging.

The basic study tool of micromechanics is continuum mechanics. The research object is composite materials or heterogeneous materials. The research method covers two scales. Through the stress analysis at microscale, one can predict the effective mechanical behavior of the composites. There are numerous methods in micromechanics to solve the effective elastic moduli of the composite. Variational approaches generate the bounds of effective properties by substituting the approximated fields into the energy bounds, which are obtained by taking volume average of the energy density [2,7–14]. Effective medium approaches take into account the effect of the occurrence of the particles on the averaged material properties and the average elastic field of the matrix, and then predict effective elastic properties of composites regardless of the particle distributions and microstructure. Examples include Mori–Tanaka's model [15–17], the self-consistent model [18,19], the differential scheme [20,21], and the generalized self-consistent model [22].

Though the above approaches utilize various techniques to investigate the interaction of the particles, they are essentially based on Eshelby's solution [5,6] for one particle filled in the infinite medium. This assumption limits higher levels of accuracy. In order to consider the direct pairwise particle interaction, other model methods have been studied. Chen and Acrivos [23] integrated the contribution of all particles on overall stress based on the solution [24] to two interacting particles, and reached the effective elastic moduli of an isotropic composite material containing randomly distributed spheres of the same size. Ju and Chen [25,26] also obtained effective elastic moduli for the same problem by collecting the contribution of all other particles on the eigenstrain of one particle. The former involves a complicated summation of a Legendre's series and depends on the boundary condition, while the latter needs more physical explanation. Torquato [27] used statistical methods to quantitatively describe random microstructures and thus study the interaction among particles.

If the microstructure of the composite is periodic, it is possible to solve the local elastic field exactly. Several analytical results for periodically distributed particulate composites were obtained recently based on a Fourier series approach. Examples of this approach include Iwakuma and Nemat-Nasser [28], Walker et al. [29], Nemat-Nasser and Hori [4]. On the other hand, Levy and Papazian [30] and Swan and Kosaka [31] used the finite element method to determine the effective response of these composites. Aboudi et al. [32] constructed a higher-order theory to consider the fluctuation of the displacement in the unit cell. Fish's group has introduced homogenization methods into finite element models for multiscale modeling [33]. Both analytical and numerical methods are based on the unit cell model with periodic loading and boundary conditions, which depends on the microstructure of the composite.

1.3 A BIG PICTURE OF MICROMECHANICS-BASED MODELING

Micromechanics encompasses continuum mechanics to microstructures of materials [3] through which one can tailor the effective material properties through manipulating the microstructure of material phases. Figure 1.3 provides a big picture of a micromechanical method. Given a composite specimen, one can apply a test load on the boundary, which is a general uniform stress or displacement. At the macroscale, the composite can be considered as a homogeneous material. For any material point at the macroscale, when it is observed at a lower scale, the microstructure can be heterogeneous, which can be indicated by an RVE. Notice that the size of an RVE should be large enough to be statistically representative of the microstructure, which will be further discussed in Chapter 6. With a well-defined microstructure and corresponding boundary conditions to the test loading, one can solve the boundary value problem for the local stress and strain fields. Taking an average of the stress and strain, one can treat them as the stress and strain on the material point at the macroscale. The correlation between the stress and strain provides the effective constitutive law for the mechanical behavior. If we change the test loading from stress or displacement to other types of load, say electric or magnetic field, this micromechanics-based framework can be extended to multiphysical problems.

In general, people use a scale description in Figure 1.4 to indicate length scales so that different branches of mechanics can be applicable. Micromechanics generally studies the materials at the scale at microns. The reason is at this scale, the microstructure of most composites can be well defined. At a larger scale, say millimeter or above, one may not be able to identify the material phases, and thus the material can be assumed to be homogeneous. Continuum mechanics are generally used to analyze the mechanical behavior. With the increase of the scale, materials are used for structures and structural mechanics may be used. For a scale lower than microns, say in nanometer or below, the material is not a continuum body

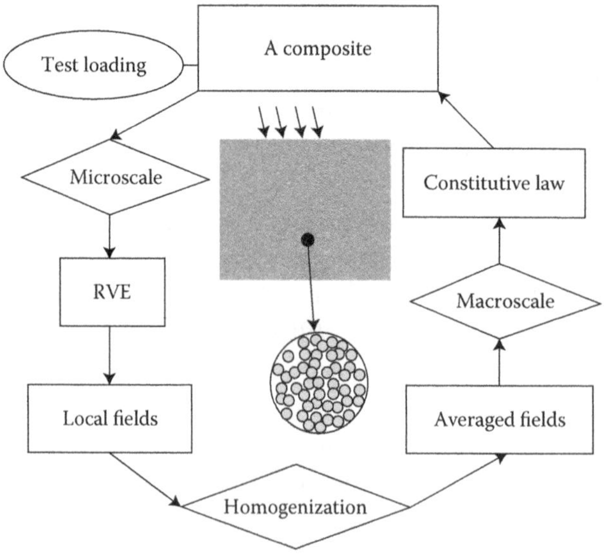

FIGURE 1.3 A micromechanics-based analysis framework.

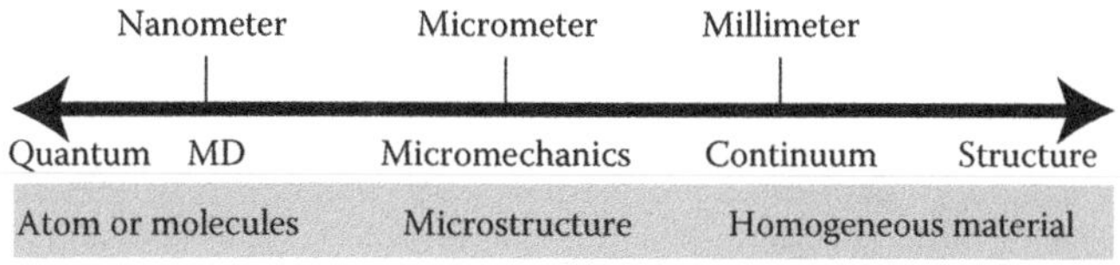

FIGURE 1.4 The length scales related to different branches of mechanics.

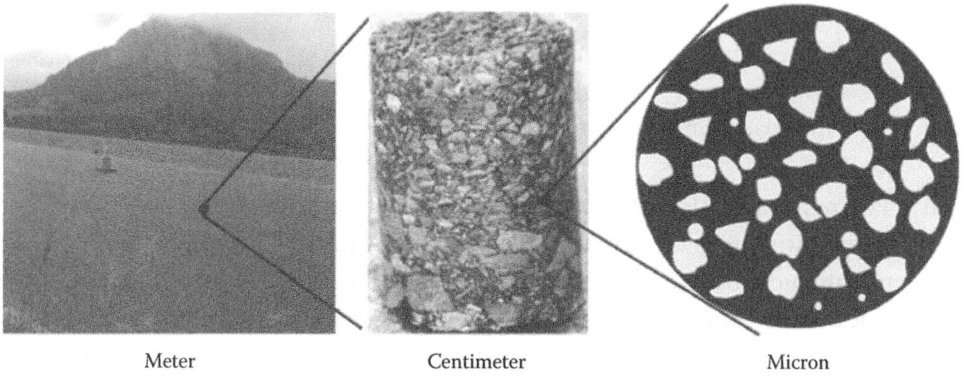

FIGURE 1.5 Applications of micromechanics to asphalt mix and mastic cross different scales.

but a discrete particle system made of atoms or molecules. Molecular dynamics (MD) and quantum mechanics may be used to study the material behavior.

This length scale description is not rigorous. The micromechanics method can be used to predict the effective material behavior based on the microstructure. The method is more essential than the actual length scale to identify a micromechanical problem. For example, Figure 1.5 indicates asphalt materials at different scales. In the meter scale, an asphalt pavement can be considered as a homogeneous material with certain mechanical properties. However, an asphalt core taken from the pavement is heterogeneous with a random microstructure of aggregate and asphalt mastic at centimeter scale. Micromechanics can be used to predict the mechanical properties of the asphalt pavement based on the microstructure and material properties of aggregate and the asphalt mastic. If asphalt mastic is considered, at the centimeter scale, it is a homogeneous material, but at the micron scale, the material is observed as mineral fillers dispersed in asphalt binder. Micromechanics can also be used to predict the mechanical properties of the asphalt mastic by the microstructure and material properties of fillers and the asphalt binder. Therefore, the same method is used at different length scales.

1.4 BASIC CONCEPTS OF MICROMECHANICS

Some basic concepts of micromechanics will be briefly introduced as follows.

1.4.1 Representative Volume Element

An RVE for a material point of a continuous mass is a material volume that is statistically representative of the mircostructure in the neighborhood of the infinitesimal material point. As

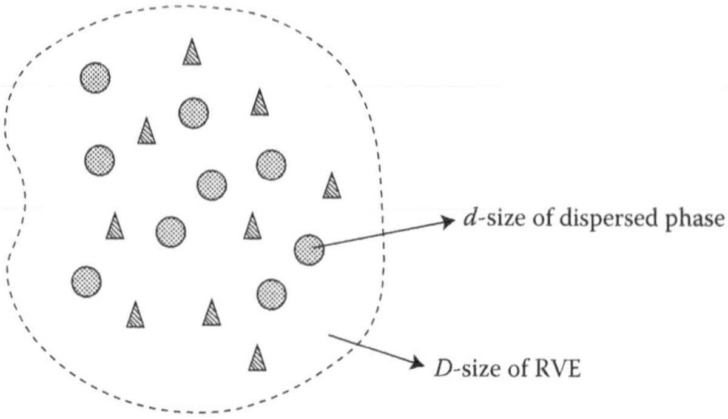

FIGURE 1.6 A material volume of a three-phase composite.

shown in Figure 1.6, an RVE bridges the macroscale material point in a statistically homogeneous composite with its microstructure and microconstituents. The size of the RVE should be large enough, such that the average material behavior of the RVE can represent the material properties of the composite. For example, Figure 1.6 illustrates a material volume of a three-phase composite. The characteristic particle size is denoted by d. The size of the RVE is shown as D. If the size D is small, the effective material properties, which is obtained by the volume average, may significantly change with the possible microstructures. Even if a statistical approach is used to consider all possible microstructures in D through an ensemble average that will be elaborated in Chapter 6, the average material properties may still not be representative of the overall composite.

As an extreme case, if D is so small as to shrink to a point even at the microscale, the effective properties will be either each dispersed phase or the matrix for each possible microstructure. Taking an ensemble average for all possible microstructure, one can see that overall averaged properties can be represented by a simple volume average, say Equation 1.1 when a displacement control test is conducted or Equation 1.2 when a stress control test is conducted [4]. As the size of D increases, the volume average properties of an RVE will change in a smaller range for all possible microstructure. Taking an ensemble average of all possible microstructure, the overall average will decrease for the displacement control test and increase for the stress control test. When D is large enough, any possible microstructure should provide almost the same average properties for either displacement or stress control tests. However, how large of D/d is sufficient is still a vital research topic because it may depend on the shape, volume fraction, size distribution, and mechanical properties of the material phases. As a rule of thumb, theoretical analysis commonly assumes $D/d \to \infty$ to avoid the convergence problems. As for numerical methods, more consideration should be taken into account for the balance of accuracy and computational cost. Typically, D/d should be higher than 10 to obtain a stable prediction by one microstructure case. If ensemble average is used, D/d could be much smaller, such as 2 [34].

1.4.2 Inclusion and Inhomogeneity

As illustrated in Figure 1.7a, a homogeneous material domain D contains a subdomain Ω that has the same mechanical properties as the rest of the region but includes a

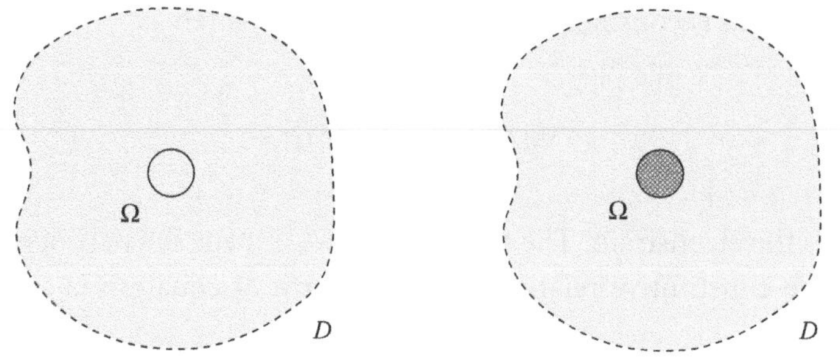

FIGURE 1.7 Illustration of an inclusion and an inhomogeneity.

nonmechanical strain, which is not caused by stress-induced elastic deformation. The subdomain Ω is called an *inclusion*. For example, within a uniform soil there is one wet part. Due to the moisture change, the soil may swell but the rest of the soil still keeps dry. If the mechanical properties of the wet and dry soil keeps uniform, the wet part can be considered as an inclusion. As another two-dimensional example, within a large steel plate there is a circular hole. A circular steel bar with a slightly larger diameter, which is cooled to a certain temperature, can perfectly fit into the hole. When temperature increases to the room temperature, the bar will experience a thermal expansion, but the steel plate does not have temperature change at all. Because both the plate and the circular bar have the same mechanical properties, the bar can be considered as an inclusion.

On the other hand, Figure 1.7b shows that a subdomain Ω has different mechanical properties from the rest of the region in the domain D. The subdomain Ω is called an *inhomogeneity*. A particle embedded in a matrix is a common example of an inhomogeneity.

Inclusion and inhomogeneity may share the same geometry of a subdomain embeded in a large domain but they have distinct meanings. An inclusion is defined by a prescribed deformation in the subdomain, whereas an inhomogeneity means the material difference in the subdomain.

1.4.3 Eigenstrain

Eigenstrain means a nonmechanical strain. The concept of eigenstrain was first introduced by Eshelby [5] and was initially named by Mura [3] to describe local inelastic strain, such as thermal expansion, phase transformation, initial strains, plastic strains, and misfit strains. In the elastic constitutive law, the eigenstrain will not contribute to the stress part. For example, general elastic stress–strain relation is written as

$$\sigma_{ij} = C_{ijkl}\epsilon_{kl} \tag{1.3}$$

where the stress $\sigma_{ij} = \sigma_{ji}$, strain $\epsilon_{kl} = \epsilon_{lk}$, and the stiffness tensor satisfies the minor symmetry $C_{ijkl} = C_{jikl} = C_{ijlk}$ and the major symmetry $C_{ijkl} = C_{klij}$. Using the minor symmetry, the stiffness tensor has 36 independent constants. The major symmetry leads to a reduction to 21 independent constants.

When an eigenstrain is considered, Equation 1.3 is rewritten as

$$\sigma_{ij} = C_{ijkl}(\epsilon_{kl} - \epsilon^*_{kl}) \tag{1.4}$$

where ϵ^*_{kl} denotes the eigenstrain. The term of $\epsilon_{kl} - \epsilon^*_{kl}$ means the part of elastic strain.

To illustrate the constitutive relation, a matrix form of equation is commonly used as follows:

$$\begin{pmatrix} \sigma_{11} \\ \sigma_{22} \\ \sigma_{33} \\ \sigma_{23} \\ \sigma_{31} \\ \sigma_{12} \end{pmatrix} = \begin{bmatrix} C_{1111} & C_{1122} & C_{1133} & C_{1123} & C_{1131} & C_{1112} \\ C_{2211} & C_{2222} & C_{2233} & C_{2223} & C_{2231} & C_{2212} \\ C_{3311} & C_{3322} & C_{3333} & C_{3323} & C_{3331} & C_{3312} \\ C_{2311} & C_{2322} & C_{2333} & C_{2323} & C_{2331} & C_{2312} \\ C_{3111} & C_{3122} & C_{3133} & C_{3123} & C_{3131} & C_{3112} \\ C_{1211} & C_{1222} & C_{1233} & C_{1223} & C_{1231} & C_{1212} \end{bmatrix} \begin{bmatrix} \epsilon_{11} - \epsilon^*_{11} \\ \epsilon_{22} - \epsilon^*_{22} \\ \epsilon_{33} - \epsilon^*_{33} \\ 2\left(\epsilon_{23} - \epsilon^*_{23}\right) \\ 2\left(\epsilon_{31} - \epsilon^*_{31}\right) \\ 2\left(\epsilon_{12} - \epsilon^*_{12}\right) \end{bmatrix} \tag{1.5}$$

1.4.4 Eshelby's Equivalent Inclusion Method

As illustrated in the left of Figure 1.8, one inhomogeneity Ω is embedded in a matrix D. When a uniform stress or strain is applied in the far field, the stress and strain in the neighborhood of the particle will be disturbed. When the particle's shape is ellipsoidal, it was found that the particle will have uniform stress and strain, which are obviously different from the applied far-field stress and strain. This feature motivated Eshelby [5] to discover an approach to solve the elastic fields, which is the so-called Eshelby's equivalent inclusion method. He replaced the inhomogeneity by an inclusion with an eigenstrain without changing the stress distribution inside and outside the inhomogeneity so that the original problem can be solved by the superposition of the two problems: the first one is that a uniform stress is applied on a uniform material, whereas the second one is that an eigenstrain is applied on the inclusion. Through appropriately selecting the eigenstrain, the final elastic field can be solved. This method is beautiful for the case of ellipsoidal particle because the eigenstrain will be a constant. However, it can be generalized to general particles or multiple particles by numerical methods in which the eigenstrain will be nonuniform over the inclusion domain.

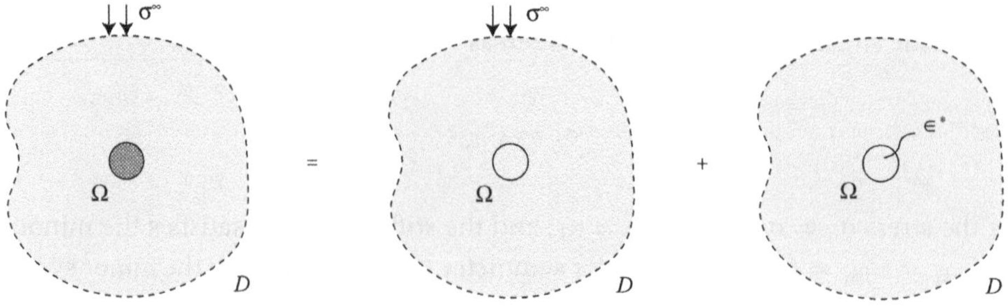

FIGURE 1.8 Schematic illustration of Eshelby's equivalent inclusion method.

1.5 CASE STUDY: HOLES SPARSELY DISTRIBUTED IN A PLATE

1.5.1 The Exact Solution to an Infinite Plate Containing a Circular Hole

Consider a finite region containing a circular hole, as shown in Figure 1.9, the shear modulus and Poisson's ratio are μ and ν. For the plane problem, the stress components σ_r, σ_θ, and $\tau_{r\theta}$ in polar coordinates can be expressed in terms of two complex potentials $\varphi(z)$ and $\psi(z)$ as [35]

$$\sigma_r + \sigma_\theta = 2\left[\Phi(z) + \overline{\Phi(z)}\right] = 2\left[\varphi'(z) + \overline{\varphi'(z)}\right] \tag{1.6}$$

$$\sigma_\theta - \sigma_r + 2i\tau_{r\theta} = 2e^{2i\theta}\left[z\Phi'(z) + \Psi(z)\right] = 2e^{2i\theta}\left[\bar{z}\varphi''(z) + \psi'(z)\right] \tag{1.7}$$

where $\Phi(z) = \varphi'(z)$, $\Psi(z) = \psi'(z)$, and the overbar denotes the complex conjugate.

And the displacements u_r and u_θ in the r and θ directions can be written as

$$2\mu\,(u_r + iu_\theta) = e^{-i\theta}\left[\kappa\varphi(z) - z\overline{\varphi'(z)} - \overline{\psi(z)}\right] \tag{1.8}$$

where κ is the Kolosov constant which has relationship with Poisson's ratio ν,

$$\kappa = \begin{cases} 3 - 4\nu & \textit{for plane strain} \\ \dfrac{3-\nu}{1+\nu} & \textit{for plane stress} \end{cases} \tag{1.9}$$

The free boundary condition along the circle edge is

$$\Phi(z) + \overline{\Phi(z)} - e^{2i\theta}\left[z\Phi'(z) + \Psi(z)\right] = F_n - iF_\tau = 0 \tag{1.10}$$

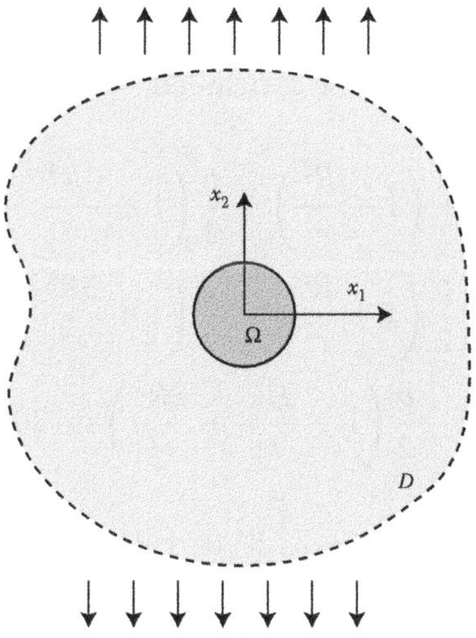

FIGURE 1.9 An infinite region containing a circular hole.

The complex potentials for the region can be expressed in terms of Fourier series for general loading at infinity,

$$\Phi(z) = \varphi'(z) = \sum \frac{a_n}{z^n} \quad (n \geq 0) \tag{1.11}$$

$$\Psi(z) = \psi'(z) = \sum \frac{b_n}{z^n} \quad (n \geq 0) \tag{1.12}$$

where

$$a_0 = \frac{\sigma_1^\infty + \sigma_2^\infty}{4}, \quad b_0 = \frac{\sigma_2^\infty - \sigma_1^\infty + 2i\tau_{12}^\infty}{2}$$

$\sigma_1^\infty, \sigma_2^\infty$, and τ_{12}^∞ are the stress components applied at infinity, and a_n and $b_n (n \geq 2)$ are the coefficients to be determined. To satisfy the free boundary conditions on the hole, one has

$$a_1 = b_1 = 0, \quad a_2 = \bar{b}_0 R^2, \quad b_2 = 2a_0 R^2, \quad b_3 = 0, \quad b_n = (n-1)a_{n-2} R^2 \tag{1.13}$$

Then the following solutions can be concluded; for uniaxial tensile case, $\sigma_x^\infty = p \neq 0$, $\sigma_y^\infty = \tau_{xy}^\infty = 0$,

$$\Phi(z) = \frac{p}{4}\left(1 - \frac{2R^2}{z^2}\right) \tag{1.14}$$

$$\Psi(z) = -\frac{p}{2}\left(1 - \frac{R^2}{z^2} + \frac{3R^4}{z^4}\right) \tag{1.15}$$

$$\varphi(z) = \frac{p}{4}\left(z + \frac{2R^2}{z}\right) \tag{1.16}$$

$$\psi(z) = -\frac{p}{2}\left(z + \frac{R^2}{z} - \frac{R^4}{z^3}\right) \tag{1.17}$$

and the corresponding stress field can be concluded,

$$\begin{cases} \sigma_r = \dfrac{p}{2}\left(1 - \dfrac{R^2}{r^2}\right) + \dfrac{p}{2}\left(1 + \dfrac{3R^4}{r^4} - \dfrac{4R^2}{r^2}\right)\cos 2\theta \\ \sigma_\theta = \dfrac{p}{2}\left(1 + \dfrac{R^2}{r^2}\right) - \dfrac{p}{2}\left(1 + \dfrac{3R^4}{r^4}\right)\cos 2\theta \\ \tau_{r\theta} = -\dfrac{p}{2}\left(1 - \dfrac{3R^4}{r^4} + \dfrac{2R^2}{r^2}\right)\sin 2\theta \end{cases} \tag{1.18}$$

$$\begin{cases} u_r = \dfrac{p}{8\mu r}\left\{(\kappa - 1)r^2 + 2R^2 + 2\left[R^2(\kappa + 1) + r^2 - \dfrac{R^4}{r^2}\right]\cos 2\theta\right\} \\ u_\theta = -\dfrac{p}{4\mu r}\left\{(\kappa - 1)R^2 + r^2 + \dfrac{R^2}{r^2}\right\}\sin 2\theta \end{cases} \tag{1.19}$$

1.5.2 Prediction of the Equivalent Property of an Infinite Plate Containing Periodic Holes

Given a plate with many sparsely distributed small holes as shown in Figure 1.10a, Young's modulus and Poisson's ratio of the solid material are E and ν.

For simplicity, only the square periodic configuration is studied. If we assume that the holes are dilute, which means that the holes are far away enough and every hole has no effect on the others, then we have $R \ll a$. To predict Young's modulus, it is assumed that the region is subjected to a uniform stress in the x_1 direction $\sigma_1^\infty = p \neq 0, \sigma_2^\infty = \tau_{12}^\infty = 0$, whereas for shear modulus, $\tau_{12}^\infty = \tau \neq 0, \sigma_1^\infty = \sigma_2^\infty = 0$, is applied at the boundary of the plate. Due to periodicity of the contribution of holes, only one unit cell, as shown in Figure 1.10b, which can represent the typical structure of the plate, the RVE, is considered. It is assumed that the RVE has the same stress/strain distribution as the corresponding area that is embedded in an infinite region. Then, the equivalent Young's modulus and Poisson's ratio are calculated by

$$E_{equiv} = \frac{\overline{\sigma_1(AB)}}{\overline{\epsilon_1(AB)}}, \quad \nu_{equiv} = -\frac{\overline{\epsilon_2(BC)}}{\overline{\epsilon_1(AB)}} \tag{1.20}$$

where $\overline{\sigma_1(AB)}$ and $\overline{\epsilon_1(AB)}$ are the average stress and strain on edge AB, and $\overline{\epsilon_2(BC)}$ is the average strain on edge BC.

$$\overline{\sigma_1(AB)} = \frac{1}{a}\int_0^a \sigma_1 dy \tag{1.21}$$

$$\overline{\epsilon_1(AB)} = \frac{\overline{u_1(AB)}}{a} = \frac{1}{a^2}\int_0^a u_1 dy \tag{1.22}$$

$$\overline{\epsilon_2(BC)} = \frac{\overline{u_2(BC)}}{a} = \frac{1}{a^2}\int_0^a u_2 dy \tag{1.23}$$

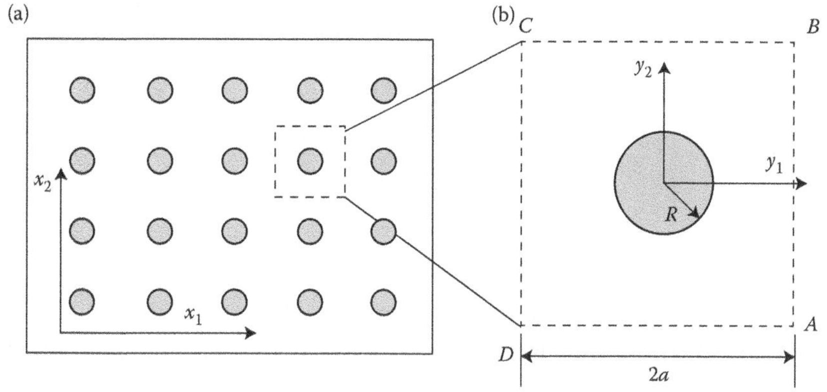

FIGURE 1.10 (a) An infinite region contains periodic holes. (b) Representation volume element.

and stress and displacement components at x_1 direction are

$$\sigma_1 = \sigma_r \cos^2\theta + \sigma_\theta \sin^2\theta - \tau \sin 2\theta \tag{1.24}$$

$$u_1 = u_r \cos\theta - u_\theta \sin\theta \tag{1.25}$$

$$u_2 = u_r \sin\theta + u_\theta \cos\theta \tag{1.26}$$

Inserting Equations 1.18,1.19,1.24, and 1.25 into Equations 1.21 and 1.22, we have

$$\overline{\sigma_1(AB)} = p\left(1 - \frac{3}{4}\frac{R^2}{a^2} + \frac{R^4}{8a^4}\right) \approx p\left(1 - \frac{3}{\pi}c\right) \tag{1.27}$$

$$\begin{aligned}\overline{\epsilon_1(BC)} &= \frac{p}{16\mu}\left[2(\kappa+1) + (\kappa\pi+4)\frac{R^2}{a^2} - \frac{R^4}{a^4}\right] \\ &\approx \frac{p}{16\mu}\left[2(\kappa+1) + \frac{4(\kappa\pi+4)}{\pi}c\right] \\ &= \frac{p(\kappa-3)}{8\mu}\left\{1 + \frac{2[4+(\kappa-2)\pi]}{\pi(\kappa-3)}c\right\}.\end{aligned} \tag{1.28}$$

$$\begin{aligned}\overline{\epsilon_2(BC)} &= \frac{p}{16\mu}\left\{2(\kappa-3) + [4+(\kappa-2)\pi]\frac{R^2}{a^2} - \frac{R^4}{a^4}\right\} \\ &\approx \frac{p}{16\mu}\left\{2(\kappa-3) + \frac{4[4+(\kappa-2)\pi]}{\pi}c\right\} \\ &= \frac{p(\kappa-3)}{8\mu}\left\{1 + \frac{2[4+(\kappa-2)\pi]}{\pi(\kappa-3)}c\right\}.\end{aligned} \tag{1.29}$$

where $c = \frac{\pi R^2}{4a^2}$ can be considered as the volume fraction of the cavities. From Equation 1.20, Young's modulus and Poisson's ratio are concluded, for plane stress problem,

$$E_{equiv} = \frac{\overline{\sigma_1(AB)}}{\overline{\epsilon_1(AB)}} = \frac{p\left(1 - \frac{3}{\pi}c\right)}{\frac{(\kappa+1)p}{8\mu}\left[1 + \frac{2(\kappa\pi+4)}{\pi(\kappa+1)}c\right]} \approx E\left[1 - \frac{3\pi + 10 - (\pi-4)\nu}{2\pi}c\right] \tag{1.30}$$

$$\nu_{equiv} = -\frac{\overline{\epsilon_2(BC)}}{\overline{\epsilon_1(AB)}} = -\frac{\frac{(\kappa+1)p}{8\mu}\left[1 + \frac{2(\kappa\pi+4)}{\pi(\kappa+1)}c\right]}{\frac{p(\kappa-3)}{8\mu}\left\{1 + \frac{2[4+(\kappa-2)\pi]}{\pi(k-3)}c\right\}} \approx \nu\left\{1 - \frac{(1+\nu)[\pi+4-(\pi-4)\nu]}{2\pi\nu}c\right\} \tag{1.31}$$

From Equations 1.30 and 1.31 one can find that the equivalent Young's modulus and Poisson's ratio decrease with the increasing of cavities volume fraction.

Notice that the above solution has been subjected to significant simplification of the original problem. As the elastic field for a single void in an infinite domain is used to simulate the elastic field for a composite with many voids, first, the interactions between voids are not considered; second, the boundary condition is not consistent between the solution and actual uniaxial loading situation; finally, the average field is calculated from the stress and displacement of the boundary of the unit cell, which may depend on the loading conditions that lead to nonunique elastic moduli. We will discuss these factors in the later chapters such as Chapter 6.

1.6 EXERCISES

1. Which of the following can be considered as composite materials? (a) ice made of salt water; (b) cable protected by a carbon fiber sheet on the surface; (c) portland cement concrete; (d) low carbon steel; (e) carbon nanotubes dispersed in an asphalt binder.

2. Classify the following cases as either an inhomogeneity and an inclusion: (a) a homogeneous cement paste with one region shrinking by drying (the moduli are assumed to be the same for different moisture contents); (b) a homogeneous polymer containing a air void; (c) a glass plate containing a glass fiber.

3. Following the case study in Section 1.5, use the classic solution of a thick wall cylinder under a uniform pressure to derive the bulk modulus for a 2D case.

4. Extend the above problem to the 3D case, and derive the bulk modulus of a porous material with air void.

CHAPTER 2

Vectors and Tensors

2.1 CARTESIAN VECTORS AND TENSORS

In this book, we use a lower case letter to represent a vector, and an upper case letter to represent a tensor. The bold font shows the symbolic form, and the italic font with index indicates the component form.

2.1.1 Summation Convention in the Index Notation

The Einstein notation or Einstein summation convention has been widely used in mechanics formulation with vectors and tensors, which can be written in terms of an index form—a letter with a subscript. For example, A_{ijk} means a third-rank tensor named A with the three free indices i, j, and k in the range of $(1, 2)$ for the two-dimensional (2D) case or $(1, 2, 3)$ for the three-dimensional (3D) case. For brevity, we use 3D in this chapter if the dimension is not specified. Mostly, the rules can be easily reduced to 2D cases.

For an expression written in terms of a high rank tensor or the multiplication of several vectors or tensors, if two indices are repeated, the summation convention will be applied in the range of $(1, 2, 3)$. For example,

$$a_i b_i = \sum_{i=1}^{3} a_i b_i = a_1 b_1 + a_2 b_2 + a_3 b_3 \tag{2.1}$$

The above expression becomes one scalar. The repeated index is also called *dummy index* because it can be replaced by other symbols without changing the outcome. For example, $a_i b_i = a_j b_j = a_k b_k$ where the repeated indices i, j, and k are dummy indices.

To uniquely define the summation, *no index can be repeated twice or higher*. For example $a_i b_i c_i$ is never allowed in the index form.

If an index is not repeated in an expression, we call it *a free index*. It determines the rank of the expression. The dimension of a tensor means the range of the index, and the rank of a tensor means the number of free indices. If an expression is in 3D and has a rank of K, it will have 3^K components.

For example, A_{ijkl} is a fourth-rank tensor and has 81 components; A_{iikl} is a second-rank tensor (k and l are free indices) and has 9 components, whereas $A_{iikk} = A_{1111} + A_{2211} + A_{3311} + A_{1122} + A_{2222} + \cdots + A_{3333}$ is a scalar of the summation of 9 terms.

2.1.2 Vector

A vector can be graphically depicted by a directed line in 2D or 3D Euclidean space. In a three-dimensional Cartesian coordinate (0, x_1, x_2, x_3) in Figure 2.1, the unit base vector along each coordinate axis is written as $\mathbf{e}_i$ ($i = 1, 2, 3$). Then an arbitrary vector $\mathbf{a}$ can be written in terms of the sum of its projection in three base vectors, that is,

$$\mathbf{a} = \sum a_i \mathbf{e}_i = a_i \mathbf{e}_i \tag{2.2}$$

where a_i is the component of $\mathbf{a}$ relative to the basis $\mathbf{e}_i$. Notice that the index notation has been used in $a_i\mathbf{e}_i$. The rules of index notation can be summarized as follows:

1. The range of the index is from 1 to n. Here n is of the dimension of the coordinate, say 3 for the case in Figure 2.1.
2. The repeated index means summation of the term over the range of the index. For example, $a_i\mathbf{e}_i = a_1\mathbf{e}_1 + a_2\mathbf{e}_2 + a_3\mathbf{e}_3$.
3. An index should NOT be repeated more than once. For example, $a_ib_i\mathbf{e}_i$ is not defined and should be avoided.

For brevity, a vector $\mathbf{a}$ can also be shortened by the index form a_i by dropping the base vector.

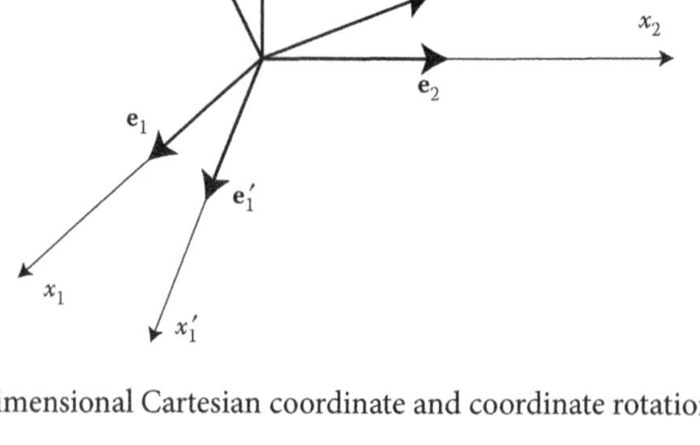

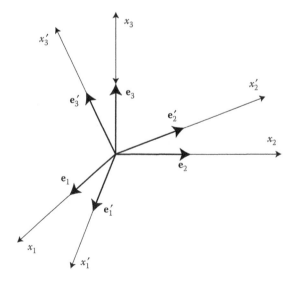

FIGURE 2.1 Three-dimensional Cartesian coordinate and coordinate rotation.

Now there is another coordinate (x'_1, x'_2, x'_3) with the base vector $\mathbf{e}'_i$ $(i = 1, 2, 3)$. The relation between $\mathbf{e}'_i$ and $\mathbf{e}_j$ can be written as

$$\mathbf{e}_i = Q_{ij}\mathbf{e}'_j \tag{2.3}$$

where $Q_{ij} = \cos(\mathbf{e}_i, \mathbf{e}'_j)$ which means the projection of vector $\mathbf{e}_i$ on the new base vector $\mathbf{e}'_j$. Q_{ij} is also called the transformation matrix. Inversely, one can also write

$$\mathbf{e}'_i = Q'_{ij}\mathbf{e}_j \tag{2.4}$$

with $Q'_{ij} = \cos(\mathbf{e}'_i, \mathbf{e}_j)$. Obviously, when Q_{ij} and Q'_{ij} are written as matrices, $\mathbf{Q}' = \mathbf{Q}^T$. Here the superscript "$T$" denotes the transpose of the matrix or the swap of the indices.

Similarly, vector $\mathbf{a}$ can be written as the sum of the new projection $\mathbf{e}'_i$,

$$\mathbf{a} = a'_i\mathbf{e}'_i \tag{2.5}$$

Using the relation of base vectors in Equation 2.3, one can write the vector in Equation 2.2 as

$$\mathbf{a} = a_i\mathbf{e}_i = a_iQ_{ij}\mathbf{e}'_j = a_jQ_{ji}\mathbf{e}'_i \tag{2.6}$$

Compare with Equation 2.5, then we have the correlation between a'_i and a_j.

$$a'_i = Q_{ji}a_j = Q^T_{ij}a_j \tag{2.7}$$

which can also be written in matrix form with the superscript T meaning the transpose of the matrix. The algebra definition of a vector can be given as an array with the components (a_1, a_2, a_3) that satisfies Equation 2.7 during the coordinate transformation.

2.1.3 Tensor

In continuum mechanics, some variables may require specification of two or more vectors for their description. For example, a stress component may be defined in a plane with a surface normal vector along a certain force orientation, which is related to the surface normal vector and the force orientation vector. It will be a second-rank tensor because two vectors are involved. In addition, as we can see from Equation 2.7, when a vector is described in two coordinates, the correlation between the components of the vector in the different coordinates can be indicated by a second-rank tensor Q_{ij}. In the Cartesian coordinate, the components of a second-rank tensor will be associated with a *dyadic product* of two base vectors, say $\mathbf{e}_i \otimes \mathbf{e}_j$, which is also called a *unit dyad*. Mathematically, a second rank tensor $\mathbf{A}$ can be written in terms of a linear combination of unit dyads, that is,

$$\mathbf{A} = A_{ij}\mathbf{e}_i \otimes \mathbf{e}_j \tag{2.8}$$

Similarly to the vector, during the coordinate transformation, the components of the tensor need to satisfy

$$A'_{ij} = Q_{mi}Q_{nj}A_{mn} = Q^T_{im}A_{mn}Q_{nj} \tag{2.9}$$

The algebra definition of a second-rank tensor can be given as a matrix A_{ij} that satisfies Equation 2.9 during the coordinate transformation.

In general, the components of a tensor may be associated with n vectors in terms of polyadic products of base vectors, say $\mathbf{e}_{i_1} \otimes \mathbf{e}_{i_2} \cdots \otimes \mathbf{e}_{i_n}$. Then Equation 2.9 becomes

$$A'_{i_1 i_2 \cdots i_n} = Q_{j_1 i_1} Q_{j_2 i_2} \cdots Q_{j_n i_n} A_{j_1 j_2 \cdots j_n} \tag{2.10}$$

When $n = 0$, the tensor becomes a scalar which does not change with coordinate transformation at all; when $n = 1$, the tensor becomes a vector; when $n = 3$, the unit polyad is called triad; when $n = 4$, the unit polyad is called tetrad. The algebra definition of a n-rank tensor can be given by Equation 2.10.

2.1.4 Special Tensors

Kronecker's delta δ_{ij} is defined as

$$\delta_{ij} = \begin{cases} 1 & i = j \\ 0 & i \neq j \end{cases} \tag{2.11}$$

which is also called the unit second-rank tensor, written as **I**. A unique feature of δ_{ij} is its substitution property: if one index of Kronecker's delta is repeated with another tensor, the index of that tensor can be replaced by the other index of Kronecker's delta. For example, $A_{ij}\delta_{jk} = A_{ik}$ and $A_{ij}\delta_{ji} = A_{ii} = A_{jj}$. For this reason, sometimes it is called the substitution tensor.

The Levi–Civita symbol, also called permutation, e_{ijk} is defined as

$$e_{ijk} = \begin{cases} 1 & \textit{even permutation of } i, j, k \\ 0 & \textit{two or more repeated index} \\ -1 & \textit{odd permutation of } i, j, k \end{cases} \tag{2.12}$$

for example, $e_{111} = 0$; $e_{123} = 1$; $e_{213} = -1$. One very useful identity of e_{ijk} is

$$e_{ijk}e_{imn} = \delta_{jm}\delta_{kn} - \delta_{jn}\delta_{km} \tag{2.13}$$

which will be proved later.

2.2 OPERATIONS OF VECTORS AND TENSORS

2.2.1 Multiplication of Vectors

Dot product, also called scalar product or inner product, is defined as

$$\mathbf{a} \cdot \mathbf{b} = |\mathbf{a}|.|\mathbf{b}| \cos(\mathbf{a}, \mathbf{b}) \tag{2.14}$$

Obviously, for a set of base vectors, their dot product can be written as

$$\mathbf{e}_i \cdot \mathbf{e}_j = \delta_{ij} \tag{2.15}$$

Equation 2.14 can be rewritten as

$$\mathbf{a} \cdot \mathbf{b} = (a_i \mathbf{e}_i) \cdot (b_j \mathbf{e}_j) = a_i b_j \delta_{ij} = a_i b_i \tag{2.16}$$

Cross product, also called vector product, is defined as

$$\mathbf{a} \times \mathbf{b} = |\mathbf{a}|.|\mathbf{b}| \sin(\mathbf{a}, \mathbf{b}) \mathbf{e}_c \tag{2.17}$$

where $\mathbf{e}_c$ is a unit vector perpendicular to the plane of **a** and **b**, positive in the direction of a right-handed screw turning **a** toward **b**. Therefore, $\mathbf{a} \times \mathbf{b} = -\mathbf{b} \times \mathbf{a}$. For a set of base vector, their cross product can be written as

$$\mathbf{e}_i \times \mathbf{e}_j = e_{ijk} \mathbf{e}_k \tag{2.18}$$

Equation 2.17 can be rewritten as

$$\mathbf{a} \times \mathbf{b} = (a_i \mathbf{e}_i) \times (b_j \mathbf{e}_j) = a_i b_j e_{ijk} \mathbf{e}_k \tag{2.19}$$

Scalar triple product, also called triple product, is defined as

$$(\mathbf{a} \times \mathbf{b}) \cdot \mathbf{c} = e_{ijk} a_i b_j c_k \tag{2.20}$$

which denotes the volume of the parallelepiped box formed by vectors **a**, **b**, and **c**, positive in the case that the directional vector of a right-handed screw turning **a** toward **b** has an angle with **c** less than $\pi/2$. It can be written in the determinant of a matrix as follows:

$$(\mathbf{a} \times \mathbf{b}) \cdot \mathbf{c} = \begin{vmatrix} a_1 & a_2 & a_3 \\ b_1 & b_2 & b_3 \\ c_1 & c_2 & c_3 \end{vmatrix} \tag{2.21}$$

The multiplication of two scalar triple products can be obtained by the multiplication of two matrices as follows:

$$\begin{aligned} [(\mathbf{a} \times \mathbf{b}) \cdot \mathbf{c}] [(\mathbf{u} \times \mathbf{v}) \cdot \mathbf{w}] &= \begin{vmatrix} a_1 & a_2 & a_3 \\ b_1 & b_2 & b_3 \\ c_1 & c_2 & c_3 \end{vmatrix} \cdot \begin{vmatrix} u_1 & u_2 & u_3 \\ v_1 & v_2 & v_3 \\ w_1 & w_2 & w_3 \end{vmatrix} \\ &= \left| \begin{pmatrix} a_1 & a_2 & a_3 \\ b_1 & b_2 & b_3 \\ c_1 & c_2 & c_3 \end{pmatrix} \cdot \begin{pmatrix} u_1 & v_1 & w_1 \\ u_2 & v_2 & w_2 \\ u_3 & v_3 & w_3 \end{pmatrix} \right| \\ &= \begin{vmatrix} \mathbf{a} \cdot \mathbf{u} & \mathbf{a} \cdot \mathbf{v} & \mathbf{a} \cdot \mathbf{w} \\ \mathbf{b} \cdot \mathbf{u} & \mathbf{b} \cdot \mathbf{v} & \mathbf{b} \cdot \mathbf{w} \\ \mathbf{c} \cdot \mathbf{u} & \mathbf{c} \cdot \mathbf{v} & \mathbf{c} \cdot \mathbf{w} \end{vmatrix} \end{aligned} \tag{2.22}$$

For a set of base vectors, their scalar triple product can be written as

$$(\mathbf{e}_i \times \mathbf{e}_j) \cdot \mathbf{e}_k = e_{ijm}\mathbf{e}_m \cdot \mathbf{e}_k = e_{ijm}\delta_{mk} = e_{ijk} \tag{2.23}$$

Therefore, substitution of Equation 2.23 into Equation 2.22 yields

$$\begin{aligned} e_{ijk}e_{pqr} = \left[(\mathbf{e}_i \times \mathbf{e}_j) \cdot \mathbf{e}_k\right]\left[(\mathbf{e}_p \times \mathbf{e}_q) \cdot \mathbf{e}_r\right] &= \begin{vmatrix} \mathbf{e}_i \cdot \mathbf{e}_p & \mathbf{e}_i \cdot \mathbf{e}_q & \mathbf{e}_i \cdot \mathbf{e}_r \\ \mathbf{e}_j \cdot \mathbf{e}_p & \mathbf{e}_j \cdot \mathbf{e}_q & \mathbf{e}_j \cdot \mathbf{e}_r \\ \mathbf{e}_k \cdot \mathbf{e}_p & \mathbf{e}_k \cdot \mathbf{e}_q & \mathbf{e}_k \cdot \mathbf{e}_r \end{vmatrix} \\ &= \begin{vmatrix} \delta_{ip} & \delta_{iq} & \delta_{ir} \\ \delta_{jp} & \delta_{jq} & \delta_{jr} \\ \delta_{kp} & \delta_{kq} & \delta_{kr} \end{vmatrix} \end{aligned} \tag{2.24}$$

Equation 2.24 implies

$$\begin{aligned} e_{ijk}e_{imn} &= \begin{vmatrix} \delta_{ii} & \delta_{im} & \delta_{in} \\ \delta_{ji} & \delta_{jm} & \delta_{jn} \\ \delta_{ki} & \delta_{km} & \delta_{kn} \end{vmatrix} = \delta_{ii}\begin{vmatrix} \delta_{jm} & \delta_{jn} \\ \delta_{km} & \delta_{kn} \end{vmatrix} - \delta_{im}\begin{vmatrix} \delta_{ji} & \delta_{jn} \\ \delta_{ki} & \delta_{kn} \end{vmatrix} + \delta_{in}\begin{vmatrix} \delta_{ji} & \delta_{jm} \\ \delta_{ki} & \delta_{km} \end{vmatrix} \\ &= \delta_{jm}\delta_{kn} - \delta_{jn}\delta_{km} \end{aligned} \tag{2.25}$$

Dyadic product, also called outer product, is defined as

$$\mathbf{a} \otimes \mathbf{b} = (a_i\mathbf{e}_i) \otimes (b_j\mathbf{e}_j) = a_ib_j\mathbf{e}_i \otimes \mathbf{e}_j \tag{2.26}$$

which makes a second-rank tensor.

2.2.2 Multiplication of Tensors

The rules of the multiplication of vectors can be easily extended to tensors as vector is indeed a first-rank tensor. Some operations of the tensors will be introduced, which will be used in the following chapters later.

Outer product of two tensors is written as

$$\mathbf{A} \otimes \mathbf{B} = (A_{ij}\mathbf{e}_i \otimes \mathbf{e}_j) \otimes (B_{mn}\mathbf{e}_m \otimes \mathbf{e}_n) = A_{ij}B_{mn}\mathbf{e}_i \otimes \mathbf{e}_j \otimes \mathbf{e}_m \otimes \mathbf{e}_n \tag{2.27}$$

which makes a fourth-rank tensor. If **B** is a vector, the above equation is reduced to

$$\mathbf{A} \otimes \mathbf{b} = (A_{ij}\mathbf{e}_i \otimes \mathbf{e}_j) \otimes (b_m\mathbf{e}_m) = A_{ij}b_m\mathbf{e}_i \otimes \mathbf{e}_j \otimes \mathbf{e}_m \tag{2.28}$$

Notice that if the order of two tensor is swapped, the results are different.

Inner product of two tensors is written as

$$\begin{aligned} \mathbf{A} \cdot \mathbf{B} &= (A_{ij}\mathbf{e}_i \otimes \mathbf{e}_j) \cdot (B_{mn}\mathbf{e}_m \otimes \mathbf{e}_n) = A_{ij}B_{mn}\mathbf{e}_i \otimes (\mathbf{e}_j \cdot \mathbf{e}_m)\,\mathbf{e}_n \\ &= A_{ij}B_{mn}\delta_{jm}\mathbf{e}_i \otimes \mathbf{e}_n = A_{ij}B_{jn}\mathbf{e}_i \otimes \mathbf{e}_n \end{aligned} \tag{2.29}$$

If **B** is a vector, the above equation is reduced to

$$\mathbf{A} \cdot \mathbf{b} = \left(A_{ij}\mathbf{e}_i \otimes \mathbf{e}_j\right) \cdot \left(b_m\mathbf{e}_m\right) = A_{ij}b_m\mathbf{e}_i\left(\mathbf{e}_j \cdot \mathbf{e}_m\right) = A_{ij}b_j\mathbf{e}_i \tag{2.30}$$

Double dot product of two tensors is

$$\mathbf{A} : \mathbf{B} = \left(A_{ij}\mathbf{e}_i \otimes \mathbf{e}_j\right) : \left(B_{mn}\mathbf{e}_m \otimes \mathbf{e}_n\right) = A_{ij}B_{mn}\left(\mathbf{e}_i \cdot \mathbf{e}_m\right)\left(\mathbf{e}_j \cdot \mathbf{e}_n\right) = A_{ij}B_{ij} \tag{2.31}$$

which makes a scalar. If **A** is a fourth-rank tensor, the last two base vectors are used in the dot product as follows:

$$\left(A_{ijkl}\mathbf{e}_i \otimes \mathbf{e}_j \otimes \mathbf{e}_k \otimes \mathbf{e}_l\right) : \left(B_{mn}\mathbf{e}_m \otimes \mathbf{e}_n\right) = A_{ijkl}B_{mn}\delta_{km}\delta_{ln}\mathbf{e}_i \otimes \mathbf{e}_j = A_{ijkl}B_{kl}\mathbf{e}_i \otimes \mathbf{e}_j \tag{2.32}$$

The double dot product of a second-rank tensor **A** with unit tensor **I** is also called the trace of **A**, that is,

$$Tr(\mathbf{A}) = \mathbf{A} : \mathbf{I} = \left(A_{ij}\mathbf{e}_i \otimes \mathbf{e}_j\right) : \left(\delta_{mn}\mathbf{e}_m \otimes \mathbf{e}_n\right) = A_{ii} \tag{2.33}$$

2.2.3 Isotropic Tensors and Stiffness Tensor

For a tensor, if the orientation of the reference coordinate system is changed, the component of the tensor generally changes through the tensor transformation rules in Equation 2.10 and thus exhibits different values. However, a certain type of tensors is called isotropic tensor, such that the values of their components always keep the same in any Cartesian reference coordinate system, that is,

$$A_{i_1 i_2 \cdots i_n} = Q_{j_1 i_1} Q_{j_2 i_2} \cdots Q_{j_n i_n} A_{j_1 j_2 \cdots j_n} \tag{2.34}$$

Rank 0—every scalar is isotropic.
Rank 1—no vector is isotropic unless all components are zero.
Rank 2—the isotropic tensor has a form of $k\delta_{ij}$.
Rank 3—the isotropic tensor has a form of ke_{ijk}.
Rank 4—the isotropic tensor has a form of $\alpha\delta_{ij}\delta_{kl} + \beta\delta_{ik}\delta_{jl} + \gamma\delta_{il}\delta_{jk}$.

For a conservative material, the stiffness tensor, a fourth-rank tensor written as C_{ijkl}, indicates the relationship between a strain tensor and a stress tensor, which is derived from a strain energy density function. It exhibits the major symmetry as

$$C_{ijkl} = C_{klij} \tag{2.35}$$

and minor symmetry as

$$C_{ijkl} = C_{jikl} = C_{ijlk} \tag{2.36}$$

For an isotropic material, the stiffness tensor should be an isotropic tensor and satisfy the symmetry conditions in Equations 2.35 and 2.36. Therefore, the general form of an isotropic stiffness tensor can be written as

$$C_{ijkl} = \lambda\delta_{ij}\delta_{kl} + \mu\left(\delta_{ik}\delta_{jl} + \delta_{il}\delta_{jk}\right) \tag{2.37}$$

where λ and μ are the Lamé constants. The compliance tensor can be considered as the inverse of the stiffness tensor. Their relationship can be described as

$$C_{ijmn}S_{mnkl} = I_{ijkl} \quad \text{or} \quad \mathbf{C} : \mathbf{S} = \mathbf{I}^{(4)} \tag{2.38}$$

where I_{ijkl} or $\mathbf{I}^{(4)}$ is a unit fourth-rank tensor with minor symmetry, which is written as

$$I_{ijkl} = \frac{1}{2}\left(\delta_{ik}\delta_{jl} + \delta_{il}\delta_{jk}\right) \tag{2.39}$$

To solve the inverse or conduct the double dot product of the fourth-rank tensor, it is convenient to define two fourth-rank isotropic tensors, $\mathbf{I}^h$ and $\mathbf{I}^d$, [18,36] as

$$I^h_{ijkl} = \frac{1}{3}\delta_{ij}\delta_{kl}; \; I^d_{ijkl} = \frac{1}{2}\left(\delta_{ik}\delta_{jl} + \delta_{il}\delta_{jk}\right) - \frac{1}{3}\delta_{ij}\delta_{kl} \tag{2.40}$$

which satisfy the following relation for the double dot product:

$$\mathbf{I}^h : \mathbf{I}^h = \mathbf{I}^h; \; \mathbf{I}^d : \mathbf{I}^d = \mathbf{I}^d; \; \mathbf{I}^h : \mathbf{I}^d = \mathbf{0}; \; \mathbf{I}^d : \mathbf{I}^h = \mathbf{0}; \quad \text{and} \quad \mathbf{I}^h + \mathbf{I}^d = \mathbf{I}^{(4)} \tag{2.41}$$

For any isotropic tensor with minor symmetry, say Equation 2.37, it can always be written in terms of the linear combination of the two tensors in Equation 2.40, that is,

$$A_{ijkl} = A^h I^h_{ijkl} + A^d I^d_{ijkl} \tag{2.42}$$

where A^h and A^d denote the coefficients of the corresponding tensors, respectively. Using Equation 2.41, one can obtain

$$\mathbf{A} : \mathbf{B} = A^h B^h \mathbf{I}^h + A^d B^d \mathbf{I}^d \tag{2.43}$$

Let $A^h B^h = 1$ and $A^d B^d = 1$. The above equation shows that **A** and **B** are inverse to each other. We can write

$$\mathbf{A}^{-1} = \frac{1}{A^h}\mathbf{I}^h + \frac{1}{A^d}\mathbf{I}^d \tag{2.44}$$

For example, the isotropic stiffness tensor in Equation 2.37 can be written as

$$\mathbf{C} = (3\lambda + 2\mu)\mathbf{I}^h + 2\mu\mathbf{I}^d \tag{2.45}$$

and then the compliance tensor **s** can be obtained by Equation 2.44 as

$$\mathbf{S} = \mathbf{C}^{-1} = \frac{1}{3\lambda + 2\mu}\mathbf{I}^h + \frac{1}{2\mu}\mathbf{I}^d \tag{2.46}$$

where $K = \frac{1}{3}(3\lambda + 2\mu)$ and μ in Equation 2.45 and Equation 2.46 are also called the bulk modulus and shear modulus, which describe an isotropic material's response to uniform pressure and shear stress, respectively.

For general conservative materials, the stiffness tensor may include 21 independent material constants. Considering symmetries of materials, the material constants will reduce. For orthotropic materials, 9 independent material constants may exist. To illustrate a general tensor with a lot of zero components, Mura's extended index notation [3] can be used. For example, an orthotropic stiffness tensor can be written as

$$C_{ijkl} = C^1_{IK}\delta_{ij}\delta_{kl} + C^2_{IJ}\left(\delta_{ik}\delta_{jl} + \delta_{il}\delta_{jk}\right) \tag{2.47}$$

where C^1_{IK} and C^2_{IJ} are symmetric 3×3 matrices. Mura's extended index notation [3] is used as follows:

1. Repeated lower case indices are summed up as usual index notation.
2. Upper case indices take on the same numbers as the corresponding lower case ones but are not summed.

For example, if the tensor in Equation 2.47 transform an isotropic strain, say $\alpha\delta_{kl}$, to a stress, the stress can be obtained as

$$\sigma_{ij} = C_{ijkl}\alpha\delta_{kl} = \alpha\left(C^1_{I1} + C^1_{I2} + C^1_{I3}\right)\delta_{ij} + 2\alpha C^2_{IJ}\delta_{ij} \tag{2.48}$$

Obviously, for one orthotropic stiffness tensor, there may exist many cases of C^1_{IK} and C^2_{IJ} because 12 material constants of C^1_{IK} and C^2_{IJ} are used to fit the 9 material constants.

2.3 CALCULUS OF VECTOR AND TENSOR FIELDS

In a domain Ω, a vector or tensor field indicates a vector or tensor function at any point $\mathbf{x}$ of Ω, which can be represented by $\mathbf{a}(\mathbf{x})$ for a vector field or $\mathbf{A}(\mathbf{x})$ for a tensor field. For dynamic problems, a vector or tensor field may change with time. For example, speed field in a fluid can be a vector field; stiffness of a graded material can be a tensor field. In this section, the calculus of tensors and vectors are introduced. To do the differential operation, the vector or tensor field is considered to be continuous and differentiable.

2.3.1 Del Operator and Operations

Del operator ∇, also called nabla or Hamilton operator, is written as $\partial_i\mathbf{e}_i$ or ∂_i, which is a vector operator. It follows the rules of vector multiplication.

2.3.1.1 Gradient

For a differentiable scalar field $\phi(\mathbf{x})$, the gradient of ϕ is defined as

$$\text{grad}(\phi) = \nabla\phi = \frac{\partial\phi}{\partial x_i}\mathbf{e}_i = \partial_i\phi\mathbf{e}_i = \boldsymbol{\phi}_{,i}\mathbf{e}_i \tag{2.49}$$

which is a vector indicating the greatest change rate and direction of the scalar field.

The gradient of a vector field $\mathbf{a}(\mathbf{x})$ is defined as

$$\text{grad}(\mathbf{a}) = \nabla \otimes \mathbf{a} = \partial_i \mathbf{e}_i \otimes a_j \mathbf{e}_j = \partial_i a_j \mathbf{e}_i \otimes \mathbf{e}_j = a_{j,i} \mathbf{e}_i \otimes \mathbf{e}_j \tag{2.50}$$

Similarly, the gradient of a tensor field $\mathbf{A}(\mathbf{x})$ is defined as

$$\text{grad}(\mathbf{A}) = \nabla \otimes \mathbf{A} = \partial_i \mathbf{e}_i \otimes \left(A_{jk} \mathbf{e}_j \otimes \mathbf{e}_k\right) = \partial_i A_{jk} \mathbf{e}_i \otimes \mathbf{e}_j \otimes \mathbf{e}_k = A_{jk,i} \mathbf{e}_i \otimes \mathbf{e}_j \otimes \mathbf{e}_k \tag{2.51}$$

Actually, the gradient of a vector or tensor field is an *outer product* of the del operator and the vector or tensor field. The rank will increase one. Following this definition, one can write

$$\mathbf{a} \otimes \nabla = \partial_j a_i \mathbf{e}_i \otimes \mathbf{e}_j = a_{i,j} \mathbf{e}_i \otimes \mathbf{e}_j$$

which is the transpose of the tensor's components defined in Equation 2.50.

2.3.1.2 Divergence

For a vector field $\mathbf{a}(\mathbf{x})$, the divergence of $\mathbf{a}$ is defined as

$$\text{div}(\mathbf{a}) = \nabla \cdot \mathbf{a} = \partial_i \mathbf{e}_i \cdot a_j \mathbf{e}_j = \partial_i a_i = a_{i,i} \tag{2.52}$$

which indicates the volume density of the outward flux of the vector field from an infinitesimal volume around the point. It is an *inner product* (*dot product*) of the del operator and the vector field. For tensor fields, the same rule is applied and the tensor's rank will reduce one. The left and right dot products of the del operator will have the same meaning for a vector field but they are different for higher-rank general tensors. For example,

$$\nabla \cdot \mathbf{A} = \partial_i \mathbf{e}_i \cdot A_{jk} \mathbf{e}_j \otimes \mathbf{e}_k = \partial_i A_{ik} \mathbf{e}_k = A_{ik,i} \mathbf{e}_k \tag{2.53}$$

whereas

$$\mathbf{A} \cdot \nabla = A_{ij} \mathbf{e}_i \otimes \mathbf{e}_j \cdot \partial_k \mathbf{e}_k = \partial_j A_{ij} \mathbf{e}_i = A_{ij,j} \mathbf{e}_i \tag{2.54}$$

2.3.1.3 Curl

For a vector field $\mathbf{a}(\mathbf{x})$, the curl of $\mathbf{a}$ is defined as

$$\text{curl}(\mathbf{a}) = \nabla \times \mathbf{a} = \partial_i \mathbf{e}_i \times a_j \mathbf{e}_j = \partial_i a_j e_{ijk} \mathbf{e}_k = a_{j,i} e_{ijk} \mathbf{e}_k \tag{2.55}$$

which describes the infinitesimal rotation of the vector field. It is a *cross product* (*vector product*) of the del operator and the vector field. For tensor fields, the same rule is applied.

From the above definition, given a vector field, one can find that $\nabla \otimes \mathbf{a} = (\mathbf{a} \otimes \nabla)^T$, $\nabla \cdot \mathbf{a} = \mathbf{a} \cdot \nabla$, and $\nabla \times \mathbf{a} = -\mathbf{a} \times \nabla$. In addition, one can see the following identities

$$\nabla \times (\nabla \phi) = \mathbf{0}, \tag{2.56}$$

$$\nabla \cdot (\nabla \times \mathbf{a}) = 0, \tag{2.57}$$

$$\nabla \times (\nabla \times \mathbf{a}) = \nabla(\nabla \cdot \mathbf{a}) - \nabla^2 \mathbf{a}, \tag{2.58}$$

and

$$\nabla \cdot (\nabla \phi) = \nabla^2 \phi = \phi_{,ii}. \tag{2.59}$$

where $\nabla^2 = \nabla \cdot \nabla = \partial_i \partial_i$ is the Laplace operator.

2.3.2 Examples

i. Given a scalar field in a 3D infinite domain as $\psi = |\mathbf{x} - \mathbf{x}^0|$, which is a distance from $\mathbf{x}$ to a known point $\mathbf{x}^0$, one can obtain

$$\nabla \psi = \frac{x_i - x_i^0}{|\mathbf{x} - \mathbf{x}^0|} \mathbf{e}_i = \mathbf{n}$$

$$\nabla \cdot \nabla \psi = \psi_{,ii} = \frac{2}{|\mathbf{x} - \mathbf{x}^0|}$$

ii. Given a scalar field in a 3D infinite domain as $\phi = 1/|\mathbf{x} - \mathbf{x}^0|$, one can obtain

$$\nabla \phi = -\frac{x_i - x_i^0}{|\mathbf{x} - \mathbf{x}^0|^3} \mathbf{e}_i = -\frac{\mathbf{n}}{|\mathbf{x} - \mathbf{x}^0|^2}$$

$$\nabla \cdot \nabla \phi = \phi_{,ii} = 0 \text{ for } \mathbf{x} - \mathbf{x}^0 \neq \mathbf{0}$$

Actually, ϕ is also called the fundamental solution of Laplace's equation in 3D as

$$\phi_{,ii} = -4\pi \delta(\mathbf{x} - \mathbf{x}^0) \tag{2.60}$$

where $\delta(\mathbf{x} - \mathbf{x}^0)$ is the Dirac function, which satisfies

$$\int_{\infty} \delta(\mathbf{x} - \mathbf{x}^0) f(\mathbf{x}) d\mathbf{x} = f(\mathbf{x}^0); \ \delta(\mathbf{x} - \mathbf{x}^0) = 0 \quad \text{for } \mathbf{x} - \mathbf{x}^0 \neq \mathbf{0}$$

iii. Given a vector field in a 3D infinite domain as $\mathbf{a} = \mathbf{x}$, which is a position vector, one can obtain

$$\nabla \otimes \mathbf{x} = \delta_{ij} \mathbf{e}_i \otimes \mathbf{e}_j$$

$$\nabla \cdot \mathbf{x} = 3$$

$$\nabla \times \mathbf{x} = \mathbf{0}$$

2.3.3 The Gauss Theorem

The Gauss theorem, also called the divergence theorem, transforms a volume integral of the divergence of a vector field $\mathbf{a}(\mathbf{x})$ with continuous derivatives in a domain Ω to the surface flux across the boundary $\partial\Omega$, that is,

$$\int_{\Omega} \nabla \cdot \mathbf{a} \, dV = \int_{\partial\Omega} \mathbf{n} \cdot \mathbf{a} \, dS \tag{2.61}$$

where dV and dS represent the volume and surface infinitesimal elements that are used for volume and surface integral. The Gauss theorem can be generalized to other operations of the del operator and other types of fields, namely scalar fields or tensor fields. Therefore, one can write the generalized Gauss as follows:

$$\int_{\Omega} \nabla * \mathbf{T}\, dV = \int_{\partial\Omega} \mathbf{n} * \mathbf{T}\, dS \tag{2.62}$$

where $*$ indicates an operation, such as inner, outer, or cross product, and $\mathbf{T}$ can be a tensor field with a certain rank, that is, 0th rank for a scalar, 1st rank for a vector, 2nd or higher for general tensors. If the order of ∇ and $\mathbf{T}$ is switched, the order of $\mathbf{n}$ and $\mathbf{T}$ will also be switched.

2.3.4 Green's Theorem and Stokes' Theorem

In a simply connected 2D domain S in the x_1–x_2 coordinate system with the boundary indicated as ∂S, U and V are continuous functions with continuous derivatives. Green's theorem reads

$$\int_{\partial S} U dx_1 + V dx_2 = \int_{S} \left(\frac{\partial V}{\partial x_1} - \frac{\partial U}{\partial x_2} \right) dx_1 dx_2 \tag{2.63}$$

where ∂S is positive along the counterclockwise direction.

Green's theorem transforms an integral over an area into a line integral over a boundary for a 2D domain. Extending it to the 3D domain, the similar equation can be obtained, which is Stokes' theorem.

For a simply connected surface S in the (x_1, x_2, x_3) coordinate system with the boundary indicated as ∂S, $\mathbf{a}$ indicates a vector field with continuous derivatives. Stokes' theorem reads

$$\int_{\partial S} \mathbf{a} \cdot d\mathbf{l} = \int_{S} (\nabla \times \mathbf{a}) \cdot \mathbf{n}\, dS \tag{2.64}$$

where the positive direction of $d\mathbf{l}$ is along the right-handed screw direction of $\mathbf{n}$, which is the surface normal direction.

2.4 POTENTIAL THEORY AND HELMHOLTZ'S DECOMPOSITION THEOREM

2.4.1 Scalar and Vector Potentials

For a simply connected domain, if a vector field $\mathbf{a}(\mathbf{x})$ satisfies $\nabla \times \mathbf{a} = \mathbf{0}$, there exists a scalar potential $\psi(x)$ such that $\mathbf{a} = \nabla\psi$. Here ψ is a continuous function, which automatically makes $\nabla \times \mathbf{a} = \mathbf{0}$ by the identity $\nabla \times \nabla\psi = \mathbf{0}$. The scalar potential typically represents the work done by a conservative force field. We can construct it as

$$\psi(\mathbf{x}) = \int_{\mathbf{x}^0}^{\mathbf{x}} \mathbf{a} \cdot d\mathbf{l} \tag{2.65}$$

Given $\mathbf{x}^0$, $\psi(\mathbf{x})$ is unique at any position $\mathbf{x}$ because the integral of Equation 2.65 is path-independent.

Similarly, if $\mathbf{a}(\mathbf{x})$ satisfies $\nabla \cdot \mathbf{a} = \mathbf{0}$, there exists a vector potential $\mathbf{b}$ such that $\mathbf{a} = \nabla \times \mathbf{b}$. Here $\mathbf{b}$ is a continuous vector field, which automatically makes $\nabla \cdot \mathbf{a} = \mathbf{0}$ by the identity $\nabla \cdot (\nabla \times \mathbf{b}) = \mathbf{0}$. The vector potential typically represents the work done by a conservative force field. We can construct $\mathbf{b}$ as

$$\begin{cases} b_1(x_1, x_2, x_3) = \int\limits_{x_3^0}^{x_3} a_2(x_1, x_2, \xi)d\xi \\ b_2(x_1, x_2, x_3) = -\int\limits_{x_3^0}^{x_3} a_1(x_1, x_2, \xi)d\xi + \int\limits_{x_1^0}^{x_1} a_3(\xi, x_2, x_3^0)d\xi \\ b_3(x_1, x_2, x_3) = 0 \end{cases} \tag{2.66}$$

where $\mathbf{x}^0$ is a fixed point in Ω. We can prove that $a_k = b_{j,i}e_{ijk}$, that is, $a_1 = b_{3,2} - b_{2,3}$, $a_2 = b_{1,3} - b_{3,1}$, and $a_3 = b_{2,1} - b_{1,2}$. Here the first two equations are obvious. The last one uses the condition of $\nabla \cdot \mathbf{a} = \mathbf{0}$, that is,

$$-\int\limits_{x_3^0}^{x_3} \left[a_{1,1}(x_1, x_2, \xi) + a_{2,2}(x_1, x_2, \xi)\right] d\xi = \int\limits_{x_3^0}^{x_3} a_{3,3}(x_1, x_2, \xi)d\xi$$

Notice that because the line integral is used here, this construction of $\mathbf{b}$ is only applicable to a convex domain. For general cases, Stevenson [37] provided further proof.

The scalar and vector potentials only exist for a conservative (irrotational) vector field and a solenoidal (incompressible) vector field, respectively. Helmholtz's decomposition theorem addresses the general cases.

2.4.2 Helmholtz's Decomposition Theorem

Helmholtz's decomposition theorem reads: For an arbitrary vector field $\mathbf{a}$ with continuous derivatives in domain Ω, it may be written in combination of a scalar potential $\psi(x)$ and a vector potential $\mathbf{b}$, such that

$$\mathbf{a} = \nabla\psi + \nabla \times \mathbf{b} \text{ with } \nabla \cdot \mathbf{b} = 0. \tag{2.67}$$

To show this, the Newtonian potential is used. Let a vector field $\mathbf{u}(\mathbf{x})$ be given as

$$\mathbf{u}(\mathbf{x}) = -\int\limits_{\Omega} \frac{1}{4\pi} \frac{\mathbf{a}(\mathbf{x}')}{|\mathbf{x} - \mathbf{x}'|} d\mathbf{x}' \tag{2.68}$$

Based on Equation 2.60, one can write

$$\mathbf{a}(\mathbf{x}) = \nabla^2\mathbf{u}(\mathbf{x}) = \nabla(\nabla \cdot \mathbf{u}) - \nabla \times (\nabla \times \mathbf{u}) \tag{2.69}$$

Define the scalar and vector potentials as

$$\psi = \nabla \cdot \mathbf{u} \quad \text{and} \quad \mathbf{b} = -\nabla \times \mathbf{u}. \tag{2.70}$$

Then, one can obtain Helmholtz's decomposition theorem.

2.5 GREEN'S IDENTITIES AND GREEN'S FUNCTIONS

2.5.1 Green's First and Second Identities

Green's identities are transformations of Gauss' theorem. Given two scalar fields u and v with continuous first and second derivatives, let $\mathbf{a} = u\nabla v$ in Gauss' theorem (2.61), the first Green's identity reads

$$\int_{\Omega} \left(\nabla u \cdot \nabla v + u\nabla^2 v\right) dV = \int_{\partial\Omega} u\nabla v \cdot \mathbf{n}\, dS = \int_{\partial\Omega} u\frac{\partial v}{\partial n}\, dS \tag{2.71}$$

where $\mathbf{n}$ is the outer normal direction of the surface $\partial\Omega$. Exchange the scalar functions in the above equation, then

$$\int_{\Omega} \left(\nabla u \cdot \nabla v + v\nabla^2 u\right) dV = \int_{\partial\Omega} v\frac{\partial u}{\partial n}\, dS \tag{2.72}$$

The difference of the above two equations yields the second Green's identity as

$$\int_{\Omega} \left(u\nabla^2 v - v\nabla^2 u\right) dV = \int_{\partial\Omega} \left(u\frac{\partial v}{\partial n} - v\frac{\partial u}{\partial n}\right) dS \tag{2.73}$$

2.5.2 Green's Function for the Laplacian

Green's function is also called the source function or influence function. It is an integral kernel to solve an inhomogeneous partial differential equation (PDE) subject to a specific boundary condition. It is typically called the fundamental solution for the PDE. For example, for an infinite domain, a scalar potential function U (electrostaticity, temperature, Newtonian gravity, and others) satisfies the following Poisson's equation:

$$\nabla^2 U = \rho(\mathbf{x})$$

where $\rho(\mathbf{x})$ is the source density. Green's fucntion can be written as

$$G(\mathbf{x}, \mathbf{x}') = -\frac{1}{4\pi|\mathbf{x} - \mathbf{x}'|}$$

where $\nabla^2 G(\mathbf{x}, \mathbf{x}') = \delta(\mathbf{x} - \mathbf{x}')$. Green's function $G(\mathbf{x}, \mathbf{x}')$ indicates the response at point $\mathbf{x}$ caused by a unit point load $\mathbf{p_0}$ located at point $\mathbf{x}'$, as shown in Figure 2.2. Consider a

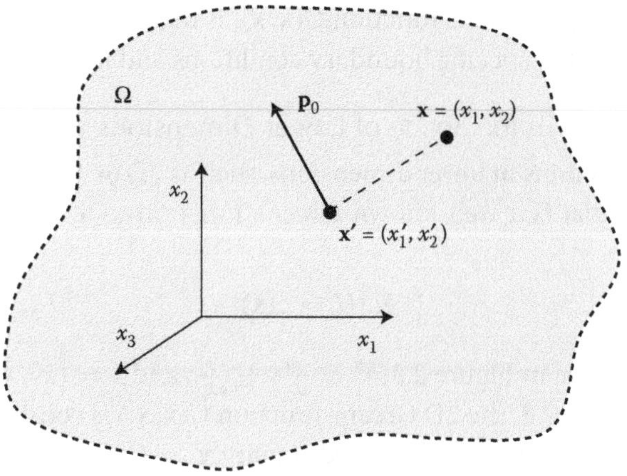

FIGURE 2.2 Green's function.

distributed source field acting on a given domain Ω, the potential at any position can be written in terms of the integral as

$$U(\mathbf{x}) = \int_{\Omega} G(\mathbf{x}, \mathbf{x}')\rho(\mathbf{x}')\, dV(\mathbf{x}') \tag{2.74}$$

From Green's second identity, one can prove the above solution as follows: Using $U(\mathbf{x})$ and $G(\mathbf{x}, \mathbf{x}')$ as u and v, respectively, one can write.

$$\begin{aligned} \int_{\Omega} (U(\mathbf{x}')\nabla^2 G(\mathbf{x}, \mathbf{x}'))dV(\mathbf{x}') &= U(\mathbf{x}) \\ &= \int_{\Omega} G(\mathbf{x}, \mathbf{x}')\nabla^2 U(\mathbf{x}')\, dV(\mathbf{x}') + \int_{\partial\Omega} \left(U(\mathbf{x}') \frac{\partial G}{\partial n} - G(\mathbf{x}, \mathbf{x}') \frac{\partial U}{\partial n} \right) dS(\mathbf{x}') \end{aligned} \tag{2.75}$$

For an infinite domain, the scalar field $U(\mathbf{x})$ in far field $\mathbf{x} \to \infty$ is zero. The second term in the above equation is eliminated. Using $\nabla^2 U = \rho(\mathbf{x})$ in the above equation, one can obtain Equation 2.74.

For a general differential equation,

$$\mathcal{L}u(\mathbf{x}) = \rho(\mathbf{x})$$

where $\mathcal{L}$ represents a differential operator, Green's function satisfies

$$\mathcal{L}G(\mathbf{x}, \mathbf{x}') = \delta(\mathbf{x} - \mathbf{x}')$$

Then, the solution is given by

$$u(\mathbf{x}) = \int_{\Omega} G(\mathbf{x}, \mathbf{x}')\rho(\mathbf{x}')dV(\mathbf{x}')$$

The physical meaning of Green's function $G(\mathbf{x}, \mathbf{x}')$ is the field value at $\mathbf{x}$ caused by a point source at $\mathbf{x}'$ considering the specific boundary conditions and initial conditions.

2.5.3 Green's Function in the Space of Lower Dimensions

There exist Green's functions in lower dimensions, such as 2D or 1D problems. For example, the logarithmic potential is a well-known Green's function to a 2D Poisson's equation as follows:

$$\nabla^2 U = \rho(\mathbf{x}) \tag{2.76}$$

Notice that in 2D space in Figure 2.3, $\nabla = \frac{\partial}{\partial x_1} + \frac{\partial}{\partial x_2}$, and $\mathbf{x} = (x_1, x_2)$.

As illustrated in Figure 2.3, the 2D Green's function $G(\mathbf{x}, \mathbf{x}')$ is considered as the response of point $\mathbf{x} = (x_1, x_2)$ caused by a force located at point $\mathbf{x}' = (x_1', x_2')$. In a 3D domain, we can imagine all points in a 2D domain are lines along the x_3 direction, where the variables along the 3D lines share the same value so that a 2D function is sufficient to describe the field. Then, the 2D Green's function can be obtained by integrating Newton's Potential in the 3rd dimension.

Then, we have

$$G(\mathbf{x}, \mathbf{x}') = \int_{-\infty}^{+\infty} -\frac{1}{4\pi|\mathbf{x} - \mathbf{x}'|} dx_3' \tag{2.77}$$

The integral $\int_{-\infty}^{+\infty} \frac{1}{|\mathbf{x}-\mathbf{x}'|} dx_3'$ may include a divergent point at $x_3 - x_3' = 0$ when $x_1 - x_1' = 0$ and $x_2 - x_2' = 0$. However, we can use the Hadamard finite-part integral [38,39] to get the solution. Rewrite the integral as

$$\int_{-\infty}^{+\infty} \frac{1}{|\mathbf{x} - \mathbf{x}'|} dx_3' = \int_{-\infty}^{+\infty} \frac{1}{\sqrt{r^{*2} + (x_3 - x_3')^2}} dx_3' = \int_{-\infty}^{+\infty} \frac{1}{\sqrt{r^{*2} + t^2}} dt \tag{2.78}$$

where

$$r^* = \sqrt{(x_1 - x_1')^2 + (x_2 - x_2')^2} \tag{2.79}$$

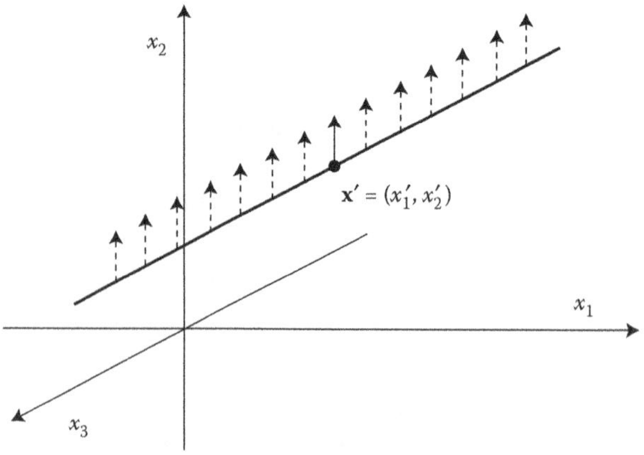

FIGURE 2.3 Dimension reduction of Green's function.

Noticing that

$$\int_{-\infty}^{+\infty} \frac{\partial}{\partial r^*} \frac{1}{\sqrt{r^{*2}+t^2}} dt = -\frac{2}{r^*} \tag{2.80}$$

if we consider the integral and differential operator is permutable, then we have

$$\frac{\partial}{\partial r'} \int_{-\infty}^{+\infty} \frac{1}{\sqrt{r^{*2}+t^2}} dt = -\frac{2}{r^*} \tag{2.81}$$

The integration of Equation 2.81 is

$$\int_{-\infty}^{+\infty} \frac{1}{\sqrt{r^{*2}+t^2}} dt = -2\ln r^* + C \tag{2.82}$$

Here $-2\ln r^*$ is the Hadamard finite part of the divergent integral $\int_{-\infty}^{+\infty} \frac{1}{\sqrt{r^{*2}+t^2}} dt$, which can be considered as the value of the integral.

Therefore, through integrating the 3D Green's function, one can obtain the 2D Green's function in Equation 2.77, simplified as

$$G(\mathbf{x}, \mathbf{x}') = \frac{1}{2\pi} \ln(\mathbf{x} - \mathbf{x}') \tag{2.83}$$

One can prove that

$$\nabla \cdot \nabla G(\mathbf{x}, \mathbf{x}') = 0 \quad \text{for } \mathbf{x} - \mathbf{x}' \neq \mathbf{0}$$

and

$$\int_{\Sigma} \nabla \cdot \nabla G(\mathbf{x}, \mathbf{x}') d\mathbf{x} = 1$$

where Σ is a closed surface including point $\mathbf{x}'$ inside.

For a 1D problem reduced from Poisson's equation, the PDE becomes a second-order ODE as follows:

$$U_{11} = \rho(x_1) \tag{2.84}$$

Green's function can be obtained simply from integrating the discontinuity function:

$$\frac{d^2 G\left(x_1 - x_1'\right)}{dx_1^2} = \delta\left(x_1 - x_1'\right) \tag{2.85}$$

Integrating it once, one can write

$$\frac{dG\left(x_1 - x_1'\right)}{dx_1} = H\left(x_1 - x_1'\right) \tag{2.86}$$

where $H(x_1 - x_1')$ is the Heaviside function, which satisfies

$$H(x - x') = \begin{cases} 1 & x \geq x'. \\ 0 & x < x'. \end{cases} \tag{2.87}$$

Integrating it once again, one can obtain the 1D Green's function as

$$G\left(x_1 - x_1'\right) = \left(x_1 - x_1'\right) H\left(x_1 - x_1'\right) \tag{2.88}$$

2.5.4 Example

Consider a simply supported straight beam with bending stiffness EI and length L is under a distributed load $q(x)$, as shown in Figure 2.4. Derive Green's function for beam bending and use it to solve deflection.

The differential equation for the beam bending is given as

$$EI\frac{d^4w}{dx^4} = q(x)$$

Green's function $G(x, x')$ can be obtained by solving the above equation with a unit concentrated load $\delta(x - x')$, which means that the total force is one acting at the point x'. By integrating $EI\frac{d^4w}{dx^4} = \delta(x - x')$, one can obtain similarly to 1D second-order ODE,

$$EI \cdot G(x - x') = \frac{(x - x')^3}{6} H(x - x') + \frac{C_1}{6}x^3 + \frac{C_2}{2}x^2 + C_3 x + C_4$$

Using the boundary conditions for $x = 0$, and $L : G(x - x') = 0$ and $G''(x - x') = 0$, respectively, one can determine the four constants as

$$C_1 = -\frac{L - x'}{L}; \; C_2 = 0; \; C_3 = \frac{(L - x')(2L - x')x'}{6L}; \; C_4 = 0$$

Then, Green's function can be written as

$$G(x - x') = \frac{1}{EI}\left[\frac{(x - x')^3}{6} H(x - x') + \frac{L - a}{6L}\left(-x^2 + 2x'L - x'^2\right)x\right]$$

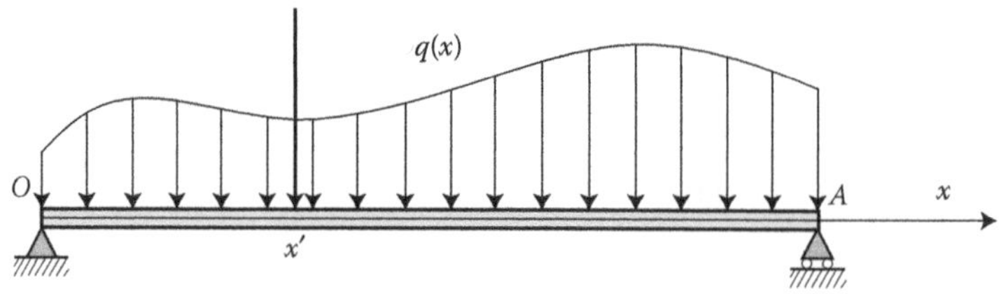

FIGURE 2.4 Straight beam under distribution loading.

Then, the beam's deflection shape function can be written as

$$w(x) = \int_{\Omega} G(x, x')q(x')dx'$$

2.6 ELASTIC EQUATIONS

2.6.1 Strain and Compatibility

In Figure 2.5, an infinitesimal line element in a deformable body is defined by two points P and A. In the reference material coordinate, their coordinates can be written in terms of $\mathbf{r}$ and $\mathbf{r} + d\mathbf{r}$, respectively. When the body is subjected to a deformation, which makes the line element PA to $\bar{P}\bar{A}$ with their coordinates written as $\mathbf{r} + \mathbf{u}$ and $\mathbf{r} + d\mathbf{r} + \mathbf{u} + d\mathbf{u}$. The new line vector can be written as

$$\bar{P}\bar{A} = d\mathbf{r} + d\mathbf{u} \tag{2.89}$$

and its length is written as

$$|\bar{P}\bar{A}| = \sqrt{(d\mathbf{r} + d\mathbf{u}) \cdot (d\mathbf{r} + d\mathbf{u})} \tag{2.90}$$

Due to the continuous deformation, the relation between $d\mathbf{r}$ and $d\mathbf{u}$ can be written as

$$d\mathbf{u} = (\mathbf{u} \otimes \nabla) \cdot d\mathbf{r} = u_{i,j} dr_j \mathbf{e}_i \tag{2.91}$$

By substitution of Equation 2.91, Equation 2.89 could be rewritten in the component form as

$$dr_i + du_i = \left(\delta_{ij} + u_{i,j}\right) dr_j \tag{2.92}$$

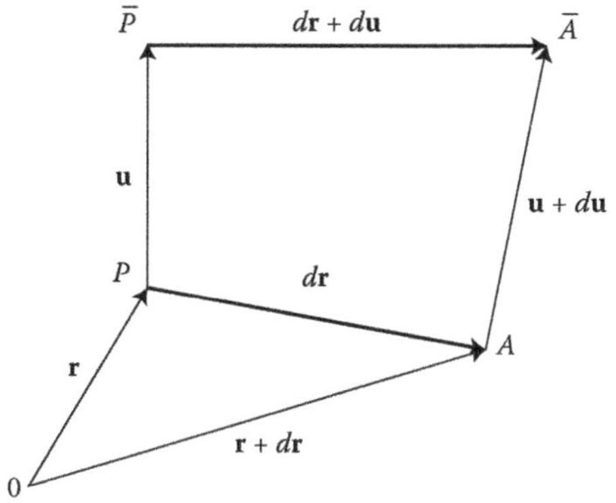

FIGURE 2.5 Deformation of an infinitesimal line element.

Then, Equation 2.90 could be rewritten as

$$\begin{aligned}\left|\bar{P}\bar{A}\right|^2 &= \left(\delta_{ij}+u_{i,j}\right)\left(\delta_{ik}+u_{i,k}\right)dr_j dr_k \\ &= \left(\delta_{jk}+u_{k,j}+u_{j,k}+u_{i,j}u_{i,k}\right)dr_j dr_k \\ &= |dr|^2 + 2G_{jk}\,|dr|^2\,\xi_j\xi_k \end{aligned} \tag{2.93}$$

where $\boldsymbol{\xi}$ indicates the directional unit vector of $d\mathbf{r}$, that is,

$$\begin{cases} dr_i = |dr|\xi_j \\ dr_k = |dr|\xi_k \end{cases} \tag{2.94}$$

and Green's strain tensor

$$G_{jk} = \frac{1}{2}\left(u_{k,j}+u_{j,k}+u_{i,j}u_{i,k}\right) \quad \text{or} \quad \mathbf{G} = \frac{1}{2}\left[\mathbf{u}\otimes\nabla + \nabla\otimes\mathbf{u} + (\nabla\otimes\mathbf{u})\cdot(\mathbf{u}\otimes\nabla)\right] \tag{2.95}$$

For small strain problem, the higher-order term of $u_{i,j}u_{i,k}$ in Equation 2.95 is negligible, thus

$$\epsilon_{ij} = \frac{1}{2}\left(u_{i,j}+u_{j,i}\right) \quad \text{or} \quad \boldsymbol{\epsilon} = \frac{1}{2}\left(\mathbf{u}\otimes\nabla + \nabla\otimes\mathbf{u}\right) \tag{2.96}$$

According to the above equations, the normal strain ϵ could be determined as

$$\epsilon = \frac{\left|d\tilde{r}\right| - |dr|}{|dr|} = \sqrt{1+2\epsilon_{jk}\xi_j\xi_k} - 1 \approx \epsilon_{jk}\xi_j\xi_k \tag{2.97}$$

Equation 2.96 is called Cauchy's strain tensor. In this textbook, strain tensor is considered as Cauchy's strain tensor unless otherwise stated. Using the identities of $\nabla\times(\nabla\otimes\mathbf{u}) = 0$ and $(\mathbf{u}\otimes\nabla)\times\nabla = 0$, one can find that Cauchy's strain tensor has to satisfy

$$\nabla\times\boldsymbol{\epsilon}\times\nabla = \mathbf{0} \quad \text{or} \quad e_{pij}e_{qkl}\epsilon_{jk,il} = 0 \tag{2.98}$$

Equation 2.98 is called compatible equation of strains, under which a unique continuous displacement field can be guaranteed.

Based on the symmetry, Equation 2.98 implies six independent equations. For example, when $p = q = 1$,

$$e_{123}e_{123}\epsilon_{32,23} + e_{132}e_{132}\epsilon_{23,23} + e_{123}e_{132}\epsilon_{33,22} + e_{132}e_{123}\epsilon_{22,33} = 0$$

that is,

$$2\epsilon_{32,23} = \epsilon_{22,33} + \epsilon_{33,22}$$

	$E =$	$\nu =$	$\lambda =$	$\mu =$	$K_0 =$
E, ν	E	ν	$\frac{E\nu}{(1+\nu)(1-2\nu)}$	$\frac{E}{2(1+\nu)}$	$\frac{E}{3(1-2\nu)}$
E, μ	E	$\frac{E}{2\mu} - 1$	$\frac{\mu(E-2\mu)}{3\mu-E}$	μ	$\frac{\mu E}{3(3\mu-E)}$
E, K_0	E	$\frac{1}{2} - \frac{E}{6K_0}$	$\frac{3K_0(3K_0-E)}{9K_0-E}$	$\frac{3K_0E}{9K_0-E}$	K_0
ν, λ	$\frac{\lambda(1+\nu)(1-2\nu)}{\nu}$	ν	λ	$\frac{\lambda(1-2\nu)}{2\nu}$	$\frac{\lambda(1+\nu)}{3\nu}$
ν, μ	$2\mu\,(1+\nu)$	ν	$\frac{2\mu\nu}{1-2\nu}$	μ	$\frac{2\mu(1+\nu)}{3(1-2\nu)}$
ν, K_0	$3K_0\,(1-2\nu)$	ν	$\frac{3K_0\nu}{1+\nu}$	$\frac{3K_0(1-2\nu)}{2(1+\nu)}$	K_0
λ, μ	$\frac{\mu(3\lambda+2\mu)}{\lambda+\mu}$	$\frac{\lambda}{2(\lambda+\mu)}$	λ	μ	$\frac{3\lambda+2\mu}{3}$
λ, K_0	$\frac{9K_0(K_0-\lambda)}{3K_0-\lambda}$	$\frac{\lambda}{3K_0-\lambda}$	λ	$\frac{3(K_0-\lambda)}{2}$	K_0
μ, K_0	$\frac{9K_0\mu}{3K_0+\mu}$	$\frac{3K_0-2\mu}{2(3K_0+\mu)}$	$\frac{3K_0-2\mu}{3}$	μ	K_0

2.6.2 Constitutive Law

In elastic problems, the constitutive law is the relation between stress and strain of a material element. When a conservative material is subjected to a given eigenstrain ϵ^*_{ij}, the stress and strain relationship reads,

$$\sigma_{ij} = C_{ijkl}\left(\epsilon_{kl} - \epsilon^*_{kl}\right) \tag{2.99}$$

where C_{ijkl} is the elasticity tensor. For an isotropic material

$$C_{ijkl} = \lambda\delta_{ij}\delta_{kl} + \mu\left(\delta_{ik}\delta_{jl} + \delta_{il}\delta_{jk}\right) \tag{2.100}$$

where λ and μ are the Lamé constants. μ is also called shear modulus or modulus of rigidity. For isotropic materials, only two independent material constants exist, but there are commonly five material constants: Young's modulus E, Poission's ratio ν, bulk modulus K_0, and the two Lamé constants. Once two of them are given, one can obtain all the other three as in the following table (for plane stress problem).

Plane strain problem: $\bar{E} = \frac{E}{1-\nu^2}$, $\bar{\nu} = \frac{\nu}{1-\nu}$, $\bar{\mu} = \frac{E}{2(1-\nu)} = \mu$

2.6.3 Equilibrium Equation

For any arbitrary domain in a static solid, the resultant force and moment should be zero to keep it in equilibrium. In Figure 2.6, in the solid, there exists a stress field σ_{ij} and a body force b_i. Taking an arbitrary domain Ω, the surface stress vector can be written as $\mathbf{t} = \mathbf{n} \cdot \boldsymbol{\sigma}$. The resultant force applied on Ω can be written as the combination of the surface force and body force as

$$\int_{\partial\Omega} \mathbf{n} \cdot \boldsymbol{\sigma} dS + \int_{\Omega} \mathbf{b} dV = 0 \tag{2.101}$$

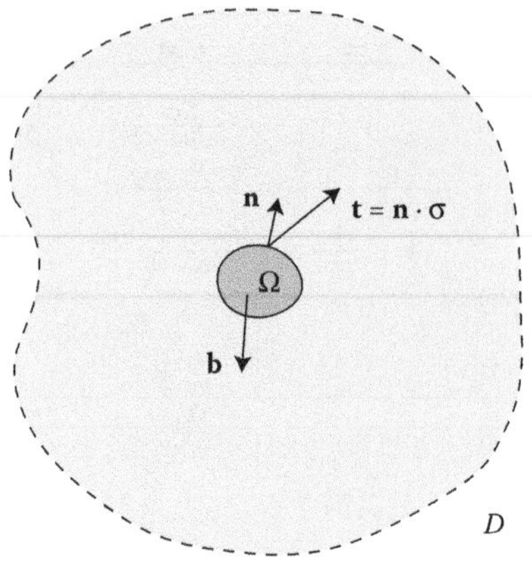

FIGURE 2.6 The equilibrium of an arbitrary domain.

According to the Gauss theorem as mentioned in Equation 2.62, Equation 2.101 becomes

$$\int_{\Omega} (\nabla \cdot \boldsymbol{\sigma} + \mathbf{b})\, dV = 0 \tag{2.102}$$

Because Ω can be an arbitrary domain considered, to make the integral in Equation 2.102 always satisfied, one should require

$$\nabla \cdot \boldsymbol{\sigma} + \mathbf{b} = 0 \;\text{ or }\; \sigma_{ij,i} + b_j = 0 \tag{2.103}$$

Equation 2.103 is the well-known equilibrium equation in linear elastic problem. In general, the body force may be considered separately, so that the equilibrium equation can often be written as a homogeneous PDE as

$$\sigma_{ij,i} = 0 \tag{2.104}$$

Because of the symmetry of stress tensor, Equation 2.104 can also be written as $\sigma_{ij,j} = 0$.

2.6.4 Governing Equations

For an isotropic material, the elastic moduli C_{ijkl} is described in Equation 2.100; the governing equation for this elastic problem could be obtained by substituting Equation 2.100 into Equation 2.99

$$\sigma_{ij} = \lambda\delta_{ij}\left(u_{k,k} - \epsilon^*_{kk}\right) + \mu\left(u_{i,j} + u_{j,i} - 2\epsilon^*_{ij}\right) \tag{2.105}$$

or

$$\sigma = \lambda\mathbf{I}\left[\nabla \cdot \mathbf{u} - tr\left(\epsilon^*\right)\right] + \mu\left(\mathbf{u} \otimes \nabla + \nabla \otimes \mathbf{u} - 2\epsilon^*\right) \tag{2.106}$$

The first derivative of stress σ_{ij} can be written as

$$\begin{aligned}\sigma_{ij,i} &= \lambda\delta_{ij}\left(u_{k,ki} - \epsilon^*_{kk,i}\right) + \mu\left(u_{i,ij} + u_{j,ii} - 2\epsilon^*_{ij,i}\right)\\ &= (\lambda + \mu)u_{i,ij} + \mu u_{j,ii} - \lambda\epsilon^*_{ii,j} - 2\mu\epsilon^*_{ij,i}\end{aligned} \tag{2.107}$$

The substitution of the above equation into Equation 2.103 provides

$$(\lambda + \mu)u_{i,ij} + \mu u_{j,ii} - \lambda\epsilon^*_{ii,j} - 2\mu\epsilon^*_{ij,i} + b_j = 0 \tag{2.108}$$

When the eigenstrain is not considered, the above equation can be reduced to

$$(\lambda + \mu)u_{i,ij} + \mu u_{j,ii} + b_j = 0 \tag{2.109}$$

or

$$(\lambda + \mu)\nabla(\nabla \cdot \mathbf{u}) + \mu\nabla^2\mathbf{u} + \mathbf{b} = 0 \tag{2.110}$$

which is Navier's equation. Let

$$\mathcal{L} = \frac{\lambda + \mu}{\mu}\nabla\nabla \cdot + \nabla^2 \tag{2.111}$$

Equation 2.110 could be rewritten as

$$\mathcal{L}\mathbf{u} = -\frac{\mathbf{b}}{\mu} \tag{2.112}$$

2.6.5 Boundary Value Problem

A linear elastic boundary value problem requires solving the elastic equations with boundary conditions. Based on the nature of the problem, different coordinates may be used to take advantage of the symmetry for the load and geometry. As long as the coordinate is orthogonal, the constitutive relation will remain the same for isotropic material, but the strain–displacement relation and the equilibrium equation may depend on the coordinate.

For example, a ball is subjected to a uniform inner pressure p with inner radius a and outer radius b as shown in Figure 2.7. Here spherical symmetry exists. Therefore, the spherical coordinate can be used. The displacement only occurs in r direction.

The strain condition is

$$\begin{cases} \epsilon_{rr} = u_{r,r} \\ \epsilon_{\theta\theta} = \epsilon_{\phi\phi} = \dfrac{u_r}{r} \end{cases} \tag{2.113}$$

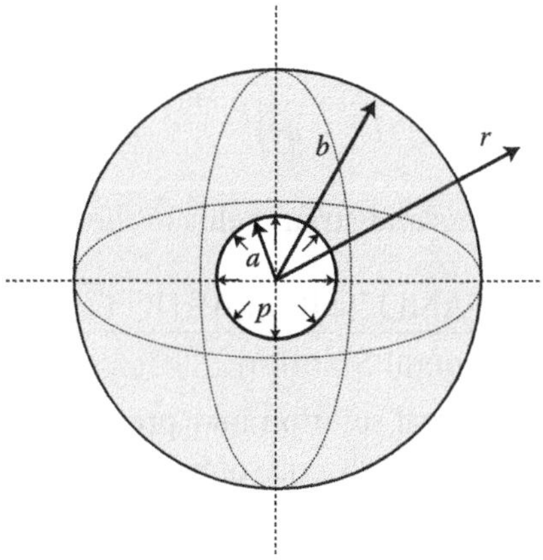

FIGURE 2.7 A ball subjected to a uniform inner pressure p.

the equilibrium equation is

$$\sigma_{rr,r} + \frac{2\sigma_{rr} - \sigma_{\theta\theta} - \sigma_{\phi\phi}}{r} = 0 \tag{2.114}$$

and the constitutive relation reads

$$\begin{cases} \sigma_{rr} = \lambda\left(\epsilon_{rr} + \epsilon_{\theta\theta} + \epsilon_{\phi\phi}\right) + 2\mu\epsilon_{rr} = (2\mu + \lambda)u_{r,r} + 2\lambda\dfrac{u_r}{r} \\ \sigma_{\theta\theta} = \lambda\left(\epsilon_{rr} + \epsilon_{\theta\theta} + \epsilon_{\phi\phi}\right) + 2\mu\epsilon_{\theta\theta} = \lambda u_{r,r} + (2\mu + 2\lambda)\dfrac{u_r}{r} \\ \sigma_{\phi\phi} = \lambda\left(\epsilon_{rr} + \epsilon_{\theta\theta} + \epsilon_{\phi\phi}\right) + 2\mu\epsilon_{\phi\phi} = \lambda u_{r,r} + (2\mu + 2\lambda)\dfrac{u_r}{r} \end{cases} \tag{2.115}$$

Substituting Equation 2.115 into Equation 2.114,

$$(2\mu + \lambda)u_{r,rr} + 2\lambda\left(\frac{u_r}{r}\right)_{,r} + 4\mu\left(\frac{u_{r,r} - \frac{u_r}{r}}{r}\right) = 0 \tag{2.116}$$

After organization, the above equation could be rewritten as

$$u_{r,rr} + \frac{2u_{r,r}}{r} - \frac{2u_r}{r^2} = 0 \tag{2.117}$$

The general solution of displacement u_r in Equation 2.117 is

$$u_r = C_1 r + \frac{C_2}{r^2} \tag{2.118}$$

According to the boundary condition

$$\begin{cases} \sigma_{rr}|_{r=a} = -p \\ \sigma_{rr}|_{r=b} = 0 \end{cases} \tag{2.119}$$

Then, the solution of displacement u could be determined as

$$u_r = \frac{a^3}{(2\mu + 3\lambda)\left(b^3 - a^3\right)} p \cdot r + \frac{a^3 b^3}{4\mu\left(b^3 - a^3\right)} \frac{p}{r^2} \tag{2.120}$$

Using it, one can obtain both stress and strain fields in the ball.

2.7 GENERAL SOLUTION AND THE ELASTIC GREEN'S FUNCTION

2.7.1 Papkovich–Neuber's General Solution

To solve elastic problems, the general solution may provide convenience for some types of problems. For Navier's equation in the absence of body force,

$$\nabla^2 \mathbf{u} + \frac{\lambda + \mu}{\mu} \nabla\left(\nabla \cdot \mathbf{u}\right) = 0 \tag{2.121}$$

where $\frac{\lambda+\mu}{\mu} = \frac{1}{1-2\nu}$, Papkovich–Neuber's general solution reads

$$\mathbf{u} = \mathbf{p} - \frac{1}{4(1-\nu)}\nabla\left(p_0 + \mathbf{x}\cdot\mathbf{p}\right) \tag{2.122}$$

where $\nabla^2\mathbf{p} = 0$, $\nabla^2 p_0 = 0$, and $\mathbf{x} = x_i\mathbf{e}_i$.

Proof: Using Helmholtz's decomposition theorem, the displacement field can be written as

$$\mathbf{u} = \nabla\phi + \nabla\times\mathbf{a} \text{ with } \nabla\cdot\mathbf{a} = 0 \tag{2.123}$$

Substitution of the above equation into Equation 2.110 yields

$$\nabla^2(\nabla\phi) + \nabla^2(\nabla\times\mathbf{a}) + \frac{1}{1-2\nu}\nabla(\nabla\cdot\nabla\phi) = 0$$

Then, it can be rewritten as

$$\nabla^2\left(\frac{2-2\nu}{1-2\nu}\nabla\phi + \nabla\times\mathbf{a}\right) = 0$$

Let

$$\mathbf{p} = \frac{2-2\nu}{1-2\nu}\nabla\phi + \nabla\times\mathbf{a} \tag{2.124}$$

then

$$\nabla^2\mathbf{p} = 0$$

Rewrite Equation 2.124 as

$$\nabla\times\mathbf{a} = \mathbf{p} - \frac{2-2\nu}{1-2\nu}\nabla\phi \tag{2.125}$$

Applying $(\nabla\cdot)$ at both sides in Equation 2.125 yields

$$\nabla\cdot\mathbf{p} = \frac{2-2\nu}{1-2\nu}\nabla^2\phi \textit{ or } \nabla^2\phi = \frac{1-2\nu}{2-2\nu}\nabla\cdot\mathbf{p} \tag{2.126}$$

where $\nabla\cdot(\nabla\times\mathbf{a}) = 0$ is used

The general solution to Equation 2.126 of ϕ is

$$\phi = \frac{1-2\nu}{4(1-\nu)}\left(\mathbf{x}\cdot\mathbf{p} + p_0\right) \text{ with } \nabla^2 p_0 = 0 \tag{2.127}$$

where p_0 is arbitrary harmonic function. Notice $\nabla^2(\mathbf{x}\cdot\mathbf{b}) = 2\nabla\cdot\mathbf{b} + \mathbf{x}\cdot\nabla^2\mathbf{b}$, Equation 2.127 can be easily verified by substituting the above equation into Equation 2.126 as follows:

$$\frac{2-2\nu}{1-2\mu}\nabla^2\phi = \frac{1}{2}\nabla^2\left(\mathbf{x}\cdot\mathbf{p} + p_0\right) = \nabla\cdot\mathbf{p} + \frac{1}{2}\nabla^2 p_0 + \mathbf{x}\cdot\nabla^2\mathbf{p} = \nabla\cdot\mathbf{p}$$

Therefore, substituting Equations 2.125 and 2.127 into Equation 2.123 yields

$$\mathbf{u} = \mathbf{p} - \frac{1}{4(1-\nu)}\nabla\left(p_0 + \mathbf{x}\cdot\mathbf{p}\right) \tag{2.128}$$

2.7.2 Kelvin's Particular Solution

Consider a concentrated force $\mathbf{P}$ acting at $\mathbf{x}'$ in an infinite domain D with an isotropic material, which can be described by a Dirac delta function as $\mathbf{P}\delta(\mathbf{x}-\mathbf{x}')$ as Figure 2.8. The governing equation can be written as

$$\mu\nabla^2\mathbf{u} + (\lambda+\mu)\nabla\left(\nabla\cdot\mathbf{u}\right) = -\mathbf{P}\delta(\mathbf{x}-\mathbf{x}') \tag{2.129}$$

Equation 2.129 could be simplified as

$$\mathcal{L}\mathbf{u} = -\frac{\mathbf{P}\delta(\mathbf{x}-\mathbf{x}')}{\mu} \tag{2.130}$$

where

$$\mathcal{L} = \nabla^2 + \frac{1}{1-2\nu}\nabla\nabla. \tag{2.131}$$

The displacement field $\mathbf{u}$ can be assumed as the same form as Papkovitch–Neuber's general solution,

$$\mathbf{u} = \mathbf{b} - \frac{1}{4(1-\nu)}\nabla\left(\phi + \mathbf{x}\cdot\mathbf{b}\right) \tag{2.132}$$

Here, $\mathbf{b}$ and ϕ are not harmonic functions any more for the nonhomogeneous term $\mathbf{P}\delta(\mathbf{x}-\mathbf{x}')$. Substituting Equation 2.132 into the left side of Equation 2.129, it could be obtained as

$$\mathcal{L}\mathbf{u} = \left(\nabla^2 + \frac{1}{1-2\nu}\nabla\nabla\cdot\right)\left[\mathbf{b} - \frac{1}{4(1-\nu)}\nabla\left(\phi + \mathbf{x}\cdot\mathbf{b}\right)\right] \tag{2.133}$$

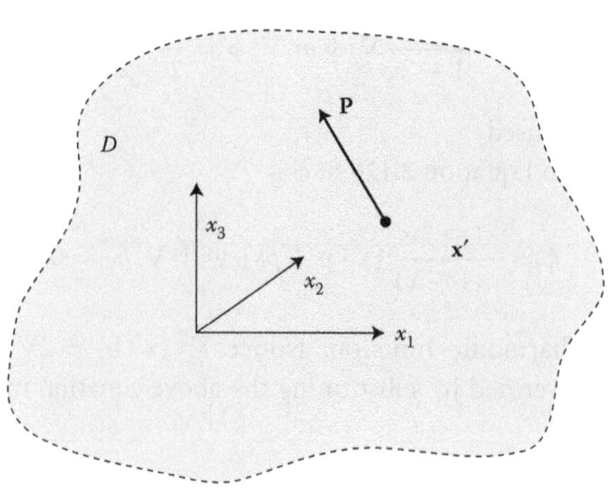

FIGURE 2.8 Concentrated force acting on an infinite domain.

which can be expanded and simplified as

$$\mathcal{L}\mathbf{u} = \nabla^2\mathbf{b} + \frac{1}{1-2\nu}\nabla\left(\nabla\cdot\mathbf{b}\right) - \frac{1}{4(1-\nu)}\nabla\left[\nabla^2\left(\phi + \mathbf{x}\cdot\mathbf{b}\right)\right] - \frac{1}{4\left(1-2\nu\right)\left(1-\nu\right)}$$
$$\times\,\nabla\left[\nabla^2\left(\phi + \mathbf{x}\cdot\mathbf{b}\right)\right] = \nabla^2\mathbf{b} - \frac{1}{2\left(1-2\nu\right)}\nabla\left(\nabla^2\phi + \mathbf{x}\cdot\nabla^2\mathbf{b}\right) \tag{2.134}$$

where $\nabla^2\left(\mathbf{x}\cdot\mathbf{b}\right) = 2\nabla\cdot\mathbf{b} + \mathbf{x}\cdot\nabla^2\mathbf{b}$ is used. Therefore, one can write

$$\boldsymbol{\nabla}^2\mathbf{b} - \frac{\mathbf{1}}{\mathbf{2(1-2\nu)}}\boldsymbol{\nabla}(\boldsymbol{\nabla}^2\phi + \mathbf{x}\cdot\boldsymbol{\nabla}^2\mathbf{b}) = -\frac{\mathbf{P}\delta(\mathbf{x}-\mathbf{x}')}{\mu} \tag{2.135}$$

In order to construct the solution for the above transformed governing equation, one can assume

$$\begin{cases} \nabla^2\mathbf{b} = -\dfrac{\mathbf{P}\delta(\mathbf{x}-\mathbf{x}')}{\mu} \\ \nabla^2\phi = -\mathbf{x}\cdot\nabla^2\mathbf{b} \end{cases} \tag{2.136}$$

For the second equation in the above, it can be written as

$$\nabla^2\phi = \frac{\mathbf{x}\cdot\mathbf{P}\delta(\mathbf{x}-\mathbf{x}')}{\mu} = \frac{\mathbf{x}'\cdot\mathbf{P}\delta(\mathbf{x}-\mathbf{x}')}{\mu} \tag{2.137}$$

so it can also be written in a Poisson's equation with Dirac delta function as the source at a given location $\mathbf{x}'$.

Thus, using Green's function of the potential problem, the solution of $\mathbf{b}$ and ϕ can be written as

$$\begin{cases} \mathbf{b} = \dfrac{1}{4\pi\mu}\dfrac{\mathbf{P}}{|\mathbf{x}-\mathbf{x}'|} \\ \phi = -\dfrac{1}{4\pi\mu}\dfrac{\mathbf{x}'\cdot\mathbf{P}}{|\mathbf{x}-\mathbf{x}'|} \end{cases} \tag{2.138}$$

The solution of displacement field can be calculated by substituting Equations 2.138 into Equation 2.132,

$$\begin{aligned} \mathbf{u}(\mathbf{x}) &= \frac{1}{4\pi\mu}\frac{\mathbf{P}}{|\mathbf{x}-\mathbf{x}'|} - \frac{1}{4(1-\nu)}\nabla\left[-\frac{1}{4\pi\mu}\frac{\mathbf{x}'\cdot\mathbf{P}}{|\mathbf{x}-\mathbf{x}'|} + \frac{1}{4\pi\mu}\frac{\mathbf{x}\cdot\mathbf{P}}{|\mathbf{x}-\mathbf{x}'|}\right] \\ &= \frac{1}{4\pi\mu}\frac{\mathbf{P}}{|\mathbf{x}-\mathbf{x}'|} - \frac{1}{4(1-\nu)}\nabla\left[\frac{1}{4\pi\mu}\frac{(\mathbf{x}-\mathbf{x}')\cdot\mathbf{P}}{|\mathbf{x}-\mathbf{x}'|}\right] \\ &= \frac{1}{16\pi\mu(1-\nu)}\left[(3-4\nu)\frac{\mathbf{P}}{|\mathbf{x}-\mathbf{x}'|} + (\mathbf{x}-\mathbf{x}')\frac{(\mathbf{x}-\mathbf{x}')\cdot\mathbf{P}}{|\mathbf{x}-\mathbf{x}'|^3}\right] \end{aligned} \tag{2.139}$$

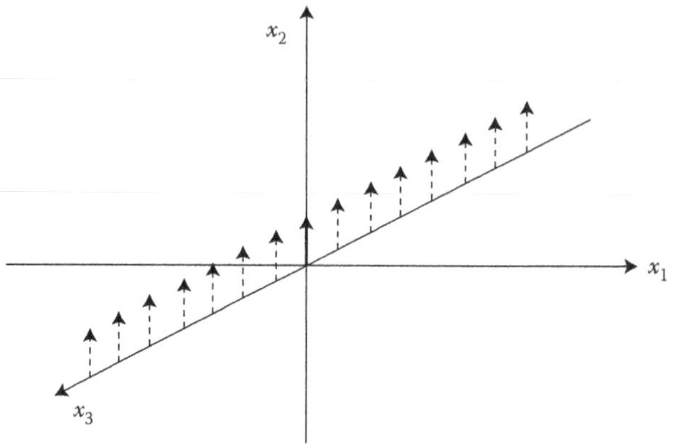

FIGURE 2.9 Dimension reduction of Kelvin's solution.

where the following equation

$$\nabla\left[\frac{(\mathbf{x}-\mathbf{x}')\cdot\mathbf{P}}{|\mathbf{x}-\mathbf{x}'|}\right]=\frac{\mathbf{P}}{|\mathbf{x}-\mathbf{x}'|}-\frac{\mathbf{n}}{|\mathbf{x}-\mathbf{x}'|^3}\mathbf{n}\cdot\mathbf{P} \tag{2.140}$$

is used to determine the displacement field of the body. Equation 2.139 can be rewritten in the index form as

$$u_i(\mathbf{x})=\frac{1}{16\pi\mu(1-\nu)}\left[(3-4\nu)\frac{\delta_{ij}}{|\mathbf{x}-\mathbf{x}'|}+\frac{\left(x_i-x_i'\right)\left(x_j-x_j'\right)}{|\mathbf{x}-\mathbf{x}'|^3}\right]P_j \tag{2.141}$$

In the infinite domain D, if body force $\mathbf{f}(\mathbf{x})$ is acting on a certain subdomain Ω, as shown in Figure 2.9, the elastic fields caused by the body force can be written by Kelvin's solution as

$$u_i(\mathbf{x})=\frac{1}{16\pi\mu(1-\nu)}\int_\Omega\left[(3-4\nu)\frac{\delta_{ij}}{|\mathbf{x}-\mathbf{x}'|}+\frac{\left(x_i-x_i'\right)\left(x_j-x_j'\right)}{|\mathbf{x}-\mathbf{x}'|^3}\right]f_j(\mathbf{x}')dV\left(\mathbf{x}'\right) \tag{2.142}$$

2.7.3 Elastic Green's Function

When a concentrated force $\mathbf{P}$ is applied at $\mathbf{x}'$ in a general elastic material, the governing equation reads

$$C_{ijkl}u_{k,li}=-P_j\delta(\mathbf{x}-\mathbf{x}') \tag{2.143}$$

The displacement at a field point $\mathbf{x}$ caused by the force at $\mathbf{x}'$ can be written in terms of the elastic Green's function as follows:

$$u_i=G_{ij}(\mathbf{x},\mathbf{x}')P_j \tag{2.144}$$

Substituting Equation 2.144 into 2.143, one can write

$$C_{ijkl}G_{km,li}P_m = -P_j\delta(\mathbf{x}-\mathbf{x}') = -\delta_{jm}P_m\delta(\mathbf{x}-\mathbf{x}') \tag{2.145}$$

Then,

$$C_{ijkl}G_{km,li} = -\delta_{jm}\delta(\mathbf{x}-\mathbf{x}') \tag{2.146}$$

For an isotropic material, the elastic Green's function can be obtained from Equation 2.141 as

$$G_{ij}(\mathbf{x},\mathbf{x}') = \frac{1}{16\pi\mu(1-\nu)}\left[(3-4\nu)\frac{\delta_{ij}}{|\mathbf{x}-\mathbf{x}'|} + \frac{\left(x_i - x_i'\right)\left(x_j - x_j'\right)}{|\mathbf{x}-\mathbf{x}'|^3}\right] \tag{2.147}$$

For the convenience of integral, another form of Green's function is also used

$$G_{ij}(\mathbf{x},\mathbf{x}') = \frac{1}{4\pi\mu}\frac{\delta_{ij}}{|\mathbf{x}-\mathbf{x}'|} - \frac{1}{16\pi\mu(1-\nu)}\frac{\partial^2\left|\mathbf{x}-\mathbf{x}'\right|}{\partial x_i \partial x_j} \tag{2.148}$$

Obviously, one can write

$$G_{ij}(\mathbf{x},\mathbf{x}') = G_{ij}\left(\mathbf{x}',\mathbf{x}\right) \tag{2.149}$$

Mathematically, $G_{ij}(\mathbf{x},\mathbf{x}')$ depends on the relative position of the field point from the source point, say $\mathbf{x}-\mathbf{x}'$. It is also written as $G_{ij}(\mathbf{x}-\mathbf{x}')$ in some cases.

For 2D case, consider a constant unit line force applied on x_3 axis, as shown in Figure 2.9, and follow the idea of Section 2.5.3.

Then, the fundamental solution of plane problem is written as

$$G_{ij}\left(\mathbf{x},\mathbf{x}'\right) = \int_{-\infty}^{+\infty} \frac{1}{16\pi\mu(1-\nu)}\left[(3-4\nu)\frac{\delta_{ij}}{\left(|\mathbf{x}-\mathbf{x}'|^2 + x_3'^2\right)^{\frac{1}{2}}} + \frac{\left(x_i - x_i'\right)\left(x_j - x_j'\right)}{\left(|\mathbf{x}-\mathbf{x}'|^2 + x_3'^2\right)^{\frac{3}{2}}}\right] dx_3' \tag{2.150}$$

where $\mathbf{x} = (x_1, x_2)$, and $i, j = 1, 2$.

Using the Hadamard finite-part integral of Equation 2.82,

$$\int_{-\infty}^{+\infty} \frac{1}{\sqrt{|\mathbf{x}-\mathbf{x}'|^2 + x_3'^2}} dx_3' = -2\ln\left|\mathbf{x}-\mathbf{x}'\right| \tag{2.151}$$

where r^* is defined in Equation 2.79. Notice that

$$\int_{-\infty}^{+\infty} \frac{1}{\left(|\mathbf{x}-\mathbf{x}'|^2 + x_3'^2\right)^{\frac{3}{2}}} dx_3' = \frac{2}{|\mathbf{x}-\mathbf{x}'|^2} \tag{2.152}$$

then Green's function of an elastic 2D problem can be written in an explicit form as

$$G_{ij}\left(\mathbf{x},\mathbf{x}'\right) = \frac{1}{8\pi\mu(1-\nu)}\left[(3-4\nu)\,\delta_{ij}\ln\left|\mathbf{x}-\mathbf{x}'\right| + \frac{\left(x_i-x_i'\right)\left(x_j-x_j'\right)}{|\mathbf{x}-\mathbf{x}'|^2}\right] \tag{2.153}$$

2.8 EXERCISES

1. Prove the following identity equations using the index notation:

$$\mathbf{a}\cdot(\mathbf{b}\times\mathbf{c}) = (\mathbf{a}\times\mathbf{b})\cdot\mathbf{c} \tag{2.154}$$

$$\mathbf{a}\times(\mathbf{b}\times\mathbf{c}) = (\mathbf{a}\cdot\mathbf{c})\,\mathbf{b} - (\mathbf{a}\cdot\mathbf{b})\,\mathbf{c} \tag{2.155}$$

$$(\mathbf{a}\times\mathbf{b})\cdot(\mathbf{c}\times\mathbf{d}) = \mathbf{a}\cdot[\mathbf{b}\times(\mathbf{c}\times\mathbf{d})] \tag{2.156}$$

2. For a rotational matrix $\mathbf{Q}$ between two Cartesian coordinates $\mathbf{x}$ and $\mathbf{x}'$ with a common origin, say $Q_{ij} = \cos(x_i, x_j')$. The angular relationship between any two Cartesian reference frames $\mathbf{x}$ and $\mathbf{x}'$ can be represented in terms of three Euler angles (ϕ,θ,ψ) as follows: (a) the first rotation is by an angle ϕ about the x_3 axis, (b) the second rotation is by an angle $\theta\in(0,\pi)$ about the x_1 axis, and (c) the third rotation is by an angle ψ about the x_3 axis again. Then, the transformation matrix can be written as

$$\mathbf{Q} = \begin{bmatrix} \cos\psi\cos\phi - \cos\theta\sin\phi\sin\psi & \cos\psi\sin\phi - \cos\theta\cos\phi\sin\psi & \sin\psi\sin\theta \\ -\sin\psi\cos\phi - \cos\theta\sin\phi\cos\psi & -\sin\psi\sin\phi + \cos\theta\cos\phi\cos\psi & \cos\psi\sin\theta \\ \sin\theta\sin\phi & -\sin\theta\cos\phi & \cos\theta \end{bmatrix}.$$

Prove

$$Q_{ik}Q_{jk} = \delta_{ij}$$

3. Prove the following tensors are isotropic for the corresponding tensor ranks:

$$A_{ij} = k\delta_{ij};\; A_{ijk} = ke_{ijk};\; A_{ijkl} = \alpha\delta_{ij}\delta_{kl} + \beta\delta_{ik}\delta_{jl} + \gamma\delta_{il}\delta_{jk}$$

4. Consider vectors $\mathbf{a} = (1,2,3)$ and $\mathbf{b} = (3,2,1)$. Determine the following quantities:

$$\mathbf{a}\cdot\mathbf{b}, \mathbf{a}\times\mathbf{b}, \mathbf{a}\otimes\mathbf{b} \tag{2.157}$$

Now, if the coordinate is rotated along the axis x_3 at 30° counterclockwise, that is, the x_1–x_2 plane is rotated 30° counterclockwise and the x_3 axis remains the same, recalculate the above quantities in the new coordinate system.

5. An orthogonal material has the stiffness tensor given in the form of $C_{ijkl} = C^1_{IK}\delta_{ij}\delta_{kl} + C^2_{IJ}\left(\delta_{ik}\delta_{jl} + \delta_{il}\delta_{jk}\right)$ with

$$\mathbf{C}^1 = \begin{bmatrix} 1 & 0.5 & 0.5 \\ 0.5 & 1.5 & 0.75 \\ 0.5 & 0.75 & 2 \end{bmatrix};\quad \mathbf{C}^2 = \begin{bmatrix} 0.5 & 0.25 & 0.25 \\ 0.25 & 0.75 & 0.5 \\ 0.25 & 0.5 & 1 \end{bmatrix}.$$

(a) Determine the components of the stiffness in the form of a 6 by 6 matrix.

(b) Calculate the components of the compliance.

(c) Write the compliance tensor in the form of $D_{ijkl} = D^1_{IK}\delta_{ij}\delta_{kl} + D^2_{IJ}\left(\delta_{ik}\delta_{jl} + \delta_{il}\delta_{jk}\right)$. Determine D^1_{IK} and D^2_{IJ} in terms of C^1_{IK} and C^2_{IJ}. Check whether the calculation in (b) is consistent with the formulation.

6. Prove the following identities:

$$\nabla \times (\nabla\phi) = \mathbf{0},$$
$$\nabla \cdot (\nabla \times \mathbf{a}) = 0,$$
$$\nabla \times (\nabla \times \mathbf{a}) = \nabla(\nabla \cdot \mathbf{a}) - \nabla^2\mathbf{a}$$

where ϕ is a scalar field and $\mathbf{a}$ a vector field.

7. Given a scalar field in a 3D infinite domain as $\psi = |\mathbf{x} - \mathbf{x}^0|$, which is a distance from $\mathbf{x}$ to a known point $\mathbf{x}^0$, derive $\nabla\nabla\psi$, which is a second-rank tensor. Similarly, for $\phi = 1/|\mathbf{x} - \mathbf{x}^0|$, derive $\nabla\nabla\phi$.

8. Given a vector field $\mathbf{a} = \mathbf{x}$ in a spherical domain: $x_1^2 + x_2^2 + x_3^2 \leq 1$, use Helmholtz's decomposition to write $\mathbf{a} = \nabla\varphi + \nabla \times \mathbf{b}$ in terms of a scalar potential φ and a vector potential $\mathbf{b}$. Find the specific form of $\mathbf{b}$ in terms of x_i.

9. Consider a scalar field $\varphi = x_1^2 + x_2^2 + x_3^2$ in a spherical domain $\Omega : x_1^2 + x_2^2 + x_3^2 \leq 1$. Confirm the following equations:

$$\int_\Omega \nabla\varphi dV = \int_{\partial\Omega} \mathbf{n}\varphi \, dS \tag{2.158}$$

$$\int_\omega \nabla^2\varphi dV = \int_{\partial\omega} n\nabla\varphi \, dS \tag{2.159}$$

10. Prove $G(\mathbf{x}, \mathbf{x}') = ln\frac{1}{|\mathbf{x}-\mathbf{x}'|}$ is Green's function for the 2D Laplace equation.

11. Derive Green's function for a cantilever beam bending.

12. Write the six equations implied by compatibility condition for strains $\nabla \times \boldsymbol{\epsilon} \times \nabla = \mathbf{0}$.

13. Solve the boundary value problem: For a solid elastic ball fully bound by an outer layer, the radius of the ball and outer surface radii are written as $R_i < R_o$, respectively. Young's moduli and Poisson's ratios of the ball and outer layer, are respectively, written as E_i and v_i and E_o and v_o. When a pressure P is applied on the outer surface, derive the displacement distribution along the radial direction in two materials.

14. The stress tensor in a semi-infinite domain $x_3 > 0$ is given by $\sigma_{ij} = (\alpha x_i x_j x_3)/|x|^5$, where $|x| = (x_i x_i)^{1/2}$ and α is a constant. Determine whether there are any body forces in the domain using the equilibrium equation.

15. In an infinite solid with material elastic constants μ and v, a concentrated force $P = (0, 0, 1)$ is applied at the origin. Calculate the surface force distribution on the sphere with radius 1 and the center at the origin. Integrate it over the surface for the resultant surface force.

CHAPTER 3

Spherical Inclusion and Inhomogeneity

3.1 SPHERICAL INCLUSION PROBLEM

Following Mura's definition [3], consider a spherical subdomain Ω in a homogeneous solid (matrix) as shown in Figure 3.1, where the subdomain is defined by the function $\Omega : x_1^2 + x_2^2 + x_3^2 \leq a^2$ with a being the radius of the sphere. The mechanical properties of the inclusion and the matrix are the same with elastic constants μ and v. An eigenstrain rate $\epsilon_{ij}^*(\mathbf{x})$ or a body force $f_i(\mathbf{x})$ is applied on Ω but is zero on D–Ω. The term of inclusion differentiates it from another counterpart—an inhomogeneity. An inhomogeneity is a subdomain in a homogeneous solid with different material properties from the matrix. In this section, we focus on the inclusion problem. The next section will address the inhomogeneity problem.

In the last chapter, when there is a distributed body force on the inclusion, the elastic field can be obtained by the integral of Kelvin's solution directly as

$$u_i(\mathbf{x}) = \int_\Omega G_{ij}(\mathbf{x}, \mathbf{x}')f_j(\mathbf{x}')d\mathbf{x}' \tag{3.1}$$

where

$$G_{ij}(\mathbf{x}, \mathbf{x}') = \frac{1}{4\pi\mu}\frac{\delta_{ij}}{|\mathbf{x} - \mathbf{x}'|} - \frac{1}{16\pi\mu(1 - \nu)}\frac{\partial^2|\mathbf{x} - \mathbf{x}'|}{\partial x_i \partial x_j} \tag{3.2}$$

When the body force is constant, Equation 3.1 can be directly written by using the identities of the integrals in Section 3.4.

If there is only a source of eigenstrain ϵ_{ij}^* applied on the subdomain Ω, the governing equation is

$$C_{ijkl}\left(\epsilon_{kl,i} - \epsilon_{kl,i}^*\right) = 0 \tag{3.3}$$

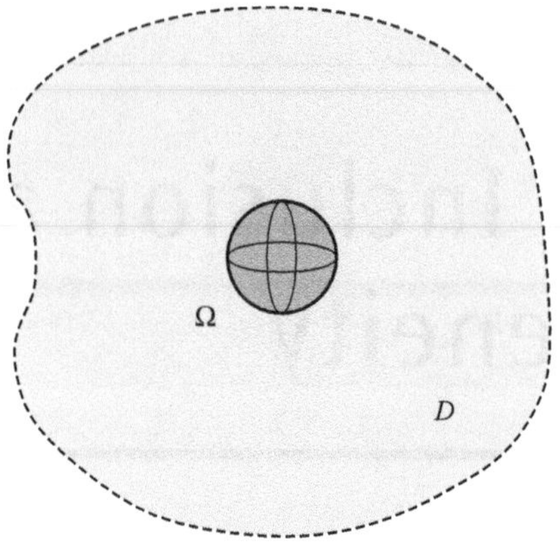

FIGURE 3.1 A spherical subdomain Ω embedded in the infinite domain D.

It can also be rewritten as

$$\mathcal{L}u_j = \frac{C_{ijkl}\epsilon^*_{kl,i}}{\mu} \tag{3.4}$$

where

$$\mathcal{L} = \nabla^2 + \frac{1}{1-2\nu}\nabla\nabla\cdot \tag{3.5}$$

According to Equation 3.4, the solution of displacement u_i can be solved in terms of Green's function as

$$\begin{aligned} u_i(\mathbf{x}) &= -\int_D \left[G_{ij}(\mathbf{x},\mathbf{x}')C_{mjkl}\frac{\partial \epsilon^*_{kl}(\mathbf{x}')}{\partial x'_m} \right] d\mathbf{x}' \\ &= -\int_D \left[\frac{\partial\left(G_{ij}C_{mjkl}\epsilon^*_{kl}\right)}{\partial x'_m} - \frac{\partial G_{ij}(\mathbf{x}')}{\partial x'_m}C_{mjkl}\epsilon^*_{kl} \right] d\mathbf{x}' \\ &= -\int_{\partial D} G_{ij}C_{mjkl}\epsilon^*_{kl}n'_m d\mathbf{x}' + \int_\Omega \frac{\partial G_{ij}}{\partial x'_m}C_{mjkl}\epsilon^*_{kl}d\mathbf{x}' \\ &= \int_\Omega \frac{\partial G_{ij}}{\partial x'_m}C_{mjkl}\epsilon^*_{kl}d\mathbf{x}' \end{aligned} \tag{3.6}$$

Using $\frac{\partial G_{ij}(x-x')}{\partial x_m} = -\frac{\partial G_{ij}(x-x')}{\partial x'_m}$, one can write

$$u_i(\mathbf{x}) = -\int_\Omega \frac{\partial G_{ij}}{\partial x_m}C_{mjkl}\epsilon^*_{kl}d\mathbf{x}' \tag{3.7}$$

Here the derivative of the Green's function is switched from variable $\mathbf{x}'$ in Equation 3.6 to $\mathbf{x}$ in Equation 3.7. It provides convenience and consistence for the derivation of strain and stress as they requires another derivative on $\mathbf{x}$ again. After the displacement field is obtained, the strain field can be solved as

$$\epsilon_{pq} = \frac{1}{2}\left(\frac{\partial u_p}{\partial x_q} + \frac{\partial u_q}{\partial x_p}\right) = -\frac{1}{2}\int_{\Omega} \left(G_{pj,mq} + G_{qj,mp}\right) C_{mjkl}\epsilon^*_{kl}d\mathbf{x}' \tag{3.8}$$

The function Γ' is used to simplify the combination of the derivatives of Green functions,

$$\Gamma'_{pqmj} = \frac{1}{2}\left(G_{pj,mq} + G_{qj,mp}\right) \tag{3.9}$$

Thus, Equation 3.8 could be rewritten as

$$\epsilon_{pq} = -\int_{\Omega} \Gamma'_{pqmj} C_{mjkl}\epsilon^*_{kl}d\mathbf{x}' \tag{3.10}$$

Due to the symmetry of the stiffness tensor $C_{mjnl} = C_{jmnl}$, one can write $\Gamma'_{pqmj}C_{mjnl} = \frac{1}{2}\left(\Gamma'_{pqmj}C_{mjnl} + \Gamma'_{pqjm}C_{mjnl}\right)$. Therefore, one can define a new tensor function

$$\Gamma_{pqmj} = \frac{1}{2}\left(\Gamma'_{pqmj} + \Gamma'_{pqjm}\right) = \frac{1}{4}\left(G_{pj,mq} + G_{qj,mp} + G_{pm,jq} + G_{qm,jp}\right) \tag{3.11}$$

where Γ is also called the modified Green's function for the eigenstrain problem, which satisfies both major symmetry and minor symmetry as follows:

$$\Gamma_{pqmj} = \Gamma_{mjqp} \text{ and } \Gamma_{pqmj} = \Gamma_{qpmj} = \Gamma_{pqjm} \tag{3.12}$$

Eventually, the strain fields could be obtained as

$$\epsilon_{ij} = -\int_{\Omega} \Gamma_{ijkl} C_{klmn}\epsilon^*_{mn}d\mathbf{x}' \tag{3.13}$$

When the eigenstrain is constant, the above equation can be directly written by using the identities of the integrals in Section 3.3 as

$$\epsilon_{ij} = -D_{ijkl}C_{klmn}\epsilon^*_{mn} \tag{3.14}$$

where D_{ijkl} can be explicitly written as

$$D_{ijkl} = \begin{cases} \dfrac{\rho^3}{60\mu\,(1-v)} \begin{bmatrix} \left(5-3\rho^2\right)\delta_{ij}\delta_{kl} - \left(5-10v+3\rho^2\right)\left(\delta_{ik}\delta_{jl}+\delta_{il}\delta_{jk}\right) \\ -15(1-\rho^2)(\delta_{ij}n_kn_l + \delta_{kl}n_in_j) \\ -15(v-\rho^2)(\delta_{ik}n_jn_l + \delta_{il}n_jn_k + \delta_{jk}n_in_l + \delta_{jl}n_in_k) \\ +15(5-7\rho^2)n_in_jn_kn_l \end{bmatrix} & \text{for } r > a \\ \dfrac{1}{30\mu\,(1-v)}\left[\delta_{ij}\delta_{kl} - (4-5v)\left(\delta_{ik}\delta_{jl}+\delta_{il}\delta_{jk}\right)\right] & \text{for } r \le a \end{cases} \tag{3.15}$$

where r, ρ, and n_i are defined as

$$\begin{cases} r = (x_i x_i)^{\frac{1}{2}} \\ \rho = \dfrac{a}{r} \\ n_i = \dfrac{x_i}{r} \end{cases} \tag{3.16}$$

Notice that the tensor $S_{ijmn} = -D_{ijkl}C_{klmn}$ for $r \le a$ is also called the Eshelby's tensor, which is a constant for any ellipsoidal domain.

Trick—How to determine the inverse of a general fourth-rank tensor D_{ijkl}?

Example: Fourth-rank tensor F_{ijkl} is written in the form of

$$F_{ijkl} = \begin{bmatrix} f_1(\delta_{ij}\delta_{kl}) + f_2(\delta_{ik}\delta_{jl} + \delta_{il}\delta_{jk}) \\ +f_3\left(\delta_{ij}n_k n_l + \delta_{kl}n_i n_j\right) \\ +f_4\left(\delta_{ik}n_j n_l + \delta_{il}n_j n_k + \delta_{jk}n_i n_l + \delta_{jl}n_i n_k\right) \\ f_5(n_i n_j n_k n_l) \end{bmatrix} \tag{3.17}$$

The inverse of F_{ijkl}, say G_{ijkl}, will have the same form and can be written as

$$G_{ijkl} = \begin{bmatrix} g_1\left(\delta_{ij}\delta_{kl}\right) + g_2\left(\delta_{ik}\delta_{jl} + \delta_{il}\delta_{jk}\right) \\ g_3\left(\delta_{ij}n_k n_l + \delta_{kl}n_i n_j\right) \\ g_4\left(\delta_{ik}n_j n_l + \delta_{il}n_j n_k + \delta_{jk}n_i n_l + \delta_{jl}n_i n_k\right) \\ g_5\left(n_i n_j n_k n_l\right) \end{bmatrix} \tag{3.18}$$

and

$$\mathbf{F} : \mathbf{G} = \mathbf{K} \tag{3.19}$$

where tensor K_{ijkl} is in the same form of F_{ijkl} and G_{ijkl}, and the coefficients are

$$\begin{aligned} k_1 &= k_3 = k_4 = k_5 = 0 \\ k_2 &= \frac{1}{2} \end{aligned} \tag{3.20}$$

then, the coefficients belonging to G_{ijkl} can be solved.

3.2 INTRODUCTION TO THE EQUIVALENT INCLUSION METHOD

Equivalent inclusion method simulates a material mismatch with a prescribed strain, namely eigenstrain, for the material subjected to a load. This section first uses a simple example to demonstrate the concept.

For a solid elastic ball fully bound to an outer layer, the radius of the ball and outer surface radii are written as $R_i < R_o$, respectively. The material constants of the ball and outer layer are, respectively, written as μ_1, λ_1 and μ_0, λ_0. When a uniform pressure P is applied on the

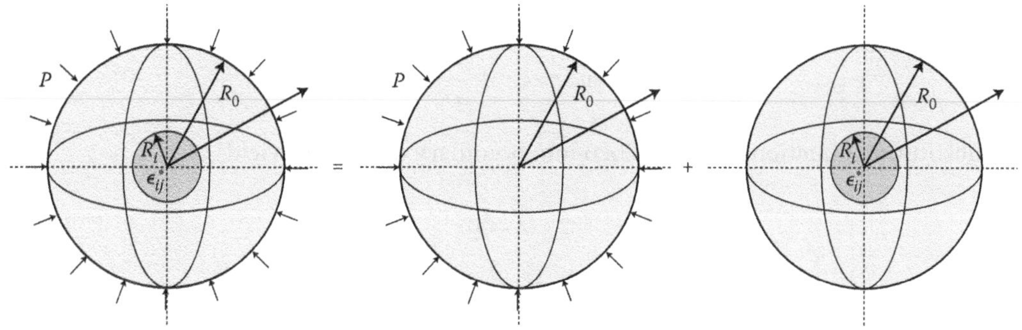

FIGURE 3.2 Layered-ball under uniform pressure P.

outer surface, the elastic fields can be solved through the equivalent inclusion method, as shown in Figure 3.2.

The ball is replaced by the outer layer material, so the material is uniform with an eigenstrain in the subdomain of the ball. The original problem could be the superposition of these two cases. First, the uniform material is subjected to a uniform pressure. The elastic field is obvious as

$$\begin{cases} \sigma_{ij} = -P\delta_{ij} \\ \epsilon_{ij} = -\dfrac{P}{3\lambda_0 + 2\mu_0}\delta_{ij} \\ u_r = -\dfrac{P}{3\lambda_0 + 2\mu_0}r \end{cases} \tag{3.21}$$

The second case, in the subdomain of the ball, that is, $r \leq R_i$, an eigenstrain of ϵ^* is included. Considering the spherical symmetry, the constitutive relation can be written as

$$\sigma_{rr} = \lambda_0 \left(u_{r,r} + 2\frac{u_r}{r} - 3\epsilon^* \right) + 2\mu_0 \left(u_{r,r} - \epsilon^* \right) \tag{3.22}$$

where $\epsilon^* = 0$ for $R_i \leq r \leq R_0$. The boundary and symmetric conditions read

$$\begin{aligned} &r = 0, \quad u_r = 0 \\ &r = R_0, \quad \sigma_{rr} = 0 \\ &r = R_i, \quad \sigma_{rr}^+ = \sigma_{rr}^-, \quad u^- = u^+ \end{aligned} \tag{3.23}$$

where "+" and "−" denote the inner and outer sides of the surface $r = R_i$. The equilibrium equation can be written in terms of the displacement field as

$$\frac{\partial^2 u_r}{\partial r^2} + \frac{2}{r}\frac{\partial u_r}{\partial r} - 2\frac{u_r}{r^2} = 0 \tag{3.24}$$

The general solution can be obtained as follows: For $r \leq R_i$,

$$u_r = Cr + \frac{D}{r^2} \tag{3.25}$$

For $R_i \leq r \leq R_o$,

$$u_r = \overline{C}r + \frac{\overline{D}}{r^2} \tag{3.26}$$

Substituting Equations 3.25 and 3.26 into boundary conditions yields,

$$D = 0 \tag{3.27}$$

$$(2\mu_0 + 3\lambda_0)\overline{C} - 4\mu_0\frac{\overline{D}}{R_o^3} = 0 \tag{3.28}$$

$$(2\mu_0 + 3\lambda_0)\,\overline{C} - 4\mu_0\frac{\overline{D}}{R_i^3} = (2\mu_0 + 3\lambda_0)\left(C - \epsilon^*\right) \tag{3.29}$$

$$CR_i = \overline{C}R_i + \frac{\overline{D}}{R_i^2} \tag{3.30}$$

Then, the combination of the last two equations yields

$$\overline{D} = \frac{2\mu_0 + 3\lambda_0}{\lambda_0 + 2\mu_0}\frac{R_i^3}{3}\epsilon^* \tag{3.31}$$

Then, one can write,

$$\overline{C} = \frac{4\mu_0}{\lambda_0 + 2\mu_0}\frac{R_i^3}{3R_o^3}\epsilon^* \tag{3.32}$$

Therefore, one can obtain

$$C = \left(\frac{4\mu_0}{\lambda_0 + 2\mu_0}\frac{R_i^3}{R_o^3} + \frac{2\mu_0 + 3\lambda_0}{\lambda_0 + 2\mu_0}\right)\frac{1}{3}\epsilon^* \tag{3.33}$$

The superposition of the two cases should make the stress equivalent over the ball. The total strain over the ball $r \leq R_i$ can be written as

$$\epsilon_{rr} = \epsilon_{\theta\theta} = \epsilon_{\varphi\varphi} = -\frac{P}{3\lambda_0 + 2\mu_0} + \left(\frac{4\mu_0}{\lambda_0 + 2\mu_0}\frac{R_i^3}{R_o^3} + \frac{2\mu_0 + 3\lambda_0}{\lambda_0 + 2\mu_0}\right)\frac{1}{3}\epsilon^*$$

The total stress for the original problem will be

$$\sigma_{rr} = (3\lambda_1 + 2\mu_1)\left[-\frac{P}{3\lambda_0 + 2\mu_0} + \left(\frac{4\mu_0}{\lambda_0 + 2\mu_0}\frac{R_i^3}{R_o^3} + \frac{2\mu_0 + 3\lambda_0}{\lambda_0 + 2\mu_0}\right)\frac{1}{3}\epsilon^*\right]$$

The combination of the two problem is

$$\sigma_{rr} = (3\lambda_0 + 2\mu_0)\left[-\frac{P}{3\lambda_0 + 2\mu_0} + \left(\frac{4\mu_0}{\lambda_0 + 2\mu_0}\frac{R_i^3}{R_o^3} + \frac{2\mu_0 + 3\lambda_0}{\lambda_0 + 2\mu_0}\right)\frac{1}{3}\epsilon^* - \epsilon^*\right]$$

Based on the equivalent inclusion condition, the above two equations should be equal, so that

$$-\frac{P}{3\lambda_0+2\mu_0}+\left(\frac{4\mu_0}{\lambda_0+2\mu_0}\frac{R_i^3}{R_o^3}+\frac{2\mu_0+3\lambda_0}{\lambda_0+2\mu_0}\right)\frac{1}{3}\epsilon^*=-\frac{3\lambda_0+2\mu_0}{3\Delta\lambda+2\Delta\mu}\epsilon^*$$

where $\Delta\lambda=\lambda_1-\lambda_0$ and $\Delta\mu=\mu_1-\mu_0$. The above equation can be rewritten as

$$\epsilon^*=\frac{1}{M}\frac{P}{3\lambda_0+2\mu_0}$$

where

$$M=\frac{1}{3}\left(\frac{4\mu_0}{\lambda_0+2\mu_0}\frac{R_i^3}{R_o^3}+\frac{2\mu_0+3\lambda_0}{\lambda_0+2\mu_0}\right)+\frac{3\lambda_0+2\mu_0}{3\Delta\lambda+2\Delta\mu}$$

Then, one can explicitly write the displacement field in terms of the following parameters:

$$\overline{D}=\frac{2\mu_0+3\lambda_0}{\lambda_0+2\mu_0}\frac{R_i^3}{3}\frac{1}{M}\frac{P}{3\lambda_0+2\mu_0} \tag{3.34}$$

Then, one can write

$$\overline{C}=\frac{4\mu_0}{\lambda_0+2\mu_0}\frac{R_i^3}{3R_o^3}\frac{1}{M}\frac{P}{3\lambda_0+2\mu_0} \tag{3.35}$$

Therefore, one can obtain

$$C=\left(\frac{4\mu_0}{\lambda_0+2\mu_0}\frac{R_i^3}{R_o^3}+\frac{2\mu_0+3\lambda_0}{\lambda_0+2\mu_0}\right)\frac{1}{3}\frac{1}{M}\frac{P}{3\lambda_0+2\mu_0} \tag{3.36}$$

From the above method, one can see that the eigenstrain is proportional to the load.

3.3 SPHERICAL INHOMOGENEITY PROBLEM

For the interior and exterior tensor D, the strain can be determined as

$$\epsilon_{ij}(x)=-D_{ijkl}(\mathbf{x})C_{klmn}\epsilon^*_{mn} \tag{3.37}$$

When $\mathbf{x}\in\Omega$, Equation 3.37 could be rewritten as

$$\epsilon'_{ij}(x)=-S_{ijmn}\epsilon^*_{mn} \tag{3.38}$$

where S_{ijkl} is the well-known Eshelby's tensor and can be written as

$$S_{ijmn}=D_{ijkl}C_{klmn}=\frac{5\nu-1}{15(1-\nu)}\delta_{ij}\delta_{mn}+\frac{4-5\nu}{15(1-\nu)}\left(\delta_{im}\delta_{jn}+\delta_{in}\delta_{jm}\right) \tag{3.39}$$

3.3.1 Eshelby's Equivalent Inclusion Method

As described before, an inclusion is subjected to an internal stress field (eigenstress) due to an eigenstrain even if the material is free of external stress. However, an inhomogeneity has different material properties from the material. It is free from any stress field unless an external load is applied. This section demonstrates Eshelby's equivalent inclusion method for an inhomogeneity Ω with stiffness C^*_{ijkl} embedded in an infinitely extended domain D with stiffness C_{ijkl}, which is subjected to a far-field stress σ^0_{ij}. Therefore, the corresponding strain ϵ^0_{ij} is induced in the far field.

In the neighborhood of the inhomogeneity, the stress field is not uniform anymore. A disturbed elastic field is assumed by material mismatch written as $\mathbf{u}'$, $\boldsymbol{\epsilon}'$, and $\boldsymbol{\sigma}'$.

The equivalent inclusion method can be illustrated as follows:

For the disturbed field u'_i, ϵ'_{ij}, and σ'_{ij} caused by the material mismatch around inhomogeneity Ω (left side in Figure 3.3), Hooke's law becomes

$$\begin{cases} \sigma^0_{ij} + \sigma'_{ij} = C^*_{ijkl}\left(\epsilon^0_{kl} + u'_{k,l}\right) & x \in \Omega \\ \sigma^0_{ij} + \sigma'_{ij} = C_{ijkl}\left(\epsilon^0_{kl} + u'_{k,l}\right) & x \in D - \Omega \end{cases} \tag{3.40}$$

Furthermore, the displacement field could also be explicitly written as

$$\tilde{u}_k = \frac{N_{ki}}{D} X_i \tag{3.41}$$

where

$$D = Det\,[\mathbf{K}] = e_{mnl} K_{m1} K_{n2} K_{l3} \tag{3.42}$$

and N_{ij} is the co-factors of the matrix K.

Consider an eigenstrain in inclusion Ω (right side in Figure 3.3), Hooke's law reads

$$\begin{cases} \sigma^0_{ij} + \sigma'_{ij} = C_{ijkl}\left(\epsilon^0_{kl} + u'_{k,l} - \epsilon^*_{kl}\right) & x \in \Omega \\ \sigma^0_{ij} + \sigma'_{ij} = C_{ijkl}\left(\epsilon^0_{kl} + u'_{k,l}\right) & x \in D - \Omega \end{cases} \tag{3.43}$$

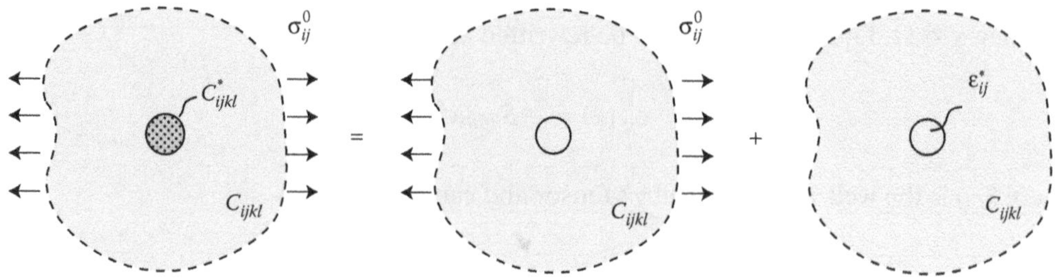

FIGURE 3.3 Eshelby's equivalent inclusion method.

The disturbed fields caused by $\boldsymbol{\epsilon}^*$ is determined as

$$\begin{cases} u'_k = -\int \dfrac{\partial G_{ki}}{\partial x_j} C_{ijmn}\epsilon^*_{mn} dV(x') \\ \epsilon'_{ij} = -D_{ijkl}C_{klmn}\epsilon^*_{mn} \end{cases} \tag{3.44}$$

The necessary and sufficient condition for the two problems in Figure 3.3 being equivalent is the equivalency of ϵ'_{ij} and σ'_{ij}

$$C^*_{ijkl}\left(\epsilon^0_{kl} + u'_{k,l}\right) = C_{ijkl}\left(\epsilon^0_{kl} + u'_{k,l} - \epsilon^*_{kl}\right) \tag{3.45}$$

or

$$C^*_{ijkl}\left(\epsilon^0_{kl} + \epsilon'_{kl}\right) = C_{ijkl}\left(\epsilon^0_{kl} + \epsilon'_{kl} - \epsilon^*_{kl}\right) \tag{3.46}$$

Thus,

$$\boldsymbol{\Delta}\mathbf{C} : \left(\boldsymbol{\epsilon}^0 + \epsilon'\right) = -\mathbf{C} : \epsilon^* \tag{3.47}$$

where

$$\boldsymbol{\Delta}\mathbf{C} = \mathbf{C}^* - \mathbf{C} \tag{3.48}$$

From Equation 3.44, the disturbed strain field $\boldsymbol{\epsilon}'$ in the inclusion is known

$$\epsilon' = -\mathbf{D}^{\Omega} : \mathbf{C} : \epsilon^* \tag{3.49}$$

where $\mathbf{D}^{\Omega}$ is the $\mathbf{D}(\mathbf{x})$ for $\mathbf{x} \in \Omega$. Multiply $\boldsymbol{\Delta}\mathbf{C}^{-1}$ on both sides and substitute Equation 3.49 into Equation 3.47, the eigenstrain can be determined as

$$\epsilon^* = \mathbf{C}^{-1} : \left(\mathbf{D}^{\Omega} - \Delta\mathbf{C}^{-1}\right)^{-1} : \epsilon^0 \tag{3.50}$$

Thus, the disturbed strain field ϵ' becomes

$$\epsilon' = -\mathbf{D}(\mathbf{x}) : \left(\mathbf{D}^{\Omega} - \Delta\mathbf{C}^{-1}\right)^{-\mathbf{1}} : \epsilon^{\mathbf{0}} \tag{3.51}$$

Therefore, all elastic fields can be written in terms of ϵ^0.

3.3.2 General Cases of Inhomogeneity with a Prescribed Eigenstrain

For the inhomogeneity already has a prescribed eigenstrain ϵ^P, such as thermal strain, the material is still subjected to a far-field strain. The disturbed elastic fields u'_i, ϵ'_{ij}, and σ'_{ij} are caused by the material mismatch around inhomogeneity Ω (left side in Figure 3.3) and the eigenstrain ϵ^p. Hooke's law becomes

$$\begin{cases} \sigma_{ij}^0 + \sigma_{ij}' = C_{ijkl}^* \left(\epsilon_{kl}^0 + u_{k,l}' - \epsilon_{kl}^p \right) & x \in \Omega \\ \sigma_{ij}^0 + \sigma_{ij}' = C_{ijkl} \left(\epsilon_{kl}^0 + u_{k,l}' \right) & x \in D - \Omega \end{cases} \tag{3.52}$$

Consider an eigenstrain in inclusion Ω (right side in Figure 3.3). Hooke's law reads

$$\begin{cases} \sigma_{ij}^0 + \sigma_{ij}' = C_{ijkl} \left(\epsilon_{kl}^0 + u_{k,l}' - \epsilon_{kl}^* - \epsilon_{kl}^p \right) & x \in \Omega \\ \sigma_{ij}^0 + \sigma_{ij}' = C_{ijkl} \left(\epsilon_{kl}^0 + u_{k,l}' \right) & x \in D - \Omega \end{cases} \tag{3.53}$$

where the total eigenstrain can be written as

$$\epsilon_{kl}^{**} = \epsilon_{kl}^* + \epsilon_{kl}^p \tag{3.54}$$

The disturbed fields caused by ϵ^{**} is determined as

$$\begin{cases} u_k' &= -\int\limits_{\Omega} \frac{\partial G_{ki}}{\partial x_j} C_{ijmn} \epsilon_{mn}^{**} dV(x') \\ \epsilon_{ij}' &= -D_{ijkl} C_{klmn} \epsilon_{mn}^{**} \end{cases} \tag{3.55}$$

The necessary and sufficient condition for the two problems in Figure 3.3 being equivalent is the equivalency of ϵ_{ij}' and σ_{ij}'

$$C_{ijkl}^* \left(\epsilon_{kl}^0 + \epsilon_{kl}' - \epsilon_{kl}^p \right) = C_{ijkl} \left(\epsilon_{kl}^0 + \epsilon_{kl}' - \epsilon_{kl}^{**} \right) \tag{3.56}$$

Thus,

$$\mathbf{\Delta C} : \left(\epsilon^0 + \epsilon' \right) = -\mathbf{C} : \epsilon^{**} + \mathbf{C}^* : \epsilon^p \tag{3.57}$$

From Equation 3.55, the disturbed strain field ϵ' in the inclusion is known

$$\epsilon' = -\mathbf{D}^{\Omega} : \mathbf{C} : \epsilon^{**} \tag{3.58}$$

Multiply $\mathbf{\Delta C}^{-1}$ on both sides and substitute Equation 3.58 into Equation 3.57, the total eigenstrain can be determined as

$$\epsilon^{**} = \mathbf{C}^{-1} : \left(\mathbf{D}^{\Omega} - \mathbf{\Delta C}^{-1} \right)^{-1} : \left[\epsilon^0 - \epsilon^p - \mathbf{\Delta C}^{-1} : \mathbf{C} : \epsilon^p \right] \tag{3.59}$$

Thus, the disturbed strain field ϵ' becomes

$$\epsilon' = -\mathbf{D}(\mathbf{x}) : \left(\mathbf{D}^{\Omega} - \mathbf{\Delta C}^{-1} \right)^{-1} : \left[\epsilon^0 - \epsilon^p - \mathbf{\Delta C}^{-1} : \mathbf{C} : \epsilon^p \right] \tag{3.60}$$

3.3.3 Interface Condition and the Uniqueness of the Solution

For $x \in \partial\Omega$, the continuity of displacement and normal stress are described as

$$\begin{cases} u_i^+ = u_i^- \\ n_i\sigma_{ij}^+ = n_i\sigma_{ij}^- \end{cases} \tag{3.61}$$

The displacement field is obtained as

$$u_k = u_k^0 - \int_\Omega G_{ki,j} C_{ijmn} \epsilon_{mn}^{**} dV(x') \tag{3.62}$$

where

$$\begin{cases} u_k^0 & \text{is continuous} \\ \int_\Omega G_{ki,j} dV(x') & \text{is continuous} \end{cases} \tag{3.63}$$

Thus, $u_i^+ = u_i^-$ satisfies. As for the continuity of stress field, the interior and exterior stress fields are

$$\begin{cases} \sigma^+ = \mathbf{C} : \left(\epsilon^0 + \epsilon'\right) \\ \sigma^- = \mathbf{C} : \left(\epsilon^0 + \epsilon' - \epsilon^{**}\right) \end{cases} \tag{3.64}$$

Then

$$\sigma^+ - \sigma^- = \mathbf{C} : \left(\mathbf{D}^\Omega - \mathbf{D}\right) : \mathbf{C} : \epsilon^{**} + \mathbf{C} : \epsilon^{**} \tag{3.65}$$

Besides, the corresponding disturbed strain fields could be obtained as

$$\begin{cases} \epsilon' = -\mathbf{D}^+ : \mathbf{C} : \epsilon^{**}, & x \in D - \Omega \\ \epsilon' = -\mathbf{D}^\Omega : \mathbf{C} : \epsilon^{**}, & \mathbf{x} \in \boldsymbol{\Omega} \end{cases} \tag{3.66}$$

By substituting Equation 3.66, Equation 3.65 becomes

$$\begin{cases} \sigma^+ = \mathbf{C}\left(\epsilon^0 - \mathbf{D}\right) : \mathbf{C} : \epsilon^{**} \\ \sigma^- = \mathbf{C}\left(\epsilon^0 - \mathbf{D}^\Omega : \mathbf{C} : \epsilon^{**} - \epsilon^{**}\right) \end{cases} \tag{3.67}$$

$$\sigma^+ - \sigma^- = \mathbf{C} : [(\mathbf{D}^\Omega - \mathbf{D}^+) : \mathbf{C} + \mathbf{I}] : \epsilon^{**}$$

$$= \mu \begin{bmatrix} \frac{2\nu}{1-\nu}\delta_{ij}\delta_{kl} + \left(\delta_{ik}\delta_{jl} + \delta_{il}\delta_{jk}\right) \\ -\frac{2\nu}{1-\nu}\left(\delta_{ij}n_k n_l + \delta_{kl}n_i n_j\right) \\ -\left(\delta_{ik}n_j n_l + \delta_{il}n_j n_k + \delta_{jk}n_i n_l + \delta_{jl}n_i n_k\right) \\ +\frac{2}{1-\nu}n_i n_j n_k n_l \end{bmatrix} \epsilon_{kl}^{**} \tag{3.68}$$

Thus,

$$n_i\left(\sigma_{ij}^{+}-\sigma_{ij}^{-}\right)=\mu\left[\begin{array}{c}\frac{2\nu}{1-\nu}n_j\delta_{kl}+\left(\delta_{jl}n_k+\delta_{jk}n_l\right)\\ -\frac{2\nu}{1-\nu}\left(n_jn_kn_l+\delta_{kl}n_j\right)\\ -\left(2n_jn_kn_l+\delta_{jk}n_l+\delta_{jl}n_k\right)\\ +\frac{2}{1-\nu}n_jn_kn_l\end{array}\right]\epsilon_{kl}^{**}$$
$$=0 \tag{3.69}$$

Therefore, the interfacial continuity conditions of both the displacement and the surface stress vector are satisfied. Based on the uniqueness of the boundary value problem, the present solution is the exact solution.

3.4 INTEGRALS OF ϕ, ψ, ϕ_P, ψ_P AND THEIR DERIVATIVES IN 3D DOMAIN

Given the following functions

$$\phi(\mathbf{x},\mathbf{x}')=\frac{1}{|\mathbf{x}-\mathbf{x}'|},\quad \psi(\mathbf{x},\mathbf{x}')=|\mathbf{x}-\mathbf{x}'|,$$

$$\phi_p(\mathbf{x},\mathbf{x}')=x_p'/|\mathbf{x}-\mathbf{x}'|,\quad \psi_p(\mathbf{x},\mathbf{x}')=x_p'|\mathbf{x}-\mathbf{x}'|$$

for any field point $\mathbf{x}$, the integral of the functions by $\mathbf{x}'$ over the spherical domain with the center at the origin can be explicitly derived, which is written as Φ, Ψ, Φ_p, and Ψ_p as follows:

$$\Phi=\begin{cases}\frac{4\pi}{3}\rho a^2 & \text{for } x>a\\ 2\pi\left(a^2-\frac{1}{3}x^2\right) & \text{for } x\le a\end{cases} \tag{3.70}$$

$$\Phi_{,i}=\begin{cases}-\frac{4\pi}{3}\rho^2 a n_i & \text{for } x>a\\ -\frac{4\pi}{3}x_i & \text{for } x\le a\end{cases} \tag{3.71}$$

$$\Phi_{,ij}=\begin{cases}-\frac{4\pi}{3}\rho^3\left(-\delta_{ij}+3n_in_j\right) & \text{for } x>a\\ -\frac{4\pi}{3}\delta_{ij} & \text{for } x\le a\end{cases} \tag{3.72}$$

$$\Phi_{p,i}=\begin{cases}-\frac{4\pi}{15}\rho^3a^2\left(-\delta_{ip}+3n_in_p\right) & \text{for } x>a\\ -\frac{2\pi}{15}\left[-\left(5a^2-3x^2\right)\delta_{ip}+6x_ix_p\right] & \text{for } x\le a\end{cases} \tag{3.73}$$

$$\Phi_{p,ij} = \begin{cases} \dfrac{4\pi}{5}\rho^4 a\left(-\delta_{ij}n_p - \delta_{ip}n_j - \delta_{jp}n_i + 5n_i n_j n_p\right) & \text{for } x > a \\ \dfrac{4\pi}{5}\left[-\delta_{ij}x_p - \delta_{ip}x_j - \delta_{jp}x_i\right] & \text{for } x \leq a \end{cases} \tag{3.74}$$

$$\Psi_{,ij} = \begin{cases} \dfrac{4\pi}{15}\rho a^2\left[\left(5-\rho^2\right)\delta_{ij} + \left(-5+3\rho^2\right)n_i n_j\right] & \text{for } x > a \\ \dfrac{4\pi}{15}\left[\left(5a^2 - x^2\right)\delta_{ij} - 2x_i x_j\right] & \text{for } x \leq a \end{cases} \tag{3.75}$$

$$\Psi_{,ijk} = \begin{cases} \dfrac{4\pi}{15}\rho^2 a\begin{bmatrix} -\left(5-3\rho^2\right)\left(\delta_{ij}n_k + \delta_{ik}n_j + \delta_{jk}n_i\right) \\ +15\left(1-\rho^2\right)n_i n_j n_k \end{bmatrix} & \text{for } x > a \\ -\dfrac{8\pi}{15}\left(\delta_{ij}x_k + \delta_{ik}x_j + \delta_{jk}x_i\right) & \text{for } x \leq a \end{cases} \tag{3.76}$$

$$\Psi_{p,ijk} = \begin{cases} \dfrac{4\pi}{105}\rho^3 a^2\begin{bmatrix} \left(7-3\rho^2\right)\left(\delta_{ij}\delta_{kp} + \delta_{ik}\delta_{jp} + \delta_{jk}\delta_{ip}\right) \\ -3\left(7-5\rho^2\right)\left(\delta_{ij}n_k n_p + \delta_{kp}n_i n_j + \delta_{ik}n_j n_p\right) \\ -3\left(7-5\rho^2\right)\left(\delta_{ip}n_j n_k + \delta_{jk}n_i n_p + \delta_{jp}n_i n_k\right) \\ +105\left(1-\rho^2\right)n_i n_j n_k n_p \end{bmatrix} & \text{for } x > a \\ \dfrac{4\pi}{105}\begin{bmatrix} \left(7a^2 - 3x^2\right)\left(\delta_{ij}\delta_{kp} + \delta_{ik}\delta_{jp} + \delta_{jk}\delta_{ip}\right) \\ -6\left(\delta_{ij}x_k x_p + \delta_{kp}x_i x_j + \delta_{ik}x_j x_p\right) \\ -6\left(\delta_{ip}x_j x_k + \delta_{jk}x_i x_p + \delta_{jp}x_i x_k\right) \end{bmatrix} & \text{for } x \leq a \end{cases} \tag{3.77}$$

$$\Psi_{p,ijkl} = \begin{cases} \dfrac{4\pi}{35}\rho^4 a\begin{bmatrix} -\left(7-5\rho^2\right)\left(\delta_{ij}\delta_{kl} + \delta_{ik}\delta_{jl} + \delta_{il}\delta_{jk}\right)n_p \\ -\left(7-5\rho^2\right)\left(\delta_{jp}\delta_{kl} + \delta_{kp}\delta_{jl} + \delta_{lp}\delta_{jk}\right)n_i \\ -\left(7-5\rho^2\right)\left(\delta_{ip}\delta_{kl} + \delta_{ik}\delta_{lp} + \delta_{il}\delta_{kp}\right)n_j \\ -\left(7-5\rho^2\right)\left(\delta_{ij}\delta_{lp} + \delta_{ip}\delta_{jl} + \delta_{il}\delta_{jp}\right)n_k \\ -\left(7-5\rho^2\right)\left(\delta_{ij}\delta_{kp} + \delta_{ik}\delta_{jp} + \delta_{ip}\delta_{jk}\right)n_l \\ +35\left(1-\rho^2\right)\left(\delta_{ij}n_k n_l n_p + \delta_{ik}n_j n_l n_p + \delta_{il}n_k n_k n_p\right) \\ +35\left(1-\rho^2\right)\left(\delta_{ip}n_j n_k n_l + \delta_{jk}n_i n_l n_p + \delta_{jl}n_i n_k n_p\right) \\ +35\left(1-\rho^2\right)\left(\delta_{jp}n_i n_k n_l + \delta_{kl}n_i n_j n_p + \delta_{kp}n_i n_j n_l\right) \\ +35\left(1-\rho^2\right)\left(\delta_{lp}n_i n_j n_k\right) - 35\left(7-9\rho^2\right)n_i n_j n_k n_l n_p \end{bmatrix} & \text{for } x > a \\ -\dfrac{8\pi}{35}\begin{bmatrix} \left(\delta_{ij}\delta_{kl} + \delta_{ik}\delta_{jl} + \delta_{il}\delta_{jk}\right)x_p \\ +\left(\delta_{jp}\delta_{kl} + \delta_{kp}\delta_{jl} + \delta_{lp}\delta_{jk}\right)x_i \\ +\left(\delta_{ip}\delta_{kl} + \delta_{ik}\delta_{lp} + \delta_{il}\delta_{kp}\right)x_j \\ \left(\delta_{ij}\delta_{lp} + \delta_{ip}\delta_{jl} + \delta_{il}\delta_{jp}\right)x_k \\ \left(\delta_{ij}\delta_{kp} + \delta_{ik}\delta_{jp} + \delta_{ip}\delta_{jk}\right)x_l \end{bmatrix} & \text{for } x \leq a \end{cases} \tag{3.78}$$

where r, ρ, and n_i are defined in Equation 3.16.

3.5 EXERCISES

1. In a homogeneous infinite domain with uniform Young's modulus E and Poisson's ratio υ, a body force b acts on a spherical domain $\Omega : x_1^2 + x_2^2 + x_3^2 \leq 1$. Derive the stress and strain fields caused by the body force.

2. In an infinite domain, a uniform eigenstrain $\varepsilon_{ij}^* = \alpha\delta_{ij}$ is distributed on a spherical domain ω. For isotropic materials with μ and ν, use Green's function technique to re-derive the displacement, stress, and strain fields. Check it by directly solving the boundary value problem using the spherical coordinate.

3. For a general eigenstrain ϵ_{ij}^*, the strain field provided in the above problem can be discontinuous along $\partial\omega$. Consider ω is a unit sphere with the center at the origin and radius 1. Calculate the stress components at the boundary point $(1/\sqrt{2}, 1/\sqrt{2}, 0)$ at the inside $\partial\omega^-$ and outside $\partial\omega^+$. Calculate the stress vector along the inside and outside of $\partial\omega$ at this point.

4. Consider a copper water pipe with an outer diameter 2.5 cm and thickness 2.5 mm. Assume when temperature $T = 0$, the stress is zero in the pipe while the water becomes ice. When the temperature drops to −25°C, calculate the stresses in the ice and the pipe along the radial direction. Here, material constants are given as follows:

	Young's Modulus (GPa)	**Poisson's Ratio**	**Coefficient of Thermal Expansion**
Copper	125	0.33	17×10^{-6}/°C
Ice	9.5	0.33	-50×10^{-6}/°C

5. One spherical air void with a radius a is embedded in a copper solid with material constants given as μ and ν. When the solid is subjected to a uniform stress σ_{11}^0 in the far field, the stress in the neighborhood of the air void is not uniform any more. Derive the stress field in the solid and calculate the stress concentration factor.

CHAPTER 4

Ellipsoidal Inclusion and Inhomogeneity

Chapter 3 introduces the spherical inclusion and inhomogeneity problems and provides an explicit close form solution for the elastic fields. Eshelby's original work focused on a general problem for ellipsoidal inclusion and inhomogeneity, which includes the solution in Chapter 3 as a special case. However, the general solution is given in terms of elliptic integrals, which is not straightforward for direct use. Therefore, this chapter will emphasize on the general methods toward the solution so that the readers can start with them for specific research problems and needs. First, the elastic Green's function for general anisotropic elastic materials is derived in the Fourier integral form. For isotropic materials, it recovers the previous solution. Using it, the elastic solutions for ellipsoidal inclusion problems with a polynomial eigenstrain and body force are provided. This method has been extended to other sources in potential flow problems. The solution is used in solving the inhomogeneity problems for multiple ellipsoidal particles embedded in the infinite domain, so particle interactions can be considered. The interface continuity are discussed.

4.1 GENERAL ELASTIC SOLUTION CAUSED BY AN EIGENSTRAIN THROUGH FOURIER INTEGRAL

4.1.1 An Eigenstrain in the Form of a Single Wave

For an elastic solid with an eigenstrain, the equilibrium equation in the absence of the body force can be written as

$$C_{ijkl}\left(u_{k,li}-\epsilon^*_{kl,i}\right)=\mathbf{0} \tag{4.1}$$

One can obtain

$$C_{ijkl}u_{k,li}=C_{ijkl}\epsilon^*_{kl,i} \tag{4.2}$$

If eigenstrain ϵ^* is given in a single wave function

$$\epsilon^*_{ij}=\tilde{\epsilon}^*_{ij}\exp\left(i\boldsymbol{\xi}\cdot\mathbf{x}\right) \tag{4.3}$$

where $\boldsymbol{\xi}$ is the wave vector with period $\frac{2\pi}{|\xi|}$, $i = \sqrt{-1}$, and $\tilde{\epsilon}^*_{ij}$ is the amplitude of the wave function. Considering the periodicity of the elastic field, the displacement can also be written in the same form

$$u_i = \tilde{u}_i \exp(i\boldsymbol{\xi} \cdot \mathbf{x}) \tag{4.4}$$

Substituting Equation 4.4 into Equation 4.2,

$$C_{ijkl}\tilde{u}_k\xi_l\xi_i \exp(i\boldsymbol{\xi} \cdot \mathbf{x}) = -C_{ijkl}\tilde{\epsilon}^*_{kl}(i\xi_i) \exp(i\boldsymbol{\xi} \cdot \mathbf{x})$$

that is,

$$C_{ijkl}\tilde{u}_k\xi_l\xi_i = -iC_{ijkl}\tilde{\epsilon}^*_{kl}\xi_i \tag{4.5}$$

Here, two tensors are introduced.

$$\begin{cases} K_{jk} = C_{ijkl}\xi_l\xi_i \\ X_j = -iC_{ijkl}\tilde{\epsilon}^*_{kl}\xi_i \end{cases} \tag{4.6}$$

By the substitution of Equation 4.6, Equation 4.5 could be rewritten as

$$K_{jk}\tilde{u}_k = X_j \tag{4.7}$$

or

$$[\tilde{\mathbf{u}}] = [\mathbf{K}]^{-1}[\mathbf{X}] \tag{4.8}$$

Substituting Equation 4.8 into Equation 4.4, we have

$$u_i = \frac{N_{ij}X_j}{D} \exp(i\boldsymbol{\xi} \cdot \mathbf{x}) \tag{4.9}$$

where N_{ij} are the cofactors of matrix $\mathbf{K}$, and D is the determinant of $\mathbf{K}$.

$$N_{ij} = \frac{1}{2}\left(e_{ikl}e_{jmn}K_{km}K_{ln}\right) \tag{4.10}$$

Using Equation 4.6, Equation 4.9 can be rewritten as

$$u_i = -iC_{jlmn}\tilde{\epsilon}^*_{mn}\xi_l N_{ij} D^{-1} \exp(i\boldsymbol{\xi} \cdot \mathbf{x})$$

Once the displacement field is obtained, the strain and stress field can be written through the strain–displacement relation and constitutive relation.

4.1.2 An Eigenstrain in the Form of Fourier Series and Fourier Integral

If the eigenstrain field is written in the Fourier series form as below

$$\epsilon^*_{ij}(\mathbf{x}) = \sum_{k=1}^{N} \tilde{\epsilon}^{*(k)}_{ij} \exp\left(ik\boldsymbol{\xi}^0 \cdot \mathbf{x}\right) \tag{4.11}$$

using the principle of superposition, then the displacement field can be written as

$$u_i(\mathbf{x}) = -i \sum_{k=1}^{N} C_{jlmn}\tilde{\epsilon}^{*(k)}_{mn} \left(k\xi^0_l\right) N_{ij}(k\boldsymbol{\xi}^0) D^{-1}(k\boldsymbol{\xi}^0) \exp\left(ik\boldsymbol{\xi}^0 \cdot \mathbf{x}\right) \tag{4.12}$$

If the basic wave vector ξ^0 is taken as an infinitesimal value, the above summation can be similarly written in terms of the Fourier integral. For any piecewisely continuous eigenstrain field $\epsilon^*_{ij}(\mathbf{x})$ in the infinite domain, in the Fourier space it can be written as

$$\tilde{\epsilon}^*_{ij}(\boldsymbol{\xi}) = \frac{1}{(2\pi)^3} \int_{-\infty}^{\infty} \epsilon^*_{ij}(\mathbf{x}') \exp\left(-i\boldsymbol{\xi} \cdot \mathbf{x}'\right) d\mathbf{x}' \tag{4.13}$$

where the integral variable $\mathbf{x}'$ is the coordinate in the real domain. Then the eigenstrain in the real domain can be inversely written as

$$\epsilon^*_{ij}(\mathbf{x}) = \int_{-\infty}^{\infty} \exp\left(i\boldsymbol{\xi} \cdot \mathbf{x}\right) \tilde{\epsilon}^*_{ij}(\boldsymbol{\xi})\, d\boldsymbol{\xi} \tag{4.14}$$

where

$$\begin{cases} d\mathbf{x} & = dx_1 dx_2 dx_3 \\ d\boldsymbol{\xi} & = d\xi_1 d\xi_2 d\xi_3 \end{cases} \tag{4.15}$$

Similarly to Equation 4.12, substituting Equation 4.14 into Equation 4.9 yields

$$u_i(\mathbf{x}) = -i \int_{-\infty}^{\infty} C_{jlmn}\tilde{\epsilon}^*_{mn}(\boldsymbol{\xi})\, \xi_l N_{ij}(\boldsymbol{\xi})\, D^{-1}(\boldsymbol{\xi}) \exp\left(i\boldsymbol{\xi} \cdot \mathbf{x}\right) d\boldsymbol{\xi} \tag{4.16}$$

Substituting Equation 4.13 into the above equation yields

$$\begin{aligned} u_i(\mathbf{x}) &= -i\frac{1}{(2\pi)^3} \int_{-\infty}^{\infty}\int_{-\infty}^{\infty} C_{jlmn}\epsilon^*_{mn}(\mathbf{x}')\, \xi_l N_{ij}(\boldsymbol{\xi})\, D^{-1}(\boldsymbol{\xi}) \exp\left(i\boldsymbol{\xi} \cdot (\mathbf{x} - \mathbf{x}')\right) d\boldsymbol{\xi}\, d\mathbf{x}' \\ &= -\frac{1}{(2\pi)^3}\frac{\partial}{\partial x_l} \int_{-\infty}^{\infty}\int_{-\infty}^{\infty} C_{jlmn}\epsilon^*_{mn}(\mathbf{x}')\, N_{ij}(\boldsymbol{\xi})\, D^{-1}(\boldsymbol{\xi}) \exp\left(i\boldsymbol{\xi} \cdot (\mathbf{x} - \mathbf{x}')\right) d\boldsymbol{\xi}\, d\mathbf{x}' \end{aligned} \tag{4.17}$$

Now introduce a tensor function G_{ij} as

$$G_{ij}(\mathbf{x}, \mathbf{x}') = \frac{1}{(2\pi)^3} \int_{-\infty}^{\infty} N_{ij}(\boldsymbol{\xi})\, D^{-1}(\boldsymbol{\xi}) \exp\left(i\boldsymbol{\xi} \cdot (\mathbf{x} - \mathbf{x}')\right) d\boldsymbol{\xi} \tag{4.18}$$

The displacement field can be written as

$$u_i(\mathbf{x}) = -\int_{-\infty}^{\infty} C_{jlmn}\epsilon^*_{mn}\left(\mathbf{x}'\right) G_{ij,l}(\mathbf{x}, \mathbf{x}')d\mathbf{x}' \tag{4.19}$$

If the eigenstrain is distributed only on an inclusion domain Ω, the above integral domain will be reduced to the inclusion Ω. The above equation is the same as Equation 3.7. However, in Chapter 3, the elastic Green's function has been explicitly derived from Kelvin's solution specifically for isotropic materials. Equation 4.18 provides a general form of elastic Green's function for any stiffness tensor $\mathbf{C}$, which will change the variables N_{ij} and D. Next Section will show that for isotropic material, the elastic Green's function is in the same form as

$$G_{ij}(\mathbf{x}, \mathbf{x}') = \frac{1}{4\pi\mu} \frac{\delta_{ij}}{|\mathbf{x} - \mathbf{x}'|} - \frac{1}{16\pi\mu(1-v)} \frac{\partial |\mathbf{x} - \mathbf{x}'|}{\partial x_i \partial x_j} = \frac{1}{4\pi\mu}\delta_{ij}\phi - \frac{1}{16\pi\mu(1-v)}\psi \tag{4.20}$$

Thus, according to Equation 4.19, the strain field yields

$$\epsilon_{ij} = \frac{1}{2}\left(u_{i,j} + u_{j,i}\right) = -\int_{-\infty}^{\infty} C_{klmn}\epsilon^*_{mn}\left(\mathbf{x}'\right) \Gamma_{ijkl}(\mathbf{x}, \mathbf{x}')d\mathbf{x}' \tag{4.21}$$

where

$$\Gamma_{ijkl}(\mathbf{x}, \mathbf{x}') = \frac{1}{2}\left(G_{ik,lj} + G_{jk,li}\right) \tag{4.22}$$

Considering $C_{klmn} = C_{lkmn}$, not changing the value of ϵ_{ij} in Equation 4.21, one can redefine

$$\Gamma_{ijkl}(\mathbf{x}, \mathbf{x}') = \frac{1}{4}\left(G_{ik,lj} + G_{jk,li} + G_{il,kj} + G_{jl,ki}\right) \tag{4.23}$$

which exhibits both major and minor symmetries as discussed before. Therefore, the stress field can be obtained:

$$\sigma_{ij} = C_{ijkl}\left(\epsilon_{kl} - \epsilon^*_{kl}\right) \tag{4.24}$$

4.1.3 Green's Function for Isotropic Materials

For a general elastic material, Green's function reads

$$G_{ij}\left(\mathbf{x}, \mathbf{x}'\right) = \frac{1}{(2\pi)^3} \int_{-\infty}^{\infty} N_{ij}(\boldsymbol{\xi})\, D^{-1}(\boldsymbol{\xi}) \exp\left[i\boldsymbol{\xi} \cdot (\mathbf{x} - \mathbf{x}')\right] d\boldsymbol{\xi} \tag{4.25}$$

where

$$\begin{cases} N_{ij} = \frac{1}{2} e_{ikl} e_{jmn} K_{km} K_{ln} \\ D = e_{mnl} K_{m1} K_{n2} K_{l3} \\ K_{ik} = C_{ijkl} \xi_j \xi_l \end{cases} \tag{4.26}$$

For an isotropic material, $C_{ijkl} = \lambda \delta_{ij} \delta_{kl} + \mu \left(\delta_{ik} \delta_{jl} + \delta_{il} \delta_{jk} \right)$, one can write $K_{ik} = (\lambda + \mu) \xi_i \xi_k + \mu \delta_{ik} \xi_j \xi_j$. Then, we can obtain

$$\begin{cases} D(\xi) &= \mu^2 (\lambda + 2\mu) \xi^6 \\ N_{ij}(\xi) &= \mu \xi^2 \left[(\lambda + 2\mu) \delta_{ij} \xi^2 - (\lambda + \mu) \xi_i \xi_j \right] \end{cases} \tag{4.27}$$

with

$$\xi^2 = \xi_k \xi_k \tag{4.28}$$

Then

$$G_{ij} \left(\mathbf{x}, \mathbf{x}' \right) = \frac{1}{(2\pi)^3} \int_{-\infty}^{\infty} \frac{\mu \left[(\lambda + 2\mu) \delta_{ij} \xi^2 - (\lambda + \mu) \xi_i \xi_j \right]}{\mu^2 (\lambda + 2\mu) \xi^4} \exp \left[i \boldsymbol{\xi} \cdot (\mathbf{x} - \mathbf{x}') \right] d\boldsymbol{\xi} \tag{4.29}$$

where $d\boldsymbol{\xi} = \xi^2 d\xi dS$ and dS is the infinitesimal surface element on the unit sphere S^2 in the $\boldsymbol{\xi}$-space, centered at the origin of the coordinate ξ_i as Figure 4.1. Let

$$s_i = \frac{\xi_i}{\xi}; \; n_i = \frac{x_i - x_i'}{r}; \; r = |\mathbf{x} - \mathbf{x}'|$$

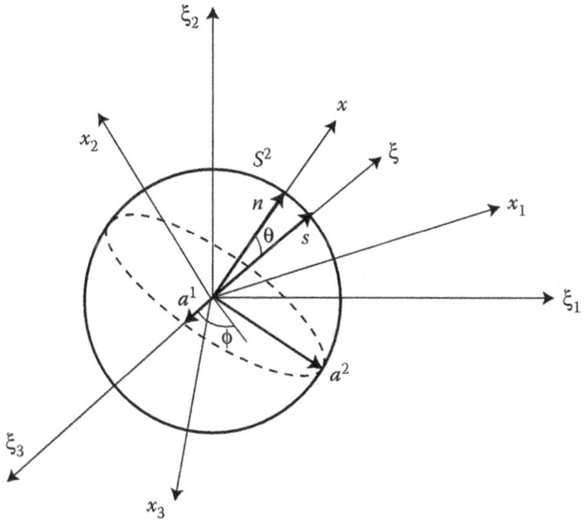

FIGURE 4.1 The unit sphere S^2 in the $\boldsymbol{\xi}$ space.

Green's function at point $\mathbf{x}$ referred to the source $\mathbf{x}'$ is written in an integral in the $\boldsymbol{\xi}$ space. For convenience, the coordinate $\mathbf{x}$ is set up with the origin at $\mathbf{x}'$ without loss of generality.

Green's function can be rewritten as

$$G_{ij}\left(\mathbf{x},\mathbf{x}'\right) = \frac{1}{(2\pi)^3} \int_{S^2}\int_0^\infty \frac{\mu\left[(\lambda+2\mu)\,\delta_{ij} - (\lambda+\mu)\,s_i s_j\right]}{\mu^2\,(\lambda+2\mu)} \exp\left(i\xi r s_k n_k\right) d\xi dS \tag{4.30}$$

Notice that

$$\int_0^\infty \exp\left(i\xi r s_k n_k\right) d\xi = \frac{1}{2}\int_{-\infty}^\infty \exp\left(i\xi r s_k n_k\right) d\xi = \pi\delta\left(r s_k n_k\right) \tag{4.31}$$

which can be proved by the Fourier transform of the Dirac Delta function inversely.

Therefore, one can write

$$G_{ij}\left(\mathbf{x},\mathbf{x}'\right) = \frac{1}{8\pi^2} \int_{S^2} \delta\left(r s_k n_k\right) \frac{\mu\left[(\lambda+2\mu)\,\delta_{ij} - (\lambda+\mu)\,s_i s_j\right]}{\mu^2\,(\lambda+2\mu)} dS \tag{4.32}$$

As shown in Figure 4.1, let the angle between $\mathbf{s}$ and $\mathbf{n}$ be written by θ, that is, $s_k n_k = \cos\theta$ and the angle on the plane Σ prependicular to $\mathbf{n}$ is written by ϕ, which is referred to a pair of orthogonal unit vectors $\mathbf{a}^1$ and $\mathbf{a}^2$ that can be arbitrarily chosen in the plane. Then, $dS = \sin\theta d\theta d\phi$. The above equation is rewritten as

$$\begin{aligned}
G_{ij}\left(\mathbf{x},\mathbf{x}'\right) &= \frac{1}{8\pi^2 r}\int_0^\pi\int_0^{2\pi} \delta(\cos\theta)\,\frac{(\lambda+2\mu)\,\delta_{ij} - (\lambda+\mu)\,s_i s_j}{\mu\,(\lambda+2\mu)} \sin\theta d\theta d\phi \\
&= -\frac{1}{8\pi^2 r}\int_0^\pi \delta(\cos\theta)\, d\cos\theta \int_0^{2\pi} \frac{(\lambda+2\mu)\,\delta_{ij} - (\lambda+\mu)\,s_i s_j}{\mu\,(\lambda+2\mu)} d\phi \\
&= -\frac{1}{8\pi^2 r} 1 \cdot \int_0^{2\pi} \frac{(\lambda+2\mu)\,\delta_{ij} - (\lambda+\mu)\,s_i s_j}{\mu\,(\lambda+2\mu)} d\phi
\end{aligned}$$

because the unit vector $\mathbf{s}$ at any angle ϕ on the plane Σ can be written as $s_i = a_i^1 \cos\phi + a_i^2 \sin\phi$ and $\int_0^{2\pi} s_i s_j d\phi = \pi\left(a_i^1 a_j^1 + a_i^2 a_j^2\right) = \pi\left(\delta_{ij} - n_i n_j\right)$. Therefore, the above equation can be written as

$$\begin{aligned}
G_{ij}(\mathbf{x},\mathbf{x}') &= -\frac{1}{8\pi^2 r}\frac{2\pi(\lambda+2\mu)\delta_{ij} - (\lambda+\mu)\,\pi\left(\delta_{ij} - n_i n_j\right)}{\mu\,(\lambda+2\mu)} \\
&= -\frac{1}{4\pi r}\frac{\delta_{ij}}{\mu} + \frac{1}{8\pi r}\frac{(\lambda+\mu)\left(\delta_{ij} - n_i n_j\right)}{\mu\,(\lambda+2\mu)}
\end{aligned}$$

4.2 ELLIPSOIDAL INCLUSION PROBLEMS

4.2.1 Ellipsoidal Inclusion with a Uniform Eigenstrain

A uniform eigenstrain is assumed on an ellipsoidal inclusion Ω as shown in Figure 4.2.

$$\Omega : \frac{x_1^2}{a_1^2} + \frac{x_2^2}{a_2^2} + \frac{x_3^2}{a_3^2} \leq 1 \tag{4.33}$$

The displacement can be written as

$$\begin{aligned} u_i &= -C_{jkmn}\epsilon_{mn}^* \int_\Omega G_{ij,k}\left(\mathbf{x}, \mathbf{x}'\right) d\mathbf{x}' \\ &= -\frac{\epsilon_{mn}^*}{8\pi(1-\nu)} \int_\Omega g_{imn}\left(\mathbf{n}\right) \frac{d\mathbf{x}'}{|\mathbf{x}-\mathbf{x}'|^2} \end{aligned} \tag{4.34}$$

where

$$g_{imn} = (1-2\nu)\left(\delta_{im}n_n + \delta_{in}n_m - \delta_{mn}n_i\right) + 3n_in_mn_n \tag{4.35}$$

which can be verified as

$$\frac{g_{imn}}{8\pi(1-\nu)\,|\mathbf{x}-\mathbf{x}'|^2} = -C_{jkmn}G_{ij,k} \tag{4.36}$$

For $\mathbf{x} \in \Omega$,

$$d\mathbf{x} = dr \cdot r^2 \cdot d\omega = dr \cdot \left|\mathbf{x}' - \mathbf{x}\right|^2 \cdot d\omega \tag{4.37}$$

As shown in Figure 4.3,
Equation 4.34 can be rewritten as

$$u_i = -\frac{\epsilon_{mn}^*}{8\pi(1-\nu)} \int_{S^2} \int_0^{r(\mathbf{n})} dr \cdot g_{imn}\left(\mathbf{n}\right) d\omega = -\frac{\epsilon_{mn}^*}{8\pi(1-\nu)} \int_{S^2} r(\mathbf{n}) \cdot g_{imn}\left(\mathbf{n}\right) d\omega \tag{4.38}$$

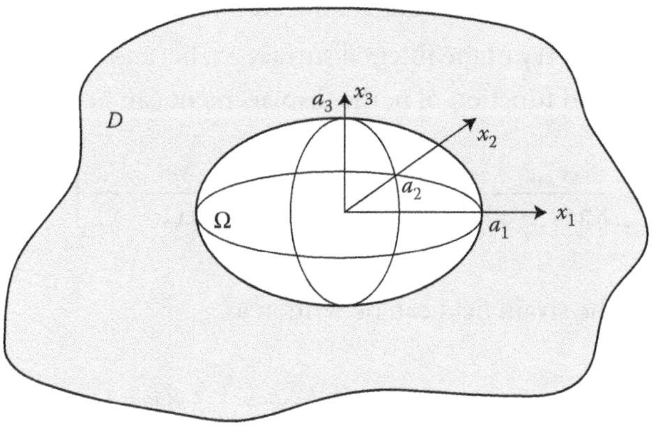

FIGURE 4.2 An ellipsoidal inclusion in the infinitely extended domain with radii a_1, a_2, and a_3.

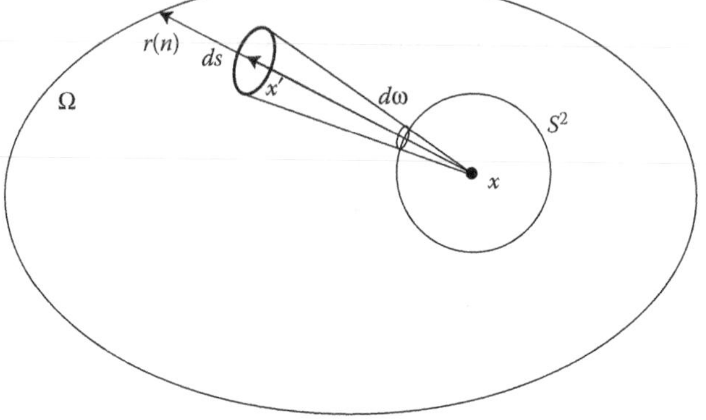

FIGURE 4.3 Integral of Green's function over an ellipsoidal inclusion.

where $r(\mathbf{n})$ is the positive root of

$$\frac{(x_i + rn_i)(x_i + rn_i)}{a_I^2} = \frac{(x_1 + rn_1)^2}{a_1^2} + \frac{(x_2 + rn_2)^2}{a_2^2} + \frac{(x_3 + rn_3)^2}{a_3^2} = 1 \tag{4.39}$$

which can be written as

$$r(\mathbf{n}) = -\frac{f}{g} + \left(\frac{f^2}{g^2} + \frac{e}{g}\right)^{1/2} \tag{4.40}$$

where

$$\begin{cases} g = \dfrac{n_i n_i}{a_I^2} \\ f = \dfrac{n_i x_i}{a_I^2} \\ e = 1 - \dfrac{x_i x_i}{a_I^2} \end{cases} \tag{4.41}$$

Notice that extended index notation is used, namely, repeated lower case indices are summed up as usual index notation, and; upper case indices take on the same numbers as the corresponding lower case ones but are not summed.

Considering the symmetry of the integral surface S^2, because $\frac{f^2}{g^2} + \frac{e}{g}$ is an even function whereas $g_{imn}(\mathbf{n})$ is an odd function of $\mathbf{n}$, the displacement can be rewritten as

$$u_i(\mathbf{x}) = \frac{\epsilon^*_{mn}}{8\pi(1-\nu)} \int_{S^2} \frac{f}{g} \cdot g_{imn}(\mathbf{n})\, d\omega = \frac{x_p \epsilon^*_{mn}}{8\pi(1-\nu)} \int_{\Sigma} \frac{\lambda_p g_{imn}}{g} d\omega \tag{4.42}$$

where $\lambda_i = \frac{n_i}{a_I^2}$. Thus, the strain field can be written as

$$\epsilon_{ij} = \frac{\epsilon^*_{mn}}{16\pi(1-\nu)} \int_{\Sigma} \frac{\lambda_i g_{jmn} + \lambda_j g_{imn}}{g} d\omega \tag{4.43}$$

Using the definition of $g_{imn}(\mathbf{n})$, the above integral can be written in terms of $\frac{n_in_j}{a_I^2 g}$, $\frac{n_in_jn_kn_l}{a_I^2 g}$, and so on. Notice that when $i \neq j$, $\frac{n_in_j}{a_I^2 g}$ is an odd function in the symmetric integral domain, so its integral must be zero. Therefore, the integral of $\frac{n_in_j}{a_I^2 g}$ can be finally written in terms of $\delta_{ij} \int_\Sigma \frac{n_I^2}{a_I^2 g} d\omega$. Similarly to the integral of $\frac{n_in_jn_kn_l}{a_I^2 g}$, we can also write it in terms of $\delta_{ij}\delta_{kl} \int_\Sigma \frac{n_I^2 n_K^2}{a_I^2 g} d\omega$ or $\delta_{ik}\delta_{jl} \int_\Sigma \frac{n_I^2 n_J^2}{a_I^2 g} d\omega$ or $\delta_{il}\delta_{jk} \int_\Sigma \frac{n_I^2 n_J^2}{a_I^2 g} d\omega$, which can be rewritten as

$$\delta_{ij}\delta_{kl} \int_\Sigma \frac{n_I^2 n_K^2}{a_I^2 g} d\omega + \delta_{ik}\delta_{jl} \int_\Sigma \frac{n_I^2 n_J^2}{a_I^2 g} d\omega + \delta_{il}\delta_{jk} \int_\Sigma \frac{n_I^2 n_J^2}{a_I^2 g} d\omega - 2\delta_{ij}\delta_{kl}\delta_{IK} \int_\Sigma \frac{n_I^4}{a_I^2 g} d\omega.$$

Eventually, we can calculate the above equation of ϵ_{ij} using the following identities of the elliptic integrals.
For interior points

$$\begin{cases} I_I = \int_\Sigma \frac{n_i^2}{a_I^2 g} d\omega = 2\pi a_1 a_2 a_3 \int_0^\infty \frac{ds}{\left(a_I^2 + s\right) \Delta(s)} \\ I_{II} = \int_\Sigma \frac{n_i^4}{a_I^4 g} d\omega = 2\pi a_1 a_2 a_3 \int_0^\infty \frac{ds}{\left(a_I^2 + s\right)^2 \Delta(s)} \\ I_{IJ} = \int_\Sigma \frac{n_i^2 n_j^2}{a_I^2 a_J^2 g} d\omega = 2\pi a_1 a_2 a_3 \int_0^\infty \frac{ds}{\left(a_I^2 + s\right)\left(a_J^2 + s\right) \Delta(s)} \end{cases} \quad \text{for } x \in \Omega \tag{4.44}$$

where

$$\Delta(s) = \sqrt{\left(a_1^2 + s\right)\left(a_2^2 + s\right)\left(a_3^2 + s\right)} \tag{4.45}$$

For exterior points

$$\begin{cases} I = 2\pi a_1 a_2 a_3 \int_\lambda^\infty \frac{ds}{\Delta(s)} \\ I_I = 2\pi a_1 a_2 a_3 \int_\lambda^\infty \frac{ds}{\left(a_I^2 + s\right)^2 \Delta(s)} \\ I_{IJ} = 2\pi a_1 a_2 a_3 \int_\lambda^\infty \frac{ds}{\left(a_I^2 + s\right)\left(a_J^2 + s\right) \Delta(s)} \end{cases} \quad \text{for } x \in D - \Omega \tag{4.46}$$

where λ is defined as the largest root of the following equation:

$$\frac{x_i x_i}{\left(a_I^2+\lambda\right)}=\frac{x_1^2}{a_1^2+\lambda}+\frac{x_2^2}{a_2^2+\lambda}+\frac{x_3^2}{a_3^2+\lambda}=1 \tag{4.47}$$

It is convenient to write Equation 4.43 as

$$\epsilon_{ij}=S_{ijmn}\epsilon_{mn}^{*} \tag{4.48}$$

and S_{ijmn} only depends on the shape of the ellipsoid, and it satisfies

$$S_{ijmn}=S_{jimn}=S_{ijnm} \tag{4.49}$$

Using Equation 4.43, some components are calculated easily,

$$\begin{cases} S_{1111}=\dfrac{3}{8\pi(1-\nu)}a^2 I_{11}+\dfrac{1-2\nu}{8\pi(1-\nu)}I_1 \\ S_{1122}=\dfrac{1}{8\pi(1-\nu)}b^2 I_{12}-\dfrac{1-2\nu}{8\pi(1-\nu)}I_1 \\ S_{1212}=\dfrac{a^2+b^2}{16\pi(1-\nu)}I_{12}+\dfrac{1-2\nu}{16\pi(1-\nu)}\left(I_1+I_2\right) \end{cases} \tag{4.50}$$

Given an exterior point of an ellipsoid, the geometric meaning of λ can be explained as follows: If the ellipsoid grows in such a manner along the three semi-axes that the elliptic radius increases from a_i to $\sqrt{a_i^2+\lambda}$ and the point is located on this new imaginary ellipsoid surface, the positive solution in Equation 4.47 can be uniquely determined by the size of the imaginary ellipsoid.

(a) For a spherical domain, that is, $a_1=a_2=a_3=a$, one can write

$$\begin{cases} I(\lambda)=4\pi a^3/\left(a^2+\lambda\right)^{1/2} \\ I_i(\lambda)=4\pi a^3/3\left(a^2+\lambda\right)^{3/2} \\ I_{\underbrace{ij\cdots k}_{n}}(\lambda)=\dfrac{4\pi a^3}{(2n+1)\,3\left(a^2+\lambda\right)^{n+1/2}} \end{cases} \tag{4.51}$$

(b) For different shapes of ellipsoids ($a_1 \neq a_2 \neq a_3$),

1. Fiber: $a_3 \to \infty$

$$\begin{cases} I_1(\lambda) = \dfrac{4\pi a_1 a_2}{a_2^2 - a_1^2}\left[\dfrac{\left(a_2^2+\lambda\right)^{1/2}}{\left(a_1^2+\lambda\right)^{1/2}} - 1\right] \\ I_2(\lambda) = \dfrac{4\pi a_1 a_2}{a_1^2 - a_2^2}\left[\dfrac{\left(a_1^2+\lambda\right)^{1/2}}{\left(a_2^2+\lambda\right)^{1/2}} - 1\right] \\ I_3(\lambda) = 0 \end{cases} \tag{4.52}$$

2. Oblate spheroid: $a_1 = a_2 > a_3$

$$\begin{cases} I(\lambda) = \dfrac{4\pi a_1^2 a_3}{\left(a_1^2 - a_3^2\right)^{1/2}}\left[\dfrac{\pi}{2} - \arctan\dfrac{\left(a_3^2+\lambda\right)^{1/2}}{\left(a_1^2-a_3^2\right)^{1/2}} - 1\right] \\ I_1(\lambda) = I_2(\lambda) = \dfrac{2\pi a_1^2 a_3\left(\arccos b - bd\right)}{\left(a_1^2-a_3^2\right)^{3/2}} \\ I_3(\lambda) = \dfrac{4\pi a_1^2 a_3}{\Delta(\lambda)} - 2I_1(\lambda) \end{cases} \tag{4.53}$$

where

$$b = \frac{\left(a_3^2+\lambda\right)^{1/2}}{\left(a_1^2+\lambda\right)^{1/2}}, \quad d = \frac{\left(a_1^2-a_3^2\right)^{1/2}}{\left(a_1^2+\lambda\right)^{1/2}}.$$

3. Prolate spheroid: $a_1 > a_2 = a_3$

$$\begin{cases} I(\lambda) = \dfrac{4\pi a_1 a_2^2}{\left(a_1^2-a_2^2\right)^{1/2}} \operatorname{arccos h}\bar{b} \\ I_1(\lambda) = \dfrac{4\pi a_1 a_2^2\left(\operatorname{arccos h}\bar{b} - \bar{d}/\bar{b}\right)}{\left(a_1^2-a_2^2\right)^{3/2}} \\ I_2(\lambda) = I_3(\lambda) = \dfrac{2\pi a_1 a_2^2\left(\operatorname{arccos h}\bar{b} - \bar{b}\bar{d}\right)}{\left(a_1^2-a_2^2\right)^{3/2}} \end{cases} \tag{4.54}$$

where

$$\bar{b} = \frac{\left(a_1^2+\lambda\right)^{1/2}}{\left(a_2^2+\lambda\right)^{1/2}}, \quad \bar{d} = \frac{\left(a_1^2-a_2^2\right)^{1/2}}{\left(a_2^2+\lambda\right)^{1/2}},$$

The details are described in book [3, pp. 93–95].

For a general ellipsoidal inclusion, Sun [40] has organized Eshelby's tensor in a compact form for both interior and exterior points of the inclusion as follows:

$$D_{ijkl} = \begin{cases} \begin{bmatrix} S^1_{IK}(\lambda)\delta_{ij}\delta_{kl} + S^2_{IJ}(\lambda)\left(\delta_{ik}\delta_{jl} + \delta_{il}\delta_{jk}\right) + S^3_I(\lambda)\delta_{ij}\hat{n}_k\hat{n}_l \\ + S^4_K(\lambda)\delta_{kl}\hat{n}_i\hat{n}_j + S^5_I(\lambda)\left(\delta_{ik}\hat{n}_j\hat{n}_l + \delta_{il}\hat{n}_j\hat{n}_k\right) \\ + S^6_J(\lambda)\left(\delta_{jk}\hat{n}_i\hat{n}_l + \delta_{jl}\hat{n}_i\hat{n}_k\right) \\ + S^7_{IJKL}(\lambda)\hat{n}_i\hat{n}_j\hat{n}_k\hat{n}_l \end{bmatrix} & \text{for } r > a \\ S^1_{IK}(0)\,\delta_{ij}\delta_{kl} + S^2_{IJ}(0)\left(\delta_{ik}\delta_{jl} + \delta_{il}\delta_{jk}\right) & \text{for } r \le a \end{cases} \tag{4.55}$$

where $\hat{\mathbf{n}}$ denotes the outward unit normal vector on the imaginary ellipsoid surface, that is,

$$\hat{n}_i = \frac{x_i}{\left(a_I^2 + \lambda\right)\Theta(\lambda)} \tag{4.56}$$

in which

$$\Theta(\lambda) = \sqrt{\Theta_i(\lambda)\Theta_i(\lambda)}$$

with

$$\Theta_i(\lambda) = \frac{x_i}{\left(a_I^2 + \lambda\right)}$$

and the tensor coefficients are provided as follows:

$$S^1_{IK} = -\frac{\nu}{2(1-\nu)}J_I(\lambda) + \frac{1}{4(1-\nu)}\left[\frac{a_I^2}{a_I^2 - a_K^2}J_I(\lambda) + \frac{a_K^2}{a_K^2 - a_I^2}J_K(\lambda)\right] \tag{4.57}$$

$$S^2_{IJ} = -\frac{1}{4}\left[J_I(\lambda) + J_J(\lambda)\right] + \frac{1}{4(1-\nu)}\left[\frac{a_I^2}{a_I^2 - a_J^2}J_I(\lambda) + \frac{a_J^2}{a_J^2 - a_I^2}J_J(\lambda)\right] \tag{4.58}$$

$$S^3_I = \frac{\rho^3(\lambda)}{2(1-\nu)}\left[1 - \rho_I^2(\lambda)\right] \tag{4.59}$$

$$S^4_K = \frac{\rho^3(\lambda)}{2(1-\nu)}\left[1 - 2\nu - \rho_K^2(\lambda)\right] \tag{4.60}$$

$$S^5_I = \frac{\rho^3(\lambda)}{2(1-\nu)}\left[\nu - \rho_I^2(\lambda)\right] \tag{4.61}$$

$$S^6_J = \frac{\rho^3(\lambda)}{2(1-\nu)}\left[\nu - \rho_J^2(\lambda)\right] \tag{4.62}$$

$$S^7_{IJKL} = \frac{\rho^3(\lambda)}{2(1-\nu)}\begin{bmatrix} 2\left(\rho_I^2(\lambda) + \rho_J^2(\lambda) + \rho_K^2(\lambda) + \rho_L^2(\lambda)\right) + \rho_m(\lambda)\rho_m(\lambda) \\ -\dfrac{4\rho_M^2(\lambda)\Theta_m(\lambda)\Theta_m(\lambda)}{\Theta^2(\lambda)} - 5 \end{bmatrix} \tag{4.63}$$

in which

$$\rho_I(\lambda) = \frac{a_I}{\sqrt{a_I^2 + \lambda}}, \ \rho(\lambda) = \sqrt[3]{\rho_1(\lambda)\rho_2(\lambda)\rho_3(\lambda)}$$

and

$$J_I(\lambda) = -\int_{\lambda}^{\infty} \frac{\rho^3(s)}{\left(a_I^2 + s\right)} ds.$$

4.2.2 Ellipsoidal Inclusion with a Polynomial Eigenstrain

For a continuous eigenstrain field over an ellipsoidal inclusion Ω, it can be approximated by Taylor's series in a polynomial form such as

$$\epsilon^*_{ij}(\mathbf{x}) = \begin{cases} E_{ij} + E_{ijk}x_k + E_{ijkl}x_k x_l + \cdots, & \mathbf{x} \in \Omega \\ 0. & \mathbf{x} \in D - \Omega. \end{cases} \tag{4.64}$$

Then, the displacement and strain fields can be written as

$$u_i = -\int_{\Omega} G_{ik,l}\left(\mathbf{x}, \mathbf{x}'\right) C_{klmn}\epsilon^*_{mn}(\mathbf{x}')d\mathbf{x}'$$

and

$$\epsilon_{ij} = -\int_{\Omega} \Gamma_{ijkl}\left(\mathbf{x}, \mathbf{x}'\right) C_{klmn}\epsilon^*_{mn}(\mathbf{x}')d\mathbf{x}'$$

For an isotropic material, we can obtain

$$C_{klmn}\epsilon^*_{mn} = \lambda\delta_{kl}\epsilon^*_{mm} + 2\mu\epsilon^*_{kl}$$

Using

$$G_{ij}\left(\mathbf{x}, \mathbf{x}'\right) = \frac{1}{4\pi\mu}\delta_{ij}\phi - \frac{1}{16\pi\mu(1-v)}\psi_{,ij},$$

the following equations can be used to straightforwardly obtain the integral of Green's function over the ellipsoidal domain Ω:

$$u_i = \frac{1}{8\pi(1-v)}\left[\begin{array}{c} \Psi_{,ikl}E_{kl} - 2v\Phi_{,i}E_{mm} - 4(1-v)\Phi_{,m}E_{im} + \Psi_{m,ikl}E_{klm} \\ -2v\Phi_{k,i}E_{mmk} - 4(1-v)\Phi_{k,m}E_{imk} + \cdots \end{array}\right] \tag{4.65}$$

and

$$\epsilon_{ij} = \frac{1}{8\pi(1-v)}\left[F_{ijkl}E_{kl} + F_{ijklm}E_{klm} + \cdots\right]$$

with

$$F_{ijkl} = \frac{1}{8\pi(1-v)}\left[\Psi_{,ijkl} - 2v\Phi_{,ij}\delta_{kl} - (1-v)\left(\Phi_{,kj}\delta_{il} + \Phi_{,ki}\delta_{jl} + \Phi_{,lj}\delta_{ik} + \Phi_{,li}\delta_{jk}\right)\right]$$

and

$$F_{ijklm} = \frac{1}{8\pi(1-v)}\left[\Psi_{m,ijkl} - 2v\Phi_{m,ij}\delta_{kl} - (1-v)\left(\Phi_{m,kj}\delta_{il} + \Phi_{m,ki}\delta_{jl} + \Phi_{m,lj}\delta_{ik} + \Phi_{m,li}\delta_{jk}\right)\right]$$

where Φ, Ψ, Φ_p, and Ψ_p denote the integral of the functions ϕ, ψ, $\phi x'_p$, and $\psi x'_p$ over an ellipsoidal inclusion Ω as

$$\begin{cases} \Phi = \int_\Omega \phi \, dx' \\ \Psi = \int_\Omega \psi \, dx' \\ \Phi_p = \int_\Omega \phi x'_p dx' \\ \Psi_p = \int_\Omega \psi x'_p dx' \end{cases} \tag{4.66}$$

In the following, the integral terms of Φ, Ψ, Φ_p, and Ψ_p and their derivative terms over an ellipsoidal domain can be derived and expressed as

$$\Phi = \frac{1}{2}\left[I(\lambda) - x_r x_r I_R(\lambda)\right] \tag{4.67}$$

$$\Phi_{,i} = -x_i I_I(\lambda) \tag{4.68}$$

$$\Phi_{,ij} = -\delta_{ij} I_I(\lambda) - x_i I_{I,j}(\lambda) \tag{4.69}$$

$$\Phi_{p,i} = \frac{1}{2} a_P^2 \left\{\delta_{ip}\left[I_P(\lambda) - x_r x_r I_{RP}(\lambda)\right] - 2x_p\left[x_i I_{IP}(\lambda)\right]\right\} \tag{4.70}$$

$$\Phi_{p,ij} = -a_P^2\left[\delta_{ip} x_j I_{JP}(\lambda) + \delta_{jp} x_i I_{IP}(\lambda) + \delta_{ij} x_p I_{IP}(\lambda) + x_i x_p I_{IP,j}(\lambda)\right] \tag{4.71}$$

$$\begin{aligned} \Psi_{,ij} = &\frac{1}{2}\delta_{ij}\left\{I(\lambda) - x_r x_r I_R(\lambda) - a_I^2\left[I_I(\lambda) - x_r x_r I_{RI}(\lambda)\right]\right\} \\ &- x_i x_j\left[I_J(\lambda) - a_I^2 I_{IJ}(\lambda)\right] \end{aligned} \tag{4.72}$$

$$\begin{aligned} \Psi_{,ijk} = &-\delta_{ij} x_k\left[I_K(\lambda) - a_I^2 I_{IK}(\lambda)\right] - \left(\delta_{ik} x_j + \delta_{jk} x_i\right)\left[I_J(\lambda) - a_I^2 I_{IJ}(\lambda)\right] \\ &- x_i x_j\left[I_J(\lambda) - a_I^2 I_{IJ}(\lambda)\right]_{,k} \end{aligned} \tag{4.73}$$

$$\Psi_{p,ijk} = a_P^2 \left\{ \begin{array}{c} \frac{1}{2}\delta_{ip}\delta_{jk}\left[I_J(\lambda) - x_r x_r I_{RJ}(\lambda)\right] \\ -\frac{1}{2}\delta_{ip}\delta_{jk}a_P^2\left[I_{JP}(\lambda) - x_r x_r I_{RJP}(\lambda)\right] \\ +\delta_{ip}x_j\left[-x_k I_{KJ}(\lambda) + a_P^2 x_k I_{KJP}(\lambda)\right] \\ +\frac{1}{2}\left(\delta_{jp}\delta_{ik} + \delta_{ij}\delta_{kp}\right)\left[I_I(\lambda) - x_r x_r I_{RI}(\lambda)\right] \\ -\frac{1}{2}\left(\delta_{jp}\delta_{ik} + \delta_{ij}\delta_{kp}\right)a_P^2\left[I_{IP}(\lambda) - x_r x_r I_{RIP}(\lambda)\right] \\ +\left(\delta_{jp}x_i + \delta_{ij}x_p\right)\left[-x_k I_{KI}(\lambda) + a_P^2 x_k I_{KIP}(\lambda)\right] \\ +\left(\delta_{ik}x_j x_p + \delta_{jk}x_i x_p + \delta_{kp}x_i x_j\right)\left[-I_{IJ}(\lambda) + a_P^2 I_{IJP}(\lambda)\right] \\ +x_i x_j x_p\left[-I_{IJ}(\lambda) + a_P^2 I_{IJP}(\lambda)\right]_{,k} \end{array} \right\} \tag{4.74}$$

$$\Psi_{p,ijkl} = a_P^2 \left\{ \begin{array}{c} \delta_{ip}\delta_{jk}x_l\left(-I_{LJ} + a_P^2 I_{LJP}\right) + \left(\delta_{ip}\delta_{jl}x_k + \delta_{ip}\delta_{kl}x_j\right)\left(-I_{KJ} + a_P^2 I_{KJP}\right) \\ +\left(\delta_{ij}\delta_{kp}x_l + \delta_{ik}\delta_{jp}x_l\right)\left(-I_{LI} + a_P^2 I_{LIP}\right) \\ +\left(\delta_{ij}\delta_{kl}x_p + \delta_{ij}\delta_{lp}x_k + \delta_{il}\delta_{jp}x_k + \delta_{jp}\delta_{kl}x_i\right)\left(-I_{KI} + a_P^2 I_{KIP}\right) \\ +\left(\delta_{ik}\delta_{jl}x_p + \delta_{ik}\delta_{lp}x_j + \delta_{il}\delta_{jk}x_p\right)\left(-I_{JI} + a_P^2 I_{JIP}\right) \\ +\left(\delta_{il}\delta_{kp}x_j + \delta_{jk}\delta_{lp}x_i + \delta_{jl}\delta_{kp}x_i\right)\left(-I_{JI} + a_P^2 I_{JIP}\right) \\ +\delta_{ip}x_j x_k\left(-I_{KJ} + a_P^2 I_{KJP}\right)_{,l} + \left(\delta_{jp}x_i x_k + \delta_{ij}x_k x_p\right)\left(-I_{KI} + a_P^2 I_{KIP}\right)_{,l} \\ +\left(\delta_{ik}x_j x_p + \delta_{jk}x_i x_p + \delta_{kp}x_i x_j\right)\left(-I_{JI} + a_P^2 I_{JIP}\right)_{,l} \\ +\left(\delta_{il}x_j x_p + \delta_{jl}x_i x_p + \delta_{lp}x_i x_j\right)\left(-I_{JI} + a_P^2 I_{JIP}\right)_{,k} \\ +x_i x_j x_p\left(-I_{JI} + a_P^2 I_{JIP}\right)_{,kl} \end{array} \right\} \tag{4.75}$$

4.2.3 Ellipsoidal Inclusion with a Body Force

Considering the body force, the equilibrium equation can be written as

$$C_{ijkl}u_{k,li} = -b_j \tag{4.76}$$

If the body force b_j is given in a single wave function

$$b_j = \widetilde{b}_j \exp\left(i\boldsymbol{\xi}\cdot\mathbf{x}\right)$$

Considering the periodicity of the elastic field, the displacement can also be written in the same form

$$u_i = \tilde{u}_i \exp\left(i\boldsymbol{\xi}\cdot\mathbf{x}\right)$$

Substituting the above two equations into Equation 4.76,

$$C_{ijkl}\tilde{u}_k\xi_l\xi_i \exp\left(i\boldsymbol{\xi}\cdot\mathbf{x}\right) = \widetilde{b}_j \exp\left(i\boldsymbol{\xi}\cdot\mathbf{x}\right)$$

that is,

$$C_{ijkl}\tilde{u}_k\xi_l\xi_i = \widetilde{b}_j$$

Considering $K_{jk} = C_{ijkl}\xi_l\xi_i$, the above equation can be rewritten as

$$K_{jk}\tilde{u}_k = \widetilde{b}_j \tag{4.77}$$

Then, one can obtain the displacement field as

$$\tilde{u}_k = \frac{N_{kj}}{D}\tilde{b}_j$$

where $D = Det\,[\mathbf{K}] = e_{mnl}K_{m1}K_{n2}K_{l3}$ and $N_{ij} = \frac{1}{2}\left(\epsilon_{ikl}\epsilon_{jmn}K_{km}K_{ln}\right)$. Then one can obtain

$$u_i = \frac{N_{ij}\tilde{b}_j}{D}\exp\left(i\boldsymbol{\xi}\cdot\mathbf{x}\right) \tag{4.78}$$

Once the displacement field is obtained, the strain and stress fields can be written through the strain–displacement relation and constitutive relation.

For any piecewisely continuous body field $b_j(\mathbf{x})$ in the infinite domain, in the Fourier space, it can be written as

$$\tilde{b}_i\left(\boldsymbol{\xi}\right) = \frac{1}{(2\pi)^3}\int_{-\infty}^{\infty} b_i\left(\mathbf{x}'\right)\exp\left(-i\boldsymbol{\xi}\cdot\mathbf{x}'\right)d\mathbf{x}' \tag{4.79}$$

where the integral variable $\mathbf{x}'$ is the coordinate in the real domain.

$$u_i(\mathbf{x}) = \int_{-\infty}^{\infty} N_{ij}\left(\boldsymbol{\xi}\right)D^{-1}\left(\boldsymbol{\xi}\right)\tilde{b}_j\exp\left(i\boldsymbol{\xi}\cdot\mathbf{x}\right)d\boldsymbol{\xi} \tag{4.80}$$

Substituting Equation 4.79 into the above equation yields

$$u_i(\mathbf{x}) = \frac{1}{(2\pi)^3}\int_{-\infty}^{\infty}\int_{-\infty}^{\infty} N_{ij}\left(\boldsymbol{\xi}\right)D^{-1}\left(\boldsymbol{\xi}\right)b_j\left(\mathbf{x}'\right)\exp\left[i\boldsymbol{\xi}\cdot\left(\mathbf{x}-\mathbf{x}'\right)\right]d\boldsymbol{\xi}d\mathbf{x}' \tag{4.81}$$

Using Green's function $G_{ij}(\mathbf{x},\mathbf{x}') = \frac{1}{(2\pi)^3}\int_{-\infty}^{\infty} N_{ij}\left(\boldsymbol{\xi}\right)D^{-1}\left(\boldsymbol{\xi}\right)\exp\left(i\boldsymbol{\xi}\cdot\left(\mathbf{x}-\mathbf{x}'\right)\right)d\boldsymbol{\xi}$, the above equation is written as

$$u_i(\mathbf{x}) = \int_{-\infty}^{\infty} G_{ij}(\mathbf{x},\mathbf{x}')b_j(\mathbf{x}')d\mathbf{x}' \tag{4.82}$$

For a distributed body force on the ellipsoidal inclusion Ω, the body force can be approximated in the polynomial form as

$$b_j(\mathbf{x}) = \begin{cases} B_j + B_{jk}x_k + \cdots, & \mathbf{x}\in\Omega \\ 0, & \mathbf{x}\in D-\Omega \end{cases} \tag{4.83}$$

Then, the displacement can be rewritten as

$$u_i(\mathbf{x}) = \int\limits_{\Omega} \left(\frac{1}{4\pi\mu}\delta_{ij}\phi - \frac{1}{16\pi\mu(1-v)}\psi_{,ij} \right) \left(B_j + B_{jk}x_k + \cdots \right) d\mathbf{x}' \tag{4.84}$$

And then one can obtain

$$u_i = \frac{1}{16\pi\mu(1-v)} \left[-\Psi_{,ij}B_j - \Psi_{k,ij}B_{jk} + 4(1-v)\left(\Phi B_j + \Phi_k B_{jk} \right) + \cdots \right]$$

Then the strain and stress fields can be obtained straightforwardly.

4.3 EQUIVALENT INCLUSION METHOD FOR ELLIPSOIDAL INHOMOGENEITIES

4.3.1 Elastic Solution for a Pair of Ellipsoidal Inhomogeneities in the Infinite Domain

For a pair of ellipsoidal inhomogeneities, Ω and $\bar{\Omega}$, with stiffness C^*_{ijkl} embedded in the infinite medium with stiffness C^0_{ijkl}, which is subjected to a uniform far-field strain ϵ^0_{ij}, set two coordinate systems, say the $\mathbf{x}$ and $\bar{\mathbf{x}}$ system, with origins indicated by point $\mathbf{0}$ and $\bar{\mathbf{0}}$, respectively, as shown in Figure 4.4. The coordinate of $\bar{\mathbf{0}}$ in the $\mathbf{x}$ coordinate system is written as $\mathbf{x}^0$. For an arbitrary point P in the domain, x^P_i and $\bar{x}^P_i$ stand for the corresponding coordinates in the two systems and their relation is expressed as

$$x^P_i = x^0_i + Q_{ij}\bar{x}^P_j \quad \text{or} \quad \bar{x}^P_j = Q_{ij}\left(x^P_i - x^0_i \right) \tag{4.85}$$

where Q_{ij} is the coordination rotational matrix with the direction cosine between the x_i axis and the $\bar{x}_j$ axis. Obviously, the eigenstrain over each particle will not be uniform because of the particle interaction. Expanding it over the corresponding local coordinate provides

$$\begin{aligned} \epsilon^*_{ij}(\mathbf{x}) &= E^I_{ij} + E^I_{ijk}x_k + \cdots \\ \bar{\epsilon}^*_{ij}(\bar{\mathbf{x}}) &= E^{II}_{ij} + E^{II}_{ijk}\bar{x}_k + \cdots \end{aligned} \tag{4.86}$$

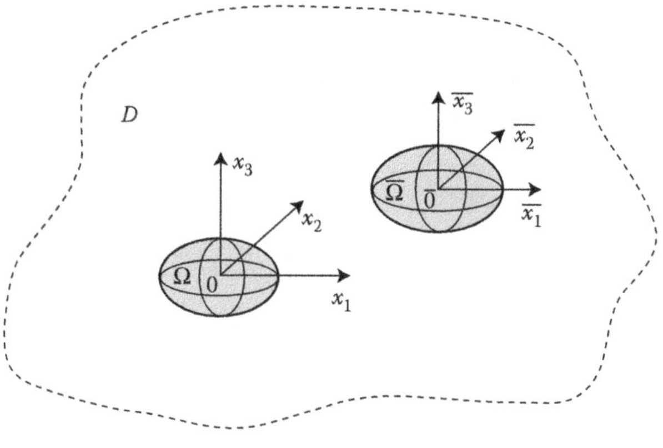

FIGURE 4.4 Interaction between two ellipsoidal particles with the local coordinates in parallel.

The strain field of any point $\mathbf{x}$ will be the superposition of the field induced by the sources of eigenstrains of the two inhomogeneities as follows:

$$\epsilon_{ij}(\mathbf{x}) = \epsilon^0_{ij} + \int\limits_{\Omega+\bar{\Omega}} \Gamma_{ijkl}\left(\mathbf{x}, \mathbf{x}'\right) C_{klmn}\epsilon^*_{mn}(\mathbf{x}')d\mathbf{x}' \tag{4.87}$$

Referred to the $\mathbf{x}$ coordinate, the induced strain can be rewritten as

$$\begin{aligned}\epsilon_{ij}(\mathbf{x}) = \epsilon^0_{ij} &+ F^I_{ijmn}(\mathbf{x})E^I_{mn} + F^I_{ijmnp}(\mathbf{x})E^I_{mnp} + F^{II}_{ijmn}(\mathbf{x})E^{II}_{mn} \\ &+ F^{II}_{ijmnp}(\mathbf{x})E^{II}_{mnp} + \cdots\end{aligned} \tag{4.88}$$

where F^I_{ijmn} and F^I_{ijmnp} are the integrals over the first particle, and F^{II}_{ijmn} and F^{II}_{ijmnp} the second particle.

Similarly, referred to the $\bar{\mathbf{x}}$ coordinate, the induced strain can be rewritten as

$$\epsilon_{ij}(\bar{\mathbf{x}}) = \epsilon^0_{ij} + F^I_{ijmn}(\bar{\mathbf{x}})E^I_{mn} + F^I_{ijmnp}(\bar{\mathbf{x}})E^I_{mnp} + F^{II}_{ijmn}(\bar{\mathbf{x}})E^{II}_{mn} + F^{II}_{ijmnp}(\bar{\mathbf{x}})E^{II}_{mnp} + \cdots \tag{4.89}$$

Based on the equivalent inclusion condition, on each inhomogeneity, one can write

$$C^*_{ijkl}\epsilon_{kl} = C^0_{ijkl}\left(\epsilon_{kl} - \epsilon^*_{kl}\right) \tag{4.90}$$

or

$$\epsilon = -\Delta\mathbf{C}^{-1} : \mathbf{C}^0 : \epsilon^* \tag{4.91}$$

with $\Delta\mathbf{C} = \mathbf{C}^* - \mathbf{C}^0$.

For the first inhomogeneity, the strain in Equation 4.88 can be expanded in the Taylor series as

$$\begin{aligned}\epsilon_{ij}(\mathbf{x}) = \epsilon^0_{ij} &+ F^I_{ijmn}(\mathbf{0})E^I_{mn} + F^I_{ijmnp}(\mathbf{0})E^I_{mnp} + F^{II}_{ijmn}(\mathbf{0})E^{II}_{mn} + F^{II}_{ijmnp}(\mathbf{0})E^{II}_{mnp} \\ &+ \left[F^I_{ijmn,r}(\mathbf{0})E^I_{mn} + F^I_{ijmnp,r}(\mathbf{0})E^I_{mnp} + F^{II}_{ijmn,r}(\mathbf{0})E^{II}_{mn} + F^{II}_{ijmnp,r}(\mathbf{0})E^{II}_{mnp}\right]x_r + \cdots\end{aligned} \tag{4.92}$$

Notice that $F^I_{ijmnp}(\mathbf{0}) = \mathbf{0}$ and $F^I_{ijmn,r}(\mathbf{0}) = \mathbf{0}$. Substituting the above equation into Equation 4.91 and comparing the coefficents at each order, one can obtain

$$\begin{cases}\epsilon^0_{ij} + F^I_{ijmn}(\mathbf{0})E^I_{mn} + F^{II}_{ijmn}(\mathbf{0})E^{II}_{mn} + F^{II}_{ijmnp}(\mathbf{0})E^{II}_{mnp} = \Delta C^{-1}_{ijkl}C^0_{klmn}E^I_{mn} \\ F^I_{ijmnp,r}(\mathbf{0})E^I_{mnp} + F^{II}_{ijmn,r}(\mathbf{0})E^{II}_{mn} + F^{II}_{ijmnp,r}(\mathbf{0})E^{II}_{mnp} = \Delta C^{-1}_{ijkl}C^0_{klmn}E^I_{mnr}\end{cases} \tag{4.93}$$

Similarly, for the second inhomogeneity, one can also write two equations as

$$\begin{cases}\epsilon^0_{ij} + F^{II}_{ijmn}(\bar{\mathbf{0}})E^{II}_{mn} + F^I_{ijmn}(\bar{\mathbf{0}})E^I_{mn} = \Delta C^{-1}_{ijkl}C^0_{klmn}E^{II}_{mn} \\ F^{II}_{ijmnp,r}(\mathbf{0})E^{II}_{mnp} + F^I_{ijmn,r}(\mathbf{0})E^{II}_{mn} + F^I_{ijmnp,r}(\mathbf{0})E^{II}_{mnp} = \Delta C^{-1}_{ijkl}C^0_{klmn}E^{II}_{mnr}\end{cases} \tag{4.94}$$

Solving the linear equation system in the above four sets of equations, the four unknown tensors E^{I}_{ij}, E^{I}_{ijk}, E^{II}_{ij}, and E^{II}_{ijk} can be determined. Then, the displacement, strain, and stress fields can be derived. The formulation can be extended to multiple particles with different properties in a matrix. The accuracy of the method can be tailored by using more terms in the polynomial form of the eigenstrain [41,42].

4.3.2 Equivalent Inclusion Method for Potential Problems of Ellipsoidal Inhomogeneities

For multiphysical material behavior of an ellipsoidal inclusion in an infinitely extended domain, the governing equation for the steady state problem can be written as a potential or Poisson's equation

$$\nabla^2 U(\mathbf{x}) = f(\mathbf{x}) \tag{4.95}$$

where U is the potential and f is a source. In the linear range, the gradient of the potential is typically proportional to a physical flow (flux). Corresponding to the elastic problem, the potential, the gradient, and the flow are mapped into displacement, strain, and stress except that in the potential problem the variables have one rank lower than the elastic problem, that is, a vector field reduces to a scalar field, and a second-rank tensor reduces to a vector field. The above equation mathematically models many important physical phenonmena as given in the following table:

Problems	**Potential**	**Proportionality Constant**	**Physical Flow (flux)**
Heat transfer	Temperature T	Thermal conductivity k	Heat flux $\mathbf{q} = -k\nabla T$
Ground water flow	Hydradic heat H	Permeability k	Water flow $\mathbf{q} = -k\nabla H$
Electrostatic field	Potential E	Permittivity ϵ	Electric flow $\mathbf{q} = -\epsilon\nabla E$
Electric conduction	Electric voltage V	Electric conductivity k	Electric current $\mathbf{I} = -k\nabla V$
Magnetic field	Potential U	Permeability μ	Induction $\mathbf{B} = -\mu\nabla U$
Gravitation	Potential ϕ	Gravitational constant G	Gravitational force $\mathbf{F} = G\nabla\phi$

For one ellipsoidal particle embedded in the infinite domain subjected to a uniform far-field load, the problem can be similarly solved by the equivalent inclusion method. Here, we use the magnetostatic problem as an example to demonstrate the method.

For an ellipsoidal particle with a magnetic permeability μ_1 embedded in a matrix with a magnetic permeability μ_0, the local magnetic field and magnetic induction satisfy the constitutive relation:

$$\begin{cases} B_i(\mathbf{x}) = \mu_1 H_i(\mathbf{x}) & x \in \Omega \\ B_i(\mathbf{x}) = \mu_0 H_i(\mathbf{x}) & x \in D - \Omega \end{cases} \tag{4.96}$$

Here, the magnetic field $H_i(\mathbf{x})$ can be written with the magnetic potential U as

$$H_i = -U_{,i} \tag{4.97}$$

When a far-field magnetic load H_i^0 is applied, disturbed fields H_i' and B_i' will be induced by the material mismatch around inhomogeneity Ω, which satisfies

$$y(n) = \begin{cases} B_i^0 + B_i'(\mathbf{x}) = \mu_1 \left(H_i^0 + H_i'(\mathbf{x})\right) & x \in \Omega \\ B_i^0 + B_i'(\mathbf{x}) = \mu_0 \left(H_i^0 + H_i'(\mathbf{x})\right) & x \in D - \Omega \end{cases} \tag{4.98}$$

Using the equivalent inclusion method, one can introduce a prescribed magnetic field, similar to eigenstrain in elastic problem, to simulate the material difference, so that the above equation is rewritten as

$$y(n) = \begin{cases} B_i^0 + B_i'(\mathbf{x}) = \mu_0 \left(H_i^0 + H_i'(\mathbf{x}) + M_i(\mathbf{x})\right) & x \in \Omega \\ B_i^0 + B_i'(\mathbf{x}) = \mu_0 \left(H_i^0 + H_i'(\mathbf{x})\right) & x \in D - \Omega \end{cases} \tag{4.99}$$

where the prescribed magnetic field $M_i(\mathbf{x})$ is also called magnetization if the matrix is a nonmagnetic material. Notice ths sign of the eigenstrain in elastic problem is "−" but here the size of the magnetization is "+", which is consistent with the traditional definition of the magnetization. Mathematically, one can consider $-M_i(\mathbf{x})$ as a counterpart to the eigenstrain. If M_i is considered a piecewise continuous function, which is zero in the matrix, the first equation in the above equations can represent the constitutive relation in the whole domain. For a steady-state magnetic field, the divergence of the magnetic induction is zero, that is,

$$B_{i,i} = 0 \tag{4.100}$$

Considering the disturbed magnetic field $H_i' = -U_{,i}$, one can write

$$U_{,ii}(\mathbf{x}) = M_{i,i}(\mathbf{x}) \tag{4.101}$$

Using Green's function technique, one can write the solution of the magnetic potential as

$$U(\mathbf{x}) = -\frac{1}{4\pi} \int_D \phi \frac{\partial M_i(\mathbf{x}')}{\partial x_i'} d\mathbf{x}' \tag{4.102}$$

Using the Gauss theorem, the above equation can be rewritten as

$$U(\mathbf{x}) = -\frac{1}{4\pi} \int_D \frac{\partial \phi}{\partial x_i} M_i(\mathbf{x}') \, d\mathbf{x}' \tag{4.103}$$

For the case of a single ellipsoidal inhomogeneity, the magnetization M_i is constant, written as M_i^0, the above equation can be rewritten as

$$U(\mathbf{x}) = -\frac{M_i^0}{4\pi} \Phi_{,i} \tag{4.104}$$

Then, the disturbed magnetic field can be written as

$$H'_i(\mathbf{x}) = \frac{M_k^0}{4\pi}\Phi_{,ik} \tag{4.105}$$

Using the equivalent inclusion method, the magnetization M_i^0 can be solved from the following equation over the inhomogeneity:

$$\mu_0\left(H_i^0 + \frac{M_k^0}{4\pi}\Phi_{,ik} + M_i^0\right) = \mu_1\left(H_i^0 + \frac{M_k^0}{4\pi}\Phi_{,ik}\right) \tag{4.106}$$

Therefore, one can obtain

$$M_i^0 = \left(\frac{\mu_0}{\mu_1 - \mu_0}\delta_{ik} - \frac{\Phi_{,ik}}{4\pi}\right)^{-1} H_k^0 \tag{4.107}$$

Then one can get magnetic potential, field, and induction straightforwardly [43,44].

For more than one inhomogeneities, the magnetization will be in a polynomial form similarly to Section 4.3.1. The same method as the last section can be used to derive the magnetic fields. The formulation can be extended to multiple particles with multiple material phases for multiphysical problems [45–47].

4.4 EXERCISES

1. Given a matrix $K_{ik}(\boldsymbol{\xi}) = C_{ijkl}\xi_j\xi_l$ with $C_{ijkl} = \lambda\delta_{ij}\delta_{kl} + \mu(\delta_{ik}\delta_{jl} + \delta_{il}\delta_{jk})$, show that: the determinant can be written as $D(\boldsymbol{\xi}) = \mu^2(\lambda + 2\mu)\xi^6$, and the cofactor matrix can be written as $N_{ij}(x) = \mu\xi^2\left[(\lambda + 2\mu)\delta_{ij}\xi^2 - (\lambda + \mu)\xi_i\xi_j\right]$, where $\xi^2 = \xi_k\xi_k$, $\frac{K_{ij}N_{jk}}{D} = \delta_{ik}$.

2. The elastic Green's function is written as

$$G_{ij}(\mathbf{x}, \mathbf{x}') = \frac{1}{(2\pi)^3}\int_{-\infty}^{\infty} \frac{N_{ij}(\xi)}{D(\xi)} \exp\left[i\xi.(\mathbf{x} - \mathbf{x}')\right] d\boldsymbol{\xi}$$

   Use the above equation to prove that $C_{ijkl}G_{km,lj}(\mathbf{x}, \mathbf{x}') = -\delta_{im}\delta(\mathbf{x} - \mathbf{x}')$. Notice that the 3D Dirac delta function can be written in the following form:

$$\delta(\mathbf{x} - \mathbf{x}') = \frac{1}{(2\pi)^3}\int_{-\infty}^{\infty} \exp\left[i\xi \cdot (\mathbf{x} - \mathbf{x}')\right] d\boldsymbol{\xi}.$$

3. Derive the Eshelby's tensor in the form of Equation 4.55 for (a) fiber: $a_1 = a_2$, $a_3 \rightarrow \infty$; (b) penny: $a_1 = a_2$, $a_3 \rightarrow 0$, and; a slit $a_2 = \infty$, $a_3 = 0$, and a_1 is a finite number.

4. A fiber with μ_1 and υ_1 included in a uniform matrix with μ_0 and υ_0 can be treated as a circular cylinder in an infinite domain with $a_1 = a_2 = a$ and $a_3 \rightarrow \infty$. Derive the stress in the cylinder subjected to a far-field stress σ_{ij}^{∞}. Notice that the integral of the modified Green's function can be found in Mura's book.

5. Consider a spheroidal particle with $a_1 = a_2 = 1$ and a_3 varying in the range of 0.2–5 embedded in an infinitely extended matrix. The particle and matrix have the material properties $\mu_1 = 2$ and $\upsilon_1 = 0.5$ and $\mu_0 = 1$ and $\upsilon_0 = 0.5$, respectively. When a body force $f_3 = 1$ is applied on the particle along the x_3 direction, illustrate the displacement along the x_1 and x_3 axes for $a_3 = 0.2, 0.5, 1, 2, 5$.

CHAPTER 5

Volume Integrals and Averages in Inclusion and Inhomogeneity Problems

CONSIDER A MICROMECHANICS-BASED FRAMEWORK, the effective material behavior is usually described by the volume average of the local fields and studied by either vectorial or energy approaches. This chapter will provide volume integrals and averages of elastic fields and strain energy for inclusion and inhomogeneity problems.

5.1 VOLUME AVERAGES OF STRESS AND STRAIN

5.1.1 Average Stress and Strain for an Inclusion Problem

For an inclusion Ω with an eigenstrain ϵ_{ij}^* embedded in a large domain D with free stress boundary condition, as shown in Figure 5.1, the stress and displacement field in the domain is continuous. The average stress can be written as

$$
\begin{aligned}
\langle\sigma_{ij}\rangle_D &= \frac{1}{V_D}\int_D \sigma_{ij}(\mathbf{x})d\mathbf{x} \\
&= \frac{1}{V_D}\int_D \sigma_{ik}(\mathbf{x})x_{j,k}d\mathbf{x} \\
&= \frac{1}{V_D}\int_D \left[\left(\sigma_{ik}(\mathbf{x})x_j\right)_{,k} - \sigma_{ik,k}(\mathbf{x})x_j\right]d\mathbf{x} \\
&= \frac{1}{V_D}\int {}_{\partial D}\sigma_{ik}(\mathbf{x})n_k x_j d\mathbf{x}
\end{aligned}
\tag{5.1}
$$

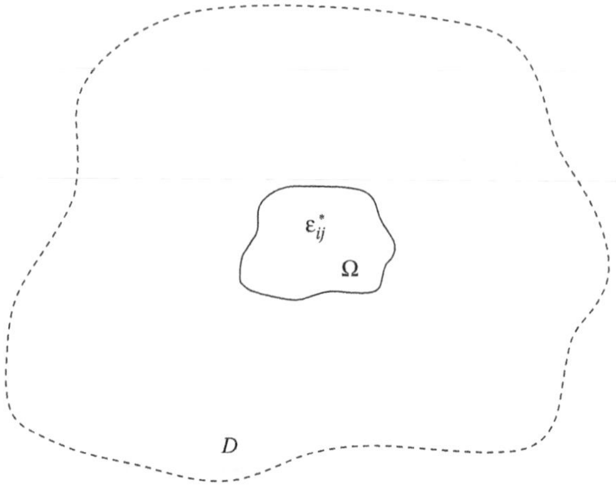

FIGURE 5.1 An inclusion Ω with an eigenstrain ϵ_{ij}^* embedded in a large domain D.

Because the stress satisfies the free stress boundary condition, the above equation yields

$$\langle \sigma_{ij} \rangle_D = 0 \tag{5.2}$$

Therefore, the eigenstrain does not produce any effective stress in the solid. The average strain can be written as

$$\begin{aligned}
\langle \epsilon_{ij} \rangle_D &= \frac{1}{V_D} \int_D \epsilon_{ij}(\mathbf{x}) d\mathbf{x} \\
&= \frac{1}{V_D} \int_D \left[\epsilon_{ij}^e(\mathbf{x}) + \epsilon_{ij}^*(\mathbf{x}) \right] d\mathbf{x} \\
&= \frac{C_{ijkl}^{-1}}{V_D} \int_D \sigma_{ij}(\mathbf{x}) d\mathbf{x} + \frac{1}{V_D} \int_D \epsilon_{ij}^*(\mathbf{x}) d\mathbf{x} \\
&= \phi \left\langle \epsilon_{ij}^* \right\rangle_\Omega
\end{aligned} \tag{5.3}$$

where ϕ is the volume fraction of the subdomain Ω to the domain D. Notice that the above two equations are general such that the inclusion can be of any shape and the overall domain can be a finite domain. For a uniform eigenstrain field over the inclusion, the above equation can be written as

$$\langle \epsilon_{ij} \rangle_D = \phi \epsilon_{ij}^* \tag{5.4}$$

5.1.2 Average Stress and Strain for an Inhomogeneity Problem

Consider an inhomogeneity problem with a stress vector $\sigma_{ij}^0 n_i$ on the boundary of ∂D, as shown in Figure 5.2 with $\mathbf{n}$ being the out-normal vector of the surface ∂D. The stiffness tensors of the inhomogeneity and the matrix are written as $\mathbf{C}^*$ and $\mathbf{C}$, respectively. One

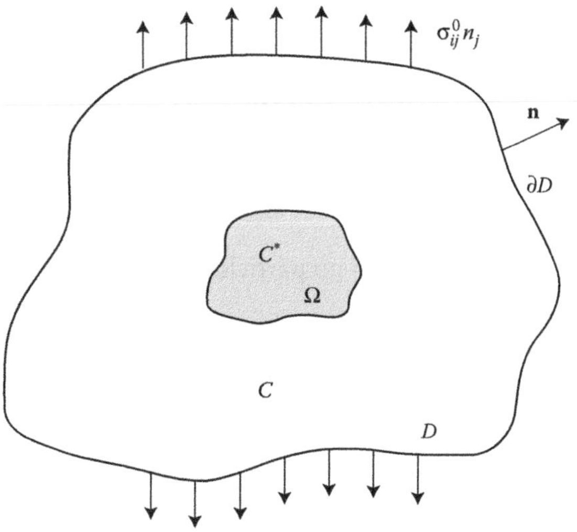

FIGURE 5.2 Inhomogeneity problem with a stress vector $\sigma_{ij}^0 n_i$ on the boundary of ∂D.

can derive the average stress and strain in the similar manner. The average stress can be written as

$$\begin{aligned}\langle\sigma_{ij}\rangle_D &= \frac{1}{V_D}\int_{\partial D}\sigma_{ik}(\mathbf{x})n_k x_j d\mathbf{x}\\ &= \frac{1}{V_D}\int_{\partial D}\sigma_{ik}^0 n_k x_j d\mathbf{x}\\ &= \frac{\sigma_{ik}^0}{V_D}V_D\delta_{jk}\\ &= \sigma_{ij}^0 \end{aligned}\tag{5.5}$$

Similarly, the average strain can be written in terms of the equivalent inclusion problem, in which the material mismatch of the inhomogeneity is simulated by an eigenstrain as

$$\begin{aligned}\langle\epsilon_{ij}\rangle_D &= \frac{C_{ijkl}^{-1}}{V_D}\int_D \sigma_{ij}(\mathbf{x})d\mathbf{x} + \frac{1}{V_D}\int_D \epsilon_{ij}^*(\mathbf{x})d\mathbf{x}\\ &= C_{ijkl}^{-1}\sigma_{kl}^0 + \phi\left\langle\epsilon_{ij}^*\right\rangle_\Omega \end{aligned}\tag{5.6}$$

where the eigenstrain satisfies the equivalent inclusion condition, such as

$$C_{ijkl}^*\epsilon_{kl}(\mathbf{x}) = C_{ijkl}\left[\epsilon_{kl}(\mathbf{x}) - \epsilon_{kl}^*(\mathbf{x})\right] \text{ for } \mathbf{x}\in\Omega \tag{5.7}$$

When D is much larger than Ω and Ω is far from the boundary, the eigenstrain can be approximated by Eshelby's solution for an ellipsoidal inhomogeneity in an infinite domain:

$$\boldsymbol{\epsilon}^* = \mathbf{C}^{-1} : \left(\mathbf{D}^\Omega - \Delta\mathbf{C}^{-1}\right)^{-1} : \mathbf{C}^{-1} : \boldsymbol{\sigma}^0 \tag{5.8}$$

Therefore, the average strain can be rewritten as

$$\langle \epsilon \rangle_D = \mathbf{C}^{-1} : \sigma^0 + \phi \mathbf{C}^{-1} : \left(\mathbf{D}^{\Omega} - \Delta \mathbf{C}^{-1} \right)^{-1} : \mathbf{C}^{-1} : \sigma^0 \tag{5.9}$$

Notice that the above equation is subjected to the following conditions:

1. Only one particle is considered, so no particle interaction is included.
2. A large domain is implied, so that the volume fraction should be very small.
3. The stress vector corresponds to a uniform stress at the boundary or a far-field stress σ^0.
4. Particle's shape is ellipsoidal to make $\mathbf{D}^{\Omega}$ as a constant tensor.

When multiple particles are considered, the eigenstrain will change as demonstrated in the last chapter. If the volume of domain D is in the same order as the volume of Ω, a finite domain must be considered. Image stresses have been used to take into account of the boundary effect, which will be introduced later in this chapter. For nonellipsoidal particles, the Tanaka–Mori theorem will be introduced.

5.1.3 Tanaka–Mori's Theorem

For an inclusion with an arbitrary shape, the stress caused by a uniform eigenstrain ϵ^*_{ij} can be written as

$$\begin{aligned} \sigma_{ij}(\mathbf{x}) &= C_{ijkl} \left[\epsilon_{kl}(\mathbf{x}) - \epsilon^*_{ij}(\mathbf{x}) \right] \\ &= -C_{ijkl} \left[\int_D \Gamma_{klmn}(\mathbf{x}, \mathbf{x}') C_{mnpq} \epsilon^*_{pq}(\mathbf{x}') d\mathbf{x}' - \epsilon_{kl}(\mathbf{x}) \right] \end{aligned} \tag{5.10}$$

The volume integral can only be explicitly obtained for a certain shape, for example ellipsoidal. In Figure 5.3, an irregular particle Ω is embedded in an ellipsoidal domain V_1 and then in another larger ellipsoidal domain V_2, which is then in the infinitely large domain D, with the relation $\Omega \subset V_1 \subset V_2 \subset D$. Tanaka and Mori [15] investigated the integral of stress over the region $V_2 - V_1$ as follows:

$$\int_{V_2 - V_1} \sigma_{ij}(\mathbf{x}) d\mathbf{x} = - \int_{V_2 - V_1} C_{ijkl} \left[\int_D \Gamma_{klmn}(\mathbf{x}, \mathbf{x}') C_{mnpq} \epsilon^*_{pq}(\mathbf{x}') d\mathbf{x}' - \epsilon_{kl}(\mathbf{x}) \right] d\mathbf{x} \tag{5.11}$$

Because the eigenstrain is uniform on Ω and zero on $D - \Omega$, the above equation can be rewritten as

$$\int_{V_2 - V_1} \sigma_{ij}(\mathbf{x}) d\mathbf{x} = -C_{ijkl} \int_{V_2 - V_1} \left[\int_{\Omega} \Gamma_{klmn}(\mathbf{x}, \mathbf{x}') d\mathbf{x}' \right] d\mathbf{x} C_{mnpq} \epsilon^*_{pq} \tag{5.12}$$

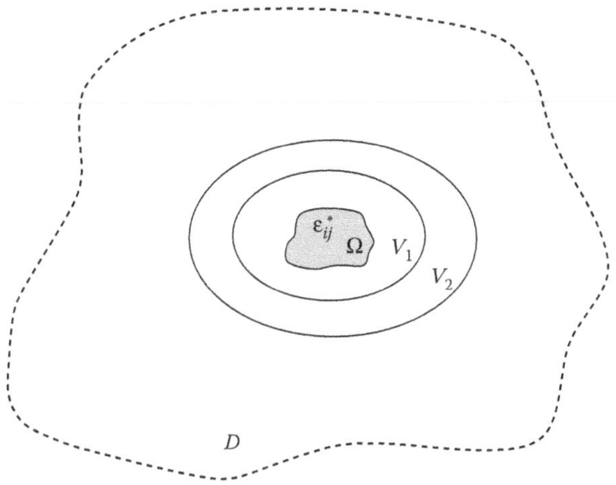

FIGURE 5.3 Tanaka–Mori's theorem.

where

$$\begin{aligned}\int_{V_2-V_1}\left[\int_{\Omega}\Gamma_{klmn}(\mathbf{x},\mathbf{x}')d\mathbf{x}'\right]d\mathbf{x} &= \int_{V_2}\left[\int_{\Omega}\Gamma_{klmn}(\mathbf{x},\mathbf{x}')d\mathbf{x}'\right]d\mathbf{x} - \int_{V_1}\left[\int_{\Omega}\Gamma_{klmn}(\mathbf{x},\mathbf{x}')d\mathbf{x}'\right]d\mathbf{x}\\ &= \int_{\Omega}\left[\int_{V_2}\Gamma_{klmn}(\mathbf{x},\mathbf{x}')d\mathbf{x}\right]d\mathbf{x}' - \int_{\Omega}\left[\int_{V_1}\Gamma_{klmn}(\mathbf{x},\mathbf{x}')d\mathbf{x}\right]d\mathbf{x}'\\ &= V_\Omega\left(D_{klmn}^{V_2} - D_{klmn}^{V_1}\right) \end{aligned} \tag{5.13}$$

Therefore, the above equation becomes

$$\int_{V_2-V_1}\sigma_{ij}(\mathbf{x})d\mathbf{x} = -V_\Omega C_{ijkl}\left(D_{klmn}^{V_2} - D_{klmn}^{V_1}\right)C_{mnpq}\epsilon_{pq}^* \tag{5.14}$$

Notice that the tensors $D_{klmn}^{V_2}$ and $D_{klmn}^{V_1}$ only depend on the shape of the domains and are not relevant to the size or location as long as $\Omega \subset V_1 \subset V_2 \subset D$. Therefore, the integral of stress in $V_2 - V_1$ is proportional to the volume of Ω and depends on the shape of the domain $V_2 - V_1$. Tanaka–Mori's theorem reads:

If V_1 and V_2 have the same shape and orientation, the average stress and strain caused by an eigenstrain over the subdomain Ω are zero. Here, the subdomain Ω can be of an arbitrary shape.

5.1.4 Image Stress and Strain for a Finite Domain

There does not exist a real infinite domain. When the size of an inhomogeneity or an inclusion is not small or it is close to the physical boundary of the matrix domain, Eshelby's solution for one particle in the infinite domain will not be directly applicable. However, one can use the concept of image stress or strain to solve for the stress or strain over the finite domain.

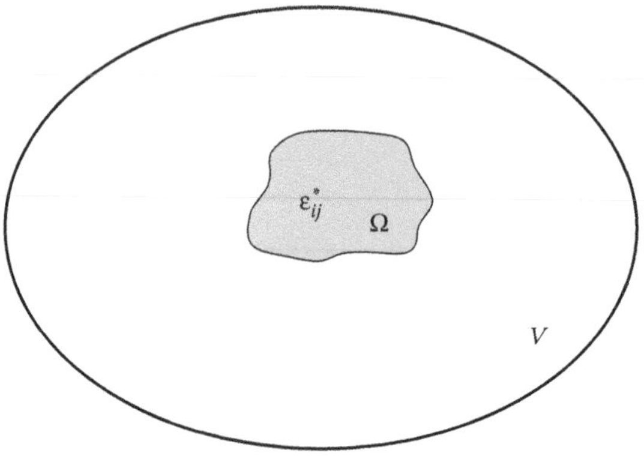

FIGURE 5.4 Image stress and strain.

Figure 5.4 illustrates that a uniform eigenstrain ϵ^*_{ij} on Ω is embedded in an ellipsoidal domain V. Although we have obtained the average stress and strain in Sections 5.1.2 and 5.1.3, for convenience to introduce the concept of the image stress and strain, we embed the domain V into an infinite domain D with the same stiffness and assume the domains Ω and V with the same shape and orientation. Let the stress caused by the eigenstrain be written as $\sigma'_{ij}(\mathbf{x})$ for the infinite domain, the volume integral of the stress over V is

$$\int_V \sigma'_{ij}(\mathbf{x})d\mathbf{x} = \int_\Omega \sigma'_{ij}(\mathbf{x})d\mathbf{x} + \int_{V-\Omega} \sigma'_{ij}(\mathbf{x})d\mathbf{x} \tag{5.15}$$

From the Tanaka–Mori theorem, the second term on the right-hand side of the above equation is zero and the above equation is reduced to

$$\int_V \sigma'_{ij}(\mathbf{x})d\mathbf{x} = \int_\Omega \sigma'_{ij}(\mathbf{x})d\mathbf{x} \tag{5.16}$$

which is obviously different from the case for free stress boundary of a finite domain in Equations 5.1 and 5.2. In the infinite domain, the stress $\sigma'_{ij}(\mathbf{x})$ at the boundary of the domain ∂V is nonzero, say $\sigma'_{ij}n_i$, but the finite domain exhibits a stress-free boundary. Because the first problem has been well studied, if a stress vector $-\sigma'_{ij}n_i$ is applied in the first problem, the superposition of Eshelby's solution with the stress field caused by the boundary stress vector $-\sigma'_{ij}n_i$ will be equivalent to the second problem. We call the stress or strain field caused by the boundary stress vector $-\sigma'_{ij}n_i$ as the image stress or strain, respectively, which is denoted by a superscript "I." One can write $\sigma_{ij}(\mathbf{x}) = \sigma'_{ij}(\mathbf{x}) + \sigma^I_{ij}(\mathbf{x})$ and $\epsilon_{ij}(\mathbf{x}) = \epsilon'_{ij}(\mathbf{x}) + \epsilon^I_{ij}(\mathbf{x})$. Here, the image stress satisfies

$$\begin{cases} \sigma^I_{ij,i}(\mathbf{x}) = 0 & \text{for } \mathbf{x} \in V. \\ \sigma^I_{ij}n_i = -\sigma'_{ij}n_i & \text{for } \mathbf{x} \in \partial V. \end{cases} \tag{5.17}$$

For some specific boundary geometries of domain V, such as semi-infinite or spherical domain, the image stress or strain can be explicitly solved. In general, it is difficult to determine the image stress or strain. However, the volume average can be derived as follows:

$$
\begin{aligned}
\left\langle \sigma_{ij}^{I} \right\rangle_V &= \frac{1}{V_V} \int\limits_V \sigma_{ij}^{I}(\mathbf{x}) d\mathbf{x} \\
&= \frac{1}{V_V} \int\limits_V \left[\left(\sigma_{k}^{I}(\mathbf{x}) x_j \right)_{,k} - \sigma_{ik,k}^{I}(\mathbf{x}) x_j \right] d\mathbf{x} \\
&= \frac{1}{V_V} \int\limits_{\partial V} \left[\sigma_{ik}^{I}(\mathbf{x}) n_k x_j \right] d\mathbf{x} \\
&= \frac{1}{V_V} \int\limits_{\partial V} \left[-\sigma_{ik}^{\prime}(\mathbf{x}) n_k x_j \right] d\mathbf{x} \\
&= -\frac{1}{V_V} \int\limits_V \left[\sigma_{ik}^{\prime}(\mathbf{x}) x_j \right]_{,k} d\mathbf{x} \\
&= -\frac{1}{V_V} \int\limits_V \sigma_{ij}^{\prime}(\mathbf{x}) d\mathbf{x} \\
&= -\frac{1}{V_V} \int\limits_\Omega \sigma_{ij}^{\prime}(\mathbf{x}) d\mathbf{x}
\end{aligned} \tag{5.18}
$$

Consider that a uniform eigenstrain ϵ_{ij}^{*} on Ω is embedded in a finite domain V. One can explicitly write the volume average of image stress and strain as follows:

$$
\left\langle \sigma_{ij}^{I} \right\rangle_V = -\frac{V_\Omega}{V_V} C_{ijkl} \left(-D_{klmn} C_{mnpq} \epsilon_{pq}^{*} - \epsilon_{kl}^{*} \right) = \phi C_{ijkl} \left(D_{klmn}^{\Omega} C_{mnpq} \epsilon_{pq}^{*} + \epsilon_{kl}^{*} \right) \tag{5.19}
$$

and

$$
\begin{aligned}
\langle \epsilon_{ij}^{I} \rangle_V &= \langle \epsilon_{ij} \rangle_V - \langle \epsilon_{ij}^{\prime} \rangle_V \\
&= \phi \epsilon_{ij}^{*} - \frac{1}{V_V} \int\limits_V \epsilon_{ij}^{\prime}(\mathbf{x}) d\mathbf{x} \\
&= \phi \epsilon_{ij}^{*} + \phi D_{ijkl}^{V} C_{klmn} \epsilon_{mn}^{*} \\
&= \phi \epsilon_{ij}^{*} - \phi S_{ijmn}^{V} \epsilon_{mn}^{*}
\end{aligned} \tag{5.20}
$$

where the Eshelby's tensor $S_{ijmn}^{V} = -D_{ijkl}^{V} C_{klmn}$ is used. For an inhomogeneity problem of finite domain, the image stress and strain can be solved in the similar process. However, the equivalent inclusion condition need to be used to solve for the eigenstrain.

5.2 VOLUME AVERAGES IN POTENTIAL PROBLEMS

The above concepts in Section 5.1 can be easily extended to potential problems. Following Section 4, we still use magnetostatic problem as an example.

5.2.1 Average Magnetic Field and Flux for an Inclusion Problem

For an inclusion Ω with a magnetization M_i and magnetic permeability μ_0 embedded in a large domain D with the same magnetic permeability μ_0 with free external magnetic flux, as shown in Figure 5.5, the magnetic flux $\mathbf{B}(\mathbf{x})$ and potential $U(\mathbf{x})$ in the domain are continuous. The average magnetic flux can be written as

$$\begin{aligned}
\langle B_i \rangle_D &= \frac{1}{V_D} \int_D B_i(\mathbf{x}) d\mathbf{x} \\
&= \frac{1}{V_D} \int_D B_j(\mathbf{x}) x_{i,j} d\mathbf{x} \\
&= \frac{1}{V_D} \int_D \left[\left(B_j(\mathbf{x}) x_i \right)_{,j} - B_{j,j}(\mathbf{x}) x_i \right] d\mathbf{x} \\
&= \frac{1}{V_D} \int {}_{\partial D} B_j(\mathbf{x}) n_j x_i d\mathbf{x}
\end{aligned} \tag{5.21}$$

where Gauss' law is used as $B_{j,j}(\mathbf{x}) = 0$. Because the magnetic flux is free on the boundary, the above equation yields

$$\langle B_i \rangle_D = 0 \tag{5.22}$$

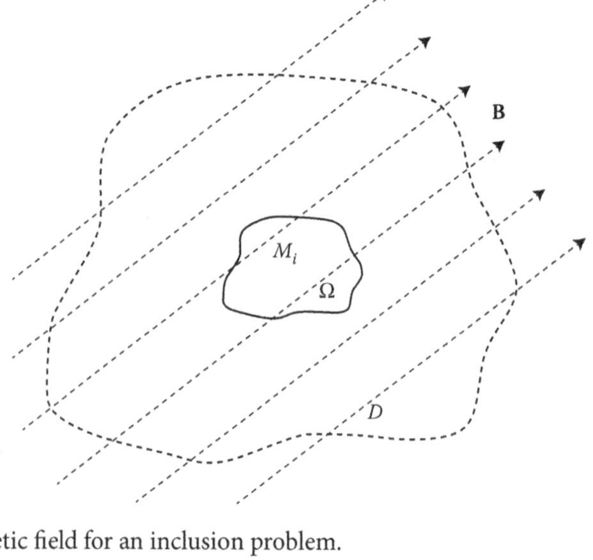

FIGURE 5.5 Magnetic field for an inclusion problem.

Therefore, the magnetization does not produce any effective magnetic flux in the material. The average magnetic field can be written as

$$\begin{aligned}
\langle H_i \rangle_D &= \frac{1}{V_D} \int_D H_i(\mathbf{x}) d\mathbf{x} \\
&= \frac{1}{V_D} \int_D \left[\frac{B_i(\mathbf{x})}{\mu_0} - M_i(\mathbf{x}) \right] d\mathbf{x} \\
&= \frac{1}{V_D} \int_D \frac{B_i(\mathbf{x})}{\mu_0} d\mathbf{x} - \frac{1}{V_D} \int_D M_i(\mathbf{x}) d\mathbf{x} \\
&= -\phi \langle M_i \rangle_\Omega
\end{aligned} \tag{5.23}$$

5.2.2 Average Magnetic Field and Flux for an Inhomogeneity Problem

Consider a spherical inhomogeneity problem with a magnetic flux $B_i^0 n_i$ on the boundary of ∂D, as shown in Figure 5.6, with $\mathbf{n}$ being the out-normal vector of the surface ∂D. The magnetic permeabilities of the inhomogeneity and the matrix are written as μ_1 and μ_0, respectively. One can derive the average magnetic field and flux in the similar manner. The average magnetic flux can be written as

$$\begin{aligned}
\langle B_i \rangle_D &= \frac{1}{V_D} \int_D B_i(\mathbf{x}) d\mathbf{x} \\
&= \frac{1}{V_D} \int_D B_j(\mathbf{x}) x_{i,j} d\mathbf{x}
\end{aligned}$$

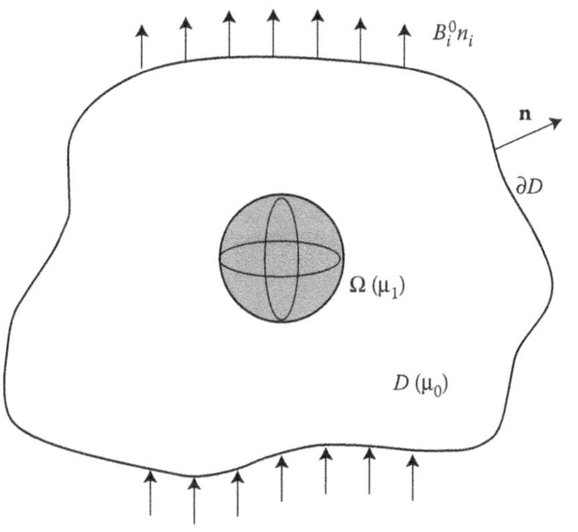

FIGURE 5.6 Spherical inhomogeneity with a magnetic flux $B_i^0 n_i$ on the boundary of ∂D.

$$= \frac{1}{V_D} \int_{\partial D} B_j(\mathbf{x}) x_i n_j d\mathbf{x} - \frac{1}{V_D} \int_{D} \left[B_{j,j}(\mathbf{x}) x_i \right] d\mathbf{x}$$

$$= B_i^0 \tag{5.24}$$

Similarly, the average magnetic field can be written in terms of the equivalent inclusion problem, in which the material mismatch of the inhomogeneity is simulated by the magnetization as

$$\langle H_i \rangle_D = \frac{1}{V_D} \int_D \left[\frac{B_i(\mathbf{x})}{\mu_0} - M_i(\mathbf{x}) \right] d\mathbf{x}$$

$$= \frac{B_i^0}{\mu_0} - \phi \langle M_i \rangle_\Omega \tag{5.25}$$

where the magnetization satisfies

$$\mu_1 H_i(\mathbf{x}) = \mu_0 \left[H_i(\mathbf{x}) + M_i(\mathbf{x}) \right] \text{ for } \mathbf{x} \in \Omega \tag{5.26}$$

When D is much larger than Ω and Ω is far from the boundary, the magnetization can be approximated by the solution for a spherical inhomogeneity in an infinite domain:

$$M_i = 3 \frac{\mu_1 - \mu_0}{\mu_1 + 3\mu_0} \frac{B_i^0}{\mu_0} \tag{5.27}$$

Therefore, the average magnetic field can be rewritten as

$$\langle H_i \rangle_D = \frac{B_i^0}{\mu_0} - 3\phi \frac{\mu_1 - \mu_0}{\mu_1 + 3\mu_0} \frac{B_i^0}{\mu_0} \tag{5.28}$$

We can obtain the similar results of Tanaka–Mori's theorem and image magnetic flux following Section 5.1.

5.3 STRAIN ENERGY IN INCLUSION AND INHOMOGENEITY PROBLEMS

5.3.1 Strain Energy for an Inclusion in an Infinite Domain

For an inclusion problem that a uniform eigenstrain ϵ_{ij}^* exists in an ellipsoidal subdomain Ω of an infinite domain D, let the elastic (displacement, strain, and stress) fields inside the inclusion ($x \in \Omega$) be denoted by a superscript Ω and the elastic fields outside the inclusion or the matrix ($x \in D - \Omega$) be denoted by a superscript M. The overall fields caused by the source ϵ_{ij}^* are denoted by a superscript T. The elastic fields can be written in the following table:

Matrix	Inhomogeneity	Comments
$\epsilon_{ij}^M = \epsilon_{ij}' + \epsilon_{ij}^0$	$\epsilon_{ij}^\Omega = \epsilon_{ij}' + \epsilon_{ij}^0$	$\epsilon_{ij}' = -D_{ijkl}(\mathbf{x})C_{klmn}\epsilon_{mn}^*$
$u_i^M = u_i' + u_i^0$	$u_i^\Omega = u_i' + u_i^0$	The rigid-body displacement is not considered
$\sigma_{ij}^M = \sigma_{ij}' + \sigma_{ij}^0$	$\sigma_{ij}^\Omega = \sigma_{ij}' + \sigma_{ij}^0$	$C_{ijkl}^*\left(\epsilon_{kl}' + \epsilon_{kl}^0\right) = C_{ijkl}\left(\epsilon_{kl}' + \epsilon_{kl}^0 - \epsilon_{kl}^*\right)$

The total elastic energy can be written as

$$\begin{aligned}
W &= \frac{1}{2}\int_{\Omega} \sigma_{ij}^\Omega \epsilon_{ij}^\Omega d\mathbf{x} + \frac{1}{2}\int_{D-\Omega} \sigma_{ij}^M \epsilon_{ij}^M d\mathbf{x} \\
&= \frac{1}{4}\int_{\Omega} \sigma_{ij}^\Omega \left(u_{i,j}^\Omega + u_{j,i}^\Omega\right) d\mathbf{x} + \frac{1}{4}\int_{D-\Omega} \sigma_{ij}^M \left(u_{i,j}^M + u_{j,i}^M\right) d\mathbf{x} \\
&= \frac{1}{2}\int_{\Omega} \sigma_{ij}^\Omega u_{j,i}^\Omega d\mathbf{x} + \frac{1}{2}\int_{D-\Omega} \sigma_{ij}^M u_{j,i}^M d\mathbf{x} \\
&= \frac{1}{2}\int_{\Omega} \left[\left(\sigma_{ij}^\Omega u_j^\Omega\right)_{,i} - \sigma_{ij,i}^\Omega u_j^\Omega\right] d\mathbf{x} + \frac{1}{2}\int_{D-\Omega} \left[\left(\sigma_{ij}^M u_j^M\right)_{,i} - \sigma_{ij,i}^M u_j^M\right] d\mathbf{x} \\
&= \frac{1}{2}\int_{\partial\Omega} \sigma_{ij}^\Omega n_i u_j^\Omega dS + \frac{1}{2}\int_{\partial D-\partial\Omega} \sigma_{ij}^M n_i u_j^M dS \\
&= \frac{1}{2}\int_{\partial\Omega} \left(\sigma_{ij}^\Omega n_i u_j^\Omega - \sigma_{ij}^M n_i u_j^M\right) dS + \frac{1}{2}\int_{\partial D} \sigma_{ij}^M n_i u_j^M dS \\
&= \frac{1}{2}\int_{\partial\Omega} \sigma_{ij}^\Omega n_i \left(u_j^\Omega - u_j^M\right) dS
\end{aligned} \tag{5.29}$$

In the last step, the stress continuity on $\partial\Omega$, that is, $\sigma_{ij}^\Omega n_i = \sigma_{ij}^M n_i$, and $\int_{\partial D} \sigma_{ij}^M n_i u_j^M dS \to 0$ as $D \to \infty$ have been used. Considering the continuity of u_i^T on $\partial\Omega$, the above equation can be written as

$$\begin{aligned}
W &= -\frac{1}{2}\int_{\partial\Omega} \sigma_{ij}^\Omega n_i \epsilon_{jk}^* x_k dS \\
&= -\frac{\epsilon_{jk}^*}{2}\int_{\partial\Omega} \sigma_{ij}^\Omega n_i x_k dS
\end{aligned} \tag{5.30}$$

Consider that a uniform eigenstrain ϵ_{ij}^* on an ellipsoidal inclusion Ω is embedded in an infinite domain D. The stress on the particle is uniform, so the above equation can be rewritten as

$$\begin{aligned} W &= -\frac{\sigma_{ij}^{\Omega}\epsilon_{jk}^*}{2}\int_{\partial\Omega} n_i x_k dS \\ &= -\frac{V_\Omega \sigma_{ij}^{\Omega}\epsilon_{jk}^*}{2}\delta_{ik} \\ &= -\frac{V_\Omega \sigma_{ij}^{\Omega}\epsilon_{ji}^*}{2} \end{aligned} \tag{5.31}$$

If the inclusion Ω is not ellipsoidal, the integral can also be written of the average stress on the inclusion using the Gauss theorem as follows:

$$W = -\frac{1}{2}V_\Omega \left\langle \sigma_{ij}^{\Omega}\right\rangle_{V_\Omega} \epsilon_{ji}^* \tag{5.32}$$

The strain energy on the inclusion can be written as

$$W^{\Omega} = \frac{1}{2}\int_{\Omega} \sigma_{ij}^{\Omega}\epsilon_{ij}^{\Omega} d\mathbf{x} = \frac{V_\Omega}{2}\sigma_{ij}^{\Omega}\left(\epsilon_{ij}^T - \epsilon_{ij}^*\right) \tag{5.33}$$

Considering the average stress $\langle\sigma_{ij}^{\Omega}\rangle_{V_\Omega} = \langle\sigma_{ij}^{T}\rangle_{V_\Omega} - \sigma_{ij}^*$, the strain energy in the matrix can be written as

$$W^M = W - W^{\Omega} = -\frac{1}{2}V_\Omega \sigma_{ij}^{\Omega}\epsilon_{ij}^T \tag{5.34}$$

The above results can also be derived by the work method that the strain energy stored in the solid must be equal to the work done to it in a reversible process. If the inclusion Ω is gradually taken out from the domain D, the strain energy in both the inclusion and the matrix will be released. This process can be divided into three steps:

1. Take the inclusion out, but keep the stress on the interface the same. The strain energy change is zero.
2. Gradually release the stress on the inclusion outer surface to zero. The strain energy change is W^{Ω}.
3. Gradually release the stress on the inner surface of the matrix to zero. The strain energy change is W^M.

5.3.2 Strain Energy for an Inclusion in a Finite Solid

Consider an inclusion problem that a uniform eigenstrain ϵ_{ij}^* exists in an ellipsoidal subdomain Ω of a finite solid D. As discussed in the last section, the elastic fields can be considered as the superposition of Eshelby's solution for one inclusion in the infinite domain and image stress/strain caused by the boundary effect. The elastic fields can be rewritten as in the following table:

Matrix	Inhomogeneity	Comments
$\epsilon_{ij}^{M} = \epsilon'_{ij} + \epsilon_{ij}^{0}$	$\epsilon_{ij}^{\Omega} = \epsilon'_{ij} + \epsilon_{ij}^{0}$	$\epsilon'_{ij} = -D_{ijkl}(\mathbf{x})C_{klmn}\epsilon_{mn}^{*}$
$u_i^{M} = u'_i + u_i^{0}$	$u_i^{\Omega} = u'_i + u_i^{0}$	The rigid-body displacement is not considered
$\sigma_{ij}^{M} = \sigma'_{ij} + \sigma_{ij}^{0}$	$\sigma_{ij}^{\Omega} = \sigma'_{ij} + \sigma_{ij}^{0}$	$C_{ijkl}^{*}\left(\epsilon'_{kl} + \epsilon_{kl}^{0}\right) = C_{ijkl}\left(\epsilon'_{kl} + \epsilon_{kl}^{0} - \epsilon_{kl}^{*}\right)$

On the boundary ∂D, the free stress boundary condition will be satisfied as

$$\sigma_{ij}^{M} n_i = 0 \tag{5.35}$$

Similarly, the total elastic energy caused by the constant eigenstrain in Ω can be written as

$$\begin{aligned} W &= \frac{1}{2}\int_{\Omega} \sigma_{ij}^{\Omega}\epsilon_{ij}^{\Omega} d\mathbf{x} + \frac{1}{2}\int_{D-\Omega} \sigma_{ij}^{M}\epsilon_{ij}^{M} d\mathbf{x} \\ &= -\frac{1}{2}\int_{\partial\Omega} \sigma_{ij}^{\Omega} n_i \epsilon_{jk}^{*} x_k dS \end{aligned} \tag{5.36}$$

Because $\sigma_{ij}^{\Omega} = \sigma_{ij}^{T} - \sigma_{ij}^{*} + \sigma_{ij}^{I}$, the strain energy can also be written in terms of two parts: W_∞ caused by $\sigma_{ij}^{T} - \sigma_{ij}^{*}$ and W_I caused by σ_{ij}^{I} as follows:

$$\begin{aligned} W_\infty &= -\frac{1}{2}\int_{\partial\Omega} \left(\sigma_{ij}^{T} - \sigma_{ij}^{*}\right) n_i \epsilon_{jk}^{*} x_k d\mathbf{x} \\ &= -\frac{V_\Omega \left(\sigma_{ij}^{T} - \sigma_{ij}^{*}\right)\epsilon_{jk}^{*}}{2}\delta_{ik} \\ &= -\frac{V_\Omega \left(\sigma_{ij}^{T} - \sigma_{ij}^{*}\right)\epsilon_{ji}^{*}}{2} \end{aligned} \tag{5.37}$$

and

$$\begin{aligned} W_I &= -\frac{1}{2}\int_{\partial\Omega} \sigma_{ij}^{I} n_i \epsilon_{jk}^{*} x_k dS \\ &= -\frac{V_\Omega \left\langle \sigma_{ij}^{I} \right\rangle_\Omega \epsilon_{ji}^{*}}{2} \end{aligned} \tag{5.38}$$

The combination of the above two equations provides

$$W = -\frac{V_\Omega\left(\sigma_{ij}^T - \sigma_{ij}^*\right)\epsilon_{ji}^*}{2} - \frac{V_\Omega\left\langle\sigma_{ij}^I\right\rangle_\Omega \epsilon_{ji}^*}{2} = -\frac{V_\Omega\left\langle\sigma_{ij}^\Omega\right\rangle_\Omega \epsilon_{ji}^*}{2} \tag{5.39}$$

which has the same form as the one for the infinite domain.

5.3.3 Strain Energy for an Inclusion with Both an Eigenstrain and an Applied Load

Using the equivalent inclusion method, an inhomogeneity problem can be treated as the superposition of an inclusion problem and a uniform stress problem. The subsection will firstly investigate an inclusion Ω embedded in a solid D subjected to a uniform eigenstrain in Ω and a surface stress vector $\sigma_{ij}^0 n_i$ on ∂D. The elastic field caused by the surface stress vector on the boundary will be a uniform stress or strain field. The elastic fields in inclusion and matrix can be rewritten as in the following table:

Matrix	**Inclusion**	**Comments**
$\epsilon_{ij}^M = \epsilon_{ij}^T + \epsilon_{ij}^I + \epsilon_{ij}^0$	$\epsilon_{ij}^\Omega = \epsilon_{ij}^T - \epsilon_{ij}^* + \epsilon_{ij}^I + \epsilon_{ij}^0$	$\epsilon_{ij}^X = \frac{1}{2}\left(u_{i,j}^X + u_{j,i}^X\right)$, $X = M, \Omega, T, I, 0$
$u_i^M = u_i^T + u_i^I + u_i^0$	$u_i^\Omega = u_i^T - \epsilon_{ij}^* x_j + u_i^I + u_i^0$	The rigid-body displacement is not considered
$\sigma_{ij}^M = \sigma_{ij}^T + \sigma_{ij}^I + \sigma_{ij}^0$	$\sigma_{ij}^\Omega = \sigma_{ij}^T - \sigma_{ij}^* + \sigma_{ij}^I + \sigma_{ij}^0$	$\sigma_{ij}^X = C_{ijkl}\epsilon_{ij}^X, X = M, \Omega, T, I, 0$

On the boundary ∂D, the stress boundary condition will be satisfied as

$$\sigma_{ij}^M n_i = \sigma_{ij}^0 n_i = t_j^0 \tag{5.40}$$

Similarly, the total elastic energy caused by the constant eigenstrain in Ω can be written as

$$W = \frac{1}{2}\int_\Omega \sigma_{ij}^\Omega \epsilon_{ij}^\Omega d\mathbf{x} + \frac{1}{2}\int_{D-\Omega} \sigma_{ij}^M \epsilon_{ij}^M d\mathbf{x} = -\frac{1}{2}\int_{\partial\Omega} \sigma_{ij}^\Omega n_i \epsilon_{jk}^* x_k dS + \frac{1}{2}\int_{\partial D} \sigma_{ij}^0 n_i u_j^M dS \tag{5.41}$$

The first term at the right-hand side of the above equation can be written as

$$-\frac{1}{2}\int_{\partial\Omega} \sigma_{ij}^\Omega n_i \epsilon_{jk}^* x_k dS = -\frac{1}{2} V_\Omega \left\langle\sigma_{ij}^\Omega\right\rangle_\Omega \epsilon_{ji}^*$$

and the second term can be written as

$$
\begin{aligned}
\frac{1}{2}\int_{\partial D} \sigma_{ij}^0 n_i u_j^M dS &= \frac{1}{2}\int_{\partial D} \sigma_{ij}^0 n_i \left(u_j^T + u_j^I + u_j^0\right) dS \\
&= \frac{1}{2}\int_{\partial D} \sigma_{ij}^0 n_i u_j^0 dS + \frac{1}{2}\int_{\partial D} \sigma_{ij}^0 n_i \left(u_j^T + u_j^I\right) dS \\
&= \frac{1}{2} V_D \sigma_{ij}^0 \epsilon_{ij}^0 + \frac{1}{2}\int_{\partial D} \sigma_{ij}^0 n_i \left(u_j^T + u_j^I\right) dS \\
&= \frac{1}{2} V_D \sigma_{ij}^0 \epsilon_{ij}^0 + \frac{1}{2}\int_{D} C_{ijkl} \epsilon_{kl}^0 \left(u_{j,i}^T + u_{j,i}^I\right) d\mathbf{x} \\
&= \frac{1}{2} V_D \sigma_{ij}^0 \epsilon_{ij}^0 + \frac{1}{2}\epsilon_{kl}^0 \int_{D} \left(\sigma_{kl}^T + \sigma_{kl}^I\right) d\mathbf{x} \\
&= \frac{1}{2} V_D \sigma_{ij}^0 \epsilon_{ij}^0 + \frac{1}{2}\epsilon_{kl}^0 \int_{D} \sigma_{ij}^* d\mathbf{x} \\
&= \frac{1}{2} V_D \sigma_{ij}^0 \epsilon_{ij}^0 + \frac{1}{2}\int_{\partial D} \sigma_{ij}^0 n_i \left(u_j^T + u_j^I\right) dS \\
&= \frac{1}{2} V_D \sigma_{ij}^0 \epsilon_{ij}^0 + \frac{1}{2} V_\Omega \left\langle \sigma_{ij}^0 \right\rangle_\Omega \epsilon_{ji}^*
\end{aligned} \tag{5.42}
$$

Therefore, one can write

$$
\begin{aligned}
W &= \frac{1}{2} V_D \sigma_{ij}^0 \epsilon_{ij}^0 + \frac{1}{2} V_\Omega \left\langle \sigma_{ij}^0 \right\rangle_\Omega \epsilon_{ji}^* - \frac{1}{2} V_\Omega \left\langle \sigma_{ij}^\Omega \right\rangle_\Omega \epsilon_{ji}^* \\
&= \frac{1}{2} V_D \sigma_{ij}^0 \epsilon_{ij}^0 - \frac{1}{2} V_\Omega \left\langle \sigma_{ij}^T - \sigma_{ij}^* + \sigma_{ij}^I \right\rangle_\Omega \epsilon_{ji}^*
\end{aligned} \tag{5.43}
$$

Actually, the first term of $\frac{1}{2} V_D \sigma_{ij}^0 \epsilon_{ij}^0$ in the above equation is corresponding to the strain energy caused by the surface stress vector $\sigma_{ij}^0 n_i$ independently applied on ∂D. The second term of $-\frac{1}{2} V_\Omega \langle \sigma_{ij}^T - \sigma_{ij}^* + \sigma_{ij}^I \rangle_\Omega \epsilon_{ji}^*$ is corresponding to the strain energy caused by the eigenstrain ϵ_{ji}^* independently applied on the inclusion, which is an internal stress field with the zero stress vector on the boundary ∂D. This is consistent with Colonettis's theorem, which reads:

> When a solid is subjected to an internal stress field and an applied stress field, the total strain energy can be expressed by the superposition of those for two individual stress states and the interaction term between them is zero.

5.3.4 Strain Energy for an Inhomogeneity Problem

Consider an inhomogeneity problem that an ellipsoidal subdomain Ω of a stiffness C^*_{ijkl} is embedded in a large domain D of a stiffness C_{ijkl}, which is subjected to an uniform far-field stress σ^0_{ij}. The local fields will be the superposition of the uniform fields, denoted by a superscript 0, and the disturbed fields, denoted by prime "′." They can be written as in the following table:

Matrix	Inhomogeneity	Comments
$\epsilon^M_{ij} = \epsilon'_{ij} + \epsilon^0_{ij}$	$\epsilon^\Omega_{ij} = \epsilon'_{ij} + \epsilon^0_{ij}$	$\epsilon'_{ij} = -D_{ijkl}(\mathbf{x})C_{klmn}\epsilon^*_{mn}$
$u^M_i = u'_i + u^0_i$	$u^\Omega_i = u'_i + u^0_i$	The rigid-body displacement is not considered
$\sigma^M_{ij} = \sigma'_{ij} + \sigma^0_{ij}$	$\sigma^\Omega_{ij} = \sigma'_{ij} + \sigma^0_{ij}$	$C^*_{ijkl}\left(\epsilon'_{kl} + \epsilon^0_{kl}\right) = C_{ijkl}\left(\epsilon'_{kl} + \epsilon^0_{kl} - \epsilon^*_{kl}\right)$

The disturbed field should satisfy the equilibrium equation

$$\sigma'_{ij,i} = 0 \tag{5.44}$$

and the stress boundary condition

$$\sigma'_{ij}n_i = 0 \tag{5.45}$$

The total elastic energy can be written as

$$W = \frac{1}{2}\int_D \left(\sigma^0_{ij} + \sigma'_{ij}\right)\left(u^0_{i,j} + u'_{i,j}\right) d\mathbf{x} \tag{5.46}$$

where

$$\begin{aligned}
\frac{1}{2}\int_D \sigma'_{ij}\left(u^0_{i,j} + u'_{i,j}\right) d\mathbf{x} &= \frac{1}{2}\int_D \left\{\left[\sigma'_{ij}\left(u^0_i + u'_i\right)\right]_{,j} - \sigma'_{ij,j}\left(u^0_i + u'_i\right)\right\} d\mathbf{x} \\
&= \frac{1}{2}\int_D \left[\sigma'_{ij}\left(u^0_i + u'_i\right)\right]_{,j} d\mathbf{x} \\
&= \frac{1}{2}\int_{\partial D} \sigma'_{ij}n_j\left(u^0_i + u'_i\right) dS \\
&= 0
\end{aligned} \tag{5.47}$$

Therefore, one can write

$$\begin{aligned}
W &= \frac{1}{2}\int_D \sigma^0_{ij}\left(u^0_{i,j} + u'_{i,j}\right) d\mathbf{x} \\
&= \frac{1}{2}V_D\sigma^0_{ij}u^0_{i,j} + \frac{1}{2}\int_D \sigma^0_{ij}u'_{i,j}d\mathbf{x}
\end{aligned} \tag{5.48}$$

where

$$\begin{aligned}\frac{1}{2}\int_D \sigma_{ij}^0 u'_{i,j} d\mathbf{x} &= \frac{1}{2}\int_D C_{ijkl}\epsilon_{kl}^0 \left(u'_{i,j} - \epsilon_{ij}^* + \epsilon_{ij}^*\right) d\mathbf{x} \\ &= \frac{1}{2}\int_D C_{ijkl}\epsilon_{kl}^0 \left(u'_{i,j} - \epsilon_{ij}^*\right) d\mathbf{x} + \frac{1}{2}\int_D C_{ijkl}\epsilon_{kl}^0 \epsilon_{ij}^* d\mathbf{x} \\ &= \frac{1}{2}\int_D u_{k,l}^0 \sigma'_{kl} d\mathbf{x} + \frac{1}{2}\int_D C_{ijkl}\epsilon_{kl}^0 \epsilon_{ij}^* d\mathbf{x} \\ &= \frac{1}{2}\int_{\partial D} \sigma'_{kl} n_l u_k^0 dS + \frac{1}{2}\int_D \sigma_{ij}^0 \epsilon_{ij}^* d\mathbf{x} \\ &= \frac{1}{2}\int_D \sigma_{ij}^0 \epsilon_{ij}^* d\mathbf{x}\end{aligned} \tag{5.49}$$

Therefore, one can obtain

$$W = \frac{1}{2} V_D \sigma_{ij}^0 u_{i,j}^0 + \frac{1}{2}\int_D \sigma_{ij}^0 \epsilon_{ij}^* d\mathbf{x} \tag{5.50}$$

For a pure elastic problem, the potential energy can be written as

$$\Pi = W - \int_{\partial D} \sigma_{ij}^0 n_i \left(u_j^0 + u'_j\right) dS \tag{5.51}$$

where the second term means the work is done by the external load on the boundary, that is,

$$\begin{aligned}\int_{\partial D} \sigma_{ij}^0 n_i \left(u_j^0 + u'_j\right) dS &= \int_D \sigma_{ij}^0 u_{j,i}^0 d\mathbf{x} + \int_D \sigma_{ij}^0 u'_{j,i} d\mathbf{x} \\ &= \int_D \sigma_{ij}^0 u_{j,i}^0 d\mathbf{x} + \int_\Omega \sigma_{ij}^0 \epsilon_{ij}^* d\mathbf{x}\end{aligned} \tag{5.52}$$

Therefore,

$$\Pi = -\frac{1}{2}\int_D \sigma_{ij}^0 u_{j,i}^0 d\mathbf{x} - \frac{1}{2}\int_\Omega \sigma_{ij}^0 \epsilon_{ij}^* d\mathbf{x} \tag{5.53}$$

Some details will be discussed in Chapter 9 later.

5.4 EXERCISES

1. Consider a spherical matrix V with a radius b and a stiffness $\mathbf{C}^0$ contains one spherical particle ω with a radius a and a stiffness $\mathbf{C}^1$. A stress vector (test load) is applied on ∂V as $t_j = \sigma^0_{ij} n_i$. Solve for the average stress and average strain over V using the image stress.

2. Extend Tanaka–Mori's theorem and image stress and strain to the potential problem; derive the image magnetic field and flux for a finite domain caused by a magnetization in a spherical inclusion.

3. Consider a large solid with stiffness $\mathbf{C}^0$ and thermal expansion coefficient α^0 including a spherical particle with stiffness $\mathbf{C}^1$ and thermal expansion coefficient α^1. Derive the strain energy in the overall solids when they are subjected to a temperature change T.

4. Consider a spherical inclusion of radius R with a uniform pure shear eigenstrain ϵ^*_{12} (other components are zero). Calculate the total elastic energy of the system as a function of R, namely $W(R)$. If a far-field stress field σ^0_{12} is applied to the solid, assuming the size of the solid is much larger than R so that Eshelby's solution is applicable, derive the total elastic energy of the system $W(R)$. Derive $dW(R)/dR$ which is related to the driven force of the inclusion growth.

CHAPTER 6

Homogenization for Effective Elasticity Based on the Energy Methods

FOR AN ENERGY conservative solid, the stress and strain relation can be described by a strain energy potential function U, such as

$$\sigma_{ij} = \frac{\partial U}{\partial \epsilon_{ij}}$$

For a linear elastic solid, the constant linearity between the stress and the strain is called the elasticity of stiffness, which can be written as

$$C_{ijkl} = \frac{\partial \sigma_{ij}}{\partial \epsilon_{kl}} = \frac{\partial^2 U}{\partial \epsilon_{ij} \partial \epsilon_{kl}}$$

The strain energy in composite materials can also be used to obtain the effective elasticity of the composites.

To obtain the effective material behavior, an RVE will be used. The volume average of stress and strain can be used to describe the constitutive behavior. This chapter will first present Hill's theorem for strain energy in terms of average stress and strain, second introduce Hill's bounds that are corresponding to the Reuss and Voigt models, then provide Hashin–Shtrikman's variational principle, and finally derive Hashin–Shtrikman's bounds.

6.1 HILL'S THEOREM

Consider an RVE with volume D. The displacement and stress fields satisfy elastic equations in D and interface continuities on ∂D and corresponding boundary conditions in terms of boundary stress vector or displacement. *Hill's theorem* [7] correlates the average elastic fields with the strain energy, which reads

$$\left\langle \sigma_{ij}\epsilon_{ij}\right\rangle_D - \left\langle \sigma_{ij}\right\rangle_D \left\langle \epsilon_{ij}\right\rangle_D = \frac{1}{V_D}\int\limits_{\partial D} \left(u_i - x_j\left\langle \epsilon_{ij}\right\rangle_D\right)\left(\sigma_{ik}n_k - \langle \sigma_{ik}\rangle_D n_k\right) d\mathbf{x} \tag{6.1}$$

Proof: The right-hand side of Equation 6.1 can be rewritten as

$$V_D \cdot RHS = \int\limits_{\partial D} \left(u_i\sigma_{ik}n_k - u_i\langle \sigma_{ik}\rangle_D n_k - x_j\left\langle \epsilon_{ij}\right\rangle_D \sigma_{ik}n_k + x_j\left\langle \epsilon_{ij}\right\rangle_D \langle \sigma_{ik}\rangle_D n_k\right) dS \tag{6.2}$$

where

$$\int\limits_{\partial D} u_i\sigma_{ik}n_k dS = \int\limits_{D} \sigma_{ik,k}u_i + \sigma_{ik}u_{i,k}d\mathbf{x} = V_D\langle \sigma_{ik}\epsilon_{ik}\rangle_D \tag{6.3}$$

$$\int\limits_{\partial D} u_i\langle \sigma_{ik}\rangle_D n_k dS = \langle \sigma_{ik}\rangle_D \int\limits_{D} u_{i,k}d\mathbf{x} = V_D\langle \sigma_{ik}\rangle_D \langle \epsilon_{ik}\rangle_D \tag{6.4}$$

$$\int\limits_{\partial D} x_j\left\langle \epsilon_{ij}\right\rangle_D \sigma_{ik}n_k dS = \left\langle \epsilon_{ij}\right\rangle_D \int\limits_{D} \left(\sigma_{ik,k}x_j + \sigma_{ik}\delta_{jk}\right) d\mathbf{x} = V_D\left\langle \epsilon_{ij}\right\rangle_D \left\langle \sigma_{ij}\right\rangle_D \tag{6.5}$$

$$\int\limits_{\partial D} x_j\left\langle \epsilon_{ij}\right\rangle_D \langle \sigma_{ik}\rangle_D n_k dS = V_D\left\langle \epsilon_{ij}\right\rangle_D \langle \sigma_{ik}\rangle_D \delta_{jk} = V_D\left\langle \epsilon_{ij}\right\rangle_D \left\langle \sigma_{ij}\right\rangle_D \tag{6.6}$$

Thus, Equation 6.2 can be written the same as the left side of Equation 6.1 multiplied by volume V_D,

$$V_D \cdot RHS = V_D\left\langle \sigma_{ij}\epsilon_{ij}\right\rangle_D - V_D\left\langle \sigma_{ij}\right\rangle_D \left\langle \epsilon_{ij}\right\rangle_D = V_D \cdot LHS \tag{6.7}$$

An RVE can be treated as a homogeneous material at the macroscale. If a stress load is applied on ∂D, that is,

$$t_j^0 = \sigma_{ij}^0 n_i \tag{6.8}$$

the average stress on the RVE can be obtained as $\left\langle \sigma_{ij}\right\rangle_D = \sigma_{ij}^0$ from Equation 5.5 in the last chapter. Then, the RHS of Equation 6.1 is zero, that is,

$$\left\langle \sigma_{ij}\epsilon_{ij}\right\rangle_D - \left\langle \sigma_{ij}\right\rangle_D \left\langle \epsilon_{ij}\right\rangle_D = 0 \tag{6.9}$$

Therefore, the average of strain energy over D is the same as the strain energy written in terms of average stress and average strain.

If the test load is provided as a displacement-control load,

$$u_i^0 = x_j\epsilon_{ij}^0 \tag{6.10}$$

The average strain field is obtained as

$$\begin{aligned}\langle \epsilon_{ij} \rangle_D &= \frac{1}{V_D} \frac{1}{2} \int_D \left(u_{i,j} + u_{j,i}\right) d\mathbf{x} \\ &= \frac{1}{V_D} \frac{1}{2} \int_{\partial D} \left(u_i n_j + u_j n_i\right) dS \\ &= \frac{1}{V_D} \frac{1}{2} \int_{\partial D} \left(\epsilon^0_{ik} x_k n_j + \epsilon^0_{jk} x_k n_i\right) dS \\ &= \epsilon^0_{ij}\end{aligned}$$

We can also write

$$\langle \sigma_{ij} \epsilon_{ij} \rangle_D = \langle \sigma_{ij} \rangle_D \langle \epsilon_{ij} \rangle_D \tag{6.11}$$

The average strain energy can be written in terms of the average stress and average strain as

$$W = \frac{1}{2V_D} \langle \sigma_{ij} \rangle_D \langle \epsilon_{ij} \rangle_D = \frac{1}{2V_D} \langle \sigma_{ij} \epsilon_{ij} \rangle_D \tag{6.12}$$

Therefore, the strain energy of an RVE with a microstructure is equivalent to that for a homogenized material point, which provides the convenience to define the effective stiffness of a heterogeneous material in a consistent manner.

6.2 HILL'S BOUNDS

The average stress and strain could be written as

$$\langle \boldsymbol{\sigma} \rangle_D = \langle \mathbf{C} \rangle_D : \langle \boldsymbol{\epsilon} \rangle_D \tag{6.13}$$

and

$$\langle \boldsymbol{\epsilon} \rangle_D = \langle \mathbf{C} \rangle_D^{-1} : \langle \boldsymbol{\sigma} \rangle_D \tag{6.14}$$

For a composite under either stress or displacement loads:

$$\langle \boldsymbol{\sigma} \rangle_D : \langle \boldsymbol{\epsilon} \rangle_D = \frac{1}{V_D} \int_D \boldsymbol{\sigma} : \boldsymbol{\epsilon} d\mathbf{x} \tag{6.15}$$

For any infinitesimal volume element at $\mathbf{x}$, whose stiffness can be written by $\mathbf{C}(\mathbf{x})$, we can define the stress caused by the average strain as

$$\tilde{\boldsymbol{\sigma}}(\mathbf{x}) = \mathbf{C}(\mathbf{x}) : \langle \boldsymbol{\epsilon} \rangle_D \tag{6.16}$$

Similarly, the strain caused by the average stress is

$$\tilde{\boldsymbol{\epsilon}}(\mathbf{x}) = \mathbf{C}(\mathbf{x}) : \langle \sigma \rangle_D \tag{6.17}$$

Therefore, in the volume element at $\mathbf{x}$, one can write

$$\sigma_{ij}\epsilon_{ij} + \left(\sigma_{ij} - \tilde{\sigma}_{ij}\right)\left(\epsilon_{ij} - \left\langle \epsilon_{ij} \right\rangle_D\right) = \tilde{\sigma}_{ij}\left\langle \epsilon_{ij} \right\rangle_D + 2\sigma_{ij}\left(\epsilon_{ij} - \left\langle \epsilon_{ij} \right\rangle_D\right) \tag{6.18}$$

because of $\tilde{\sigma}_{ij}\epsilon_{ij} = \sigma_{ij}\langle \epsilon_{ij} \rangle_D$. Because $\left(\sigma_{ij} - \tilde{\sigma}_{ij}\right)\left(\epsilon_{ij} - \left\langle \epsilon_{ij} \right\rangle_D\right) = C_{ijkl}(\mathbf{x})\left(\epsilon_{kl} - \langle \epsilon_{kl} \rangle_D\right)$ $\left(\epsilon_{ij} - \left\langle \epsilon_{ij} \right\rangle_D\right)$ is always nonnegative, we can obtain

$$\sigma_{ij}\epsilon_{ij} \leq \tilde{\sigma}_{ij}\left\langle \epsilon_{ij} \right\rangle_D + 2\sigma_{ij}\left(\epsilon_{ij} - \left\langle \epsilon_{ij} \right\rangle_D\right) \tag{6.19}$$

Because $\int_D \sigma_{ij}\left(\epsilon_{ij} - \langle \epsilon_{ij} \rangle_D\right) d\mathbf{x} = 0$ from Equation 6.11, we can write

$$V_D \left\langle C_{ijkl} \right\rangle_D \langle \epsilon_{kl} \rangle_D \langle \epsilon_{ij} \rangle_D \leq \int_D C_{ijkl}(\mathbf{x}) \langle \epsilon_{kl} \rangle_D \langle \epsilon_{ij} \rangle_D d\mathbf{x} \tag{6.20}$$

that is,

$$\left\langle C_{ijkl} \right\rangle_D \leq \frac{1}{V_D} \int_D C_{ijkl}(\mathbf{x}) d\mathbf{x} \tag{6.21}$$

or

$$\left\langle C_{ijkl} \right\rangle_D \leq \sum_{r=0}^{n} C_{ijkl}^{r} \phi^{r} \tag{6.22}$$

where the superscript r means the material phase. Generally, $r = 0$ means the matrix. And ϕ^r means the volume fraction of the material phase. Obviously, $\sum_{r=0}^{n} \phi^r = 1$.

Similarly, one can also write

$$\sigma_{ij}\epsilon_{ij} + \left(\sigma_{ij} - \left\langle \sigma_{ij} \right\rangle_D\right)\left(\epsilon_{ij} - \tilde{\epsilon}_{ij}\right) = \left\langle \sigma_{ij} \right\rangle_D \tilde{\epsilon}_{ij} + 2\left(\sigma_{ij} - \left\langle \sigma_{ij} \right\rangle_D\right)\epsilon_{ij} \tag{6.23}$$

because of $\sigma_{ij}\tilde{\epsilon}_{ij} = \left\langle \sigma_{ij} \right\rangle_D \epsilon_{ij}$. Because

$$\left(\sigma_{ij} - \left\langle \sigma_{ij} \right\rangle_D\right)\left(\epsilon_{ij} - \tilde{\epsilon}_{ij}\right) = C_{ijkl}^{-1}(\mathbf{x})\left(\sigma_{kl} - \langle \sigma_{kl} \rangle_D\right)\left(\sigma_{ij} - \left\langle \sigma_{ij} \right\rangle_D\right)$$

is always nonnegative, we can obtain

$$\sigma_{ij}\epsilon_{ij} \leq \left\langle \sigma_{ij} \right\rangle_D \tilde{\epsilon}_{ij} + 2\left(\sigma_{ij} - \left\langle \sigma_{ij} \right\rangle_D\right)\epsilon_{ij} \tag{6.24}$$

Because $\int_D \left(\sigma_{ij} - \left\langle \sigma_{ij} \right\rangle_D\right)\epsilon_{ij} d\mathbf{x} = 0$ from Equation 6.11, we can write

$$V_D \left\langle C_{ijkl}^{-1} \right\rangle_D \langle \sigma_{kl} \rangle_D \left\langle \sigma_{ij} \right\rangle_D \leq \int_D C_{ijkl}^{-1}(\mathbf{x}) \langle \sigma_{kl} \rangle_D \left\langle \sigma_{ij} \right\rangle_D d\mathbf{x} \tag{6.25}$$

that is,

$$\left\langle C^{-1}_{ijkl} \right\rangle_D \leq \frac{1}{V_D} \int\limits_D C^{-1}_{ijkl}(\mathbf{x})d\mathbf{x} \tag{6.26}$$

or

$$\langle C_{ijkl} \rangle_D \geq \left[\sum_{r=0}^{n} \left(C^{r}_{ijkl} \right)^{-1} \phi^r \right]^{-1} \tag{6.27}$$

Therefore, one can obtain

$$\left[\sum_{r=0}^{n} \left(C^{r}_{ijkl} \right)^{-1} \phi^r \right]^{-1} \leq \langle C_{ijkl} \rangle_D \leq \sum_{r=0}^{n} C^{r}_{ijkl} \phi^r \tag{6.28}$$

or

$$\left[\sum_{r=0}^{n} \left(\mathbf{C}^{r} \right)^{-1} \phi^r \right]^{-1} \leq \langle \mathbf{C} \rangle_D \leq \sum_{r=0}^{n} \phi^r \mathbf{C}_r \tag{6.29}$$

For isotropic materials, the stiffness tensor for each material phase can be written in terms of bulk and shear moduli and the bounds for the effective stiffness tensor can be straightforwardly obtained. Using the two isotropic tensors $\mathbf{I}^h$ and $\mathbf{I}^d$ which are defined in Equation 2.40, the stiffness and compliance tensor of the two phases can be written as

$$\mathbf{C}^i = 3k^i\mathbf{I}^h + 2\mu^i\mathbf{I}^d, \quad \mathbf{S}_i = \mathbf{C}^{i-1} = \frac{1}{3k^i}\mathbf{I}^h + \frac{1}{2\mu^i}\mathbf{I}^d \quad (i = 0, 1) \tag{6.30}$$

where k_i and μ_i are the bulk and shear moduli of the two phases.

Thus,

$$\begin{aligned} \left(\sum_{r=0}^{n} \mathbf{C}_r^{-1} \phi^r \right)^{-1} &= \left[\left(\frac{\phi^0}{3k^0} + \frac{\phi^1}{3k^1} \right) \mathbf{I}^h + \left(\frac{\phi^0}{2\mu^0} + \frac{\phi^0}{2\mu^1} \right) \mathbf{I}^d \right]^{-1} \\ &= \frac{3k^0k^1}{\phi^0k^1 + \phi^1k^0}\mathbf{I}^h + \frac{2\mu^0\mu^1}{\phi^0\mu^1 + \phi^1\mu^0}\mathbf{I}^d \end{aligned} \tag{6.31}$$

$$\sum_{r=0}^{n} \phi^r \mathbf{C}^r = 3\left(\phi^0k^0 + \phi^1k^1\right)\mathbf{I}^h + 2\left(\phi^0\mu^0 + \phi^1\mu^1\right)\mathbf{I}^d \tag{6.32}$$

Inserting Equations 6.31 and 6.32 back into Equation 6.29, we have

$$\begin{aligned} \frac{3k^0k^1}{\phi^0k^1 + \phi^1k^0}\mathbf{I}^h + \frac{2\mu^0\mu^1}{\phi^0\mu^1 + \phi^1\mu^0}\mathbf{I}^d &\leq \langle \mathbf{C} \rangle_D = 3\langle k \rangle_D \mathbf{I}^h + 2\langle \mu \rangle_D \mathbf{I}^d \\ &\leq 3\left(\phi^0k^0 + \phi^1k^1\right)\mathbf{I}^h + 2\left(\phi^0\mu^0 + \phi^1\mu^1\right)\mathbf{I}^d \end{aligned} \tag{6.33}$$

Then, the bound of effective bulk modulus $\langle k \rangle_D$ and effective shear modulus $\langle k \rangle_D$ will be given as

$$\frac{k^0 k^1}{\phi^0 k^1 + \phi^1 k^0} \leq \langle k \rangle_D \leq \phi^0 k^0 + \phi^1 k^1 \tag{6.34}$$

$$\frac{\mu^0 \mu^1}{\phi^0 \mu^1 + \phi^1 \mu^0} \leq \langle \mu \rangle_D \leq \phi^0 \mu^0 + \phi^1 \mu^1 \tag{6.35}$$

Actually, the two bounds correspond to the Voigt model [48] and the Reuss model [49], respectively. They can also be obtained by the classical variational principles.

6.3 CLASSICAL VARIATIONAL PRINCIPLES

For a general boundary value problem in domain D with stress boundary for $\sigma_{ij} n_i = t_j^0$ on ∂D_σ and displacement boundary for $u_i = u_i^0$ on ∂D_u, the admissible displacement need to satisfy the displacement boundary condition and be sufficiently smooth with C^1 continuity. Based on the classic variational principle, among all admissible displacement states in the solid, the actual displacement field always minimizes the potential energy in the material system, that is,

$$\Pi(u_i) = \frac{1}{2} \int_D u_{i,j} \sigma_{ij} d\mathbf{x} - \int_{\partial D_\sigma} u_i t_i^0 dS \tag{6.36}$$

reaches the minimum for the elastic solution, where no virtual work can be done on the displacment boundary. Given the boundary conditions for a specific material, the elastic field can be uniquely determined.

Given an RVE, if the test load is provided as a displacement-control load for $\mathbf{x} \in \partial D$,

$$u_i = x_j \epsilon_{ij}^0 \tag{6.37}$$

where no stress boundary is considered. The average strain can be written as

$$\langle \epsilon_{ij} \rangle_D = \frac{1}{V_D} \int_D \frac{1}{2} \left(u_{i,j} + u_{j,i} \right) d\mathbf{x} = \frac{1}{2V_D} \int_{\partial D} \left(u_i n_j + u_j n_i \right) dS = \epsilon_{ij}^0 \tag{6.38}$$

One can assume an admissible displacement field for $\mathbf{x} \in D$ has the same form as that on the boundary, that is, $\hat{u}_i = x_j \epsilon_{ij}^0$, we can obtain

$$\hat{\epsilon}(\mathbf{x}) = \epsilon^0$$

Then, the stress field will be

$$\hat{\sigma}(\mathbf{x}) = \mathbf{C}(x) : \epsilon^0 \tag{6.39}$$

The potential energy caused by the above admissible stress can be written as

$$\Pi(\hat{\sigma}_{ij}) = \frac{1}{2}\int_D \epsilon_{ij}^0 C_{ijkl}(\mathbf{x})\epsilon_{kl}^0 d\mathbf{x}$$

Notice that the second term in Equation 6.36 disappears due to no stress boundary.

Similarly, the potential energy caused by the actual stress can be obtained as a homogenized material as

$$\Pi\left(\langle\sigma_{ij}\rangle_D\right) = \frac{1}{2}\langle C_{ijkl}\rangle_D \langle\epsilon_{kl}\rangle_D \langle\epsilon_{ij}\rangle_D$$
$$= \frac{1}{2}\langle C_{ijkl}\rangle_D \epsilon_{ij}^0\epsilon_{kl}^0$$

The potential energy caused by this elastic field provides a larger value than the actual one, so one can obtain

$$\langle C_{ijkl}\rangle_D \epsilon_{ij}^0\epsilon_{kl}^0 \le \frac{1}{V_D}\int_D C_{ijkl}(\mathbf{x})d\mathbf{x}\epsilon_{kl}^0\epsilon_{ij}^0 \tag{6.40}$$

Because $\boldsymbol{\epsilon}^0$ is an arbitrary tensor, one can obtain

$$\langle C_{ijkl}\rangle_D \le \frac{1}{V_D}\int_D C_{ijkl}(\mathbf{x})d\mathbf{x} \tag{6.41}$$

or

$$\langle C_{ijkl}\rangle_D \le \sum_{r=0}^{n} C_{ijkl}^r\phi^r \tag{6.42}$$

On the other hand, one can write the complementary energy in term of admissible stresses, which is continuous (C^1) and satisfies equilibrium equation and stress boundary conditions. Among all admissible stress states in the solid, the actual stress field always minimizes the complementary energy in the material system, that is,

$$\Pi(\sigma_{ij}) = \frac{1}{2}\int_D C_{ijkl}^{-1}\sigma_{kl}\sigma_{ij}d\mathbf{x} - \int_{\partial D_u} u_i^0\sigma_{ij}n_i dS \tag{6.43}$$

reaches the minimum for the elastic solution, where no virtual work can be done on the stress boundary.

Now consider that a stress vector is applied on the boundary of the RVE, that is,

$$t_j^0 = \sigma_{ij}^0 n_i \tag{6.44}$$

One can obtain the average stress as

$$\langle \sigma_{ij} \rangle_D = \sigma^0_{ij} \tag{6.45}$$

One can assume an admissible stress field as

$$\hat{\sigma}(\mathbf{x}) = \boldsymbol{\sigma}^0$$

Then, the corresponding strain field can be

$$\hat{\epsilon}(\mathbf{x}) = \mathbf{C}^{-1}(\mathbf{x}) : \boldsymbol{\sigma}^0 \tag{6.46}$$

Using the minimum complementary energy theorem, one can obtain

$$\left\langle C^{-1}_{ijkl} \right\rangle_D \langle \sigma_{kl} \rangle_D \left\langle \sigma_{ij} \right\rangle_D \leq \left[\frac{1}{V_D} \int\limits_D C^{-1}_{ijkl}(\mathbf{x}) d\mathbf{x} \right] \sigma^0_{kl} \sigma^0_{ij} \tag{6.47}$$

Similarly, one can obtain

$$\left\langle C^{-1}_{ijkl} \right\rangle_D \leq \frac{1}{V_D} \int\limits_D C^{-1}_{ijkl}(\mathbf{x}) d\mathbf{x} \tag{6.48}$$

or

$$\left\langle C_{ijkl} \right\rangle_D \geq \left[\sum_{r=0}^{n} \left(C^r_{ijkl} \right)^{-1} \phi^r \right]^{-1} \tag{6.49}$$

6.4 HASHIN–SHTRIKMAN'S VARIATIONAL PRINCIPLE

Hill's bounds are based on a uniform stress or strain field of the average fields on all material phases, so the range of the bounds is quite large. Therefore, its applications to general composites are very limited. If the stress or strain can be given in a more accurate manner, the bounds can be tighter and thus can provide more meaningful prediction. Hashin–Shtrikman's variational principle provides a way to improve the prediction of the average stress or strain in each material phase [8,9]. It will be used to provide Hashin–Shtrikman's bounds in the next section.

Consider a heterogeneous material with stiffness $C_{ijkl}(\mathbf{x})$ is subjected to a displacement boundary condition

$$u^0_i = \epsilon^0_{ij} x_j, \quad \text{for } \mathbf{x} \in \partial D \tag{6.50}$$

Consider this material is obtained from a reference material with stiffness C^0_{ijkl} by replacing the material phase at $\mathbf{x}$ from C^0_{ijkl} to $C_{ijkl}(\mathbf{x})$, so that

$$C_{ijkl}(\mathbf{x}) = C^0_{ijkl} + C^d_{ijkl}(\mathbf{x}) \tag{6.51}$$

where $C^d_{ijkl}(\mathbf{x})$ represents the change of the stiffness.

Then, the constitutive law reads

$$\boldsymbol{\sigma}(\mathbf{x}) = \mathbf{C}^0 : \boldsymbol{\epsilon}(\mathbf{x}) + \mathbf{C}^d : \boldsymbol{\epsilon}(\mathbf{x}) = \mathbf{C}^0 : \boldsymbol{\epsilon}(\mathbf{x}) + \boldsymbol{\tau}(\mathbf{x}) \tag{6.52}$$

where $\boldsymbol{\tau}$ is the so-called polarization stress and is defined as

$$\boldsymbol{\tau}(\mathbf{x}) = \mathbf{C}^d(\mathbf{x}) : \boldsymbol{\epsilon}(\mathbf{x}) \tag{6.53}$$

The equilibrium equation can be written as

$$C^0_{ijkl} u_{k,li} + \tau_{ij,i} = 0 \tag{6.54}$$

The above equation can be further rewritten as

$$C^0_{ijkl} u_{k,li} + f_j = 0 \tag{6.55}$$

where $f_j = \tau_{ij,i}$. The solution of the above equation can be written in terms of Green's function as

$$u_i(\mathbf{x}) = \int\limits_D G_{mi}(\mathbf{x}, \mathbf{x}') f_m(\mathbf{x}') d\mathbf{x}' - \int\limits_{\partial D} C^0_{mjkl} \frac{\partial G_{ki}}{\partial x'_l} \epsilon^0_{mn} x'_n n_j dS(\mathbf{x}') \tag{6.56}$$

where Green's function for the above boundary value problem satisfies

$$C^0_{ijkl} \frac{\partial^2 G_{km}}{\partial x_l \partial x_j}(\mathbf{x}, \mathbf{x}') + \delta_{im}\delta(\mathbf{x} - \mathbf{x}') = 0, \mathbf{x}' \in D \tag{6.57}$$

and

$$G_{km}(\mathbf{x}, \mathbf{x}') = 0, \ \mathbf{x}' \in \partial D \tag{6.58}$$

Notice that the above Green's function is different from Green's function for an infinite domain because the boundary condition is enforced. For example, in the infinite domain, Green's function only depends on $\mathbf{x} - \mathbf{x}'$, whereas in the finite domain, Green's function depends on both the source $\mathbf{x}'$ and the field point $\mathbf{x}$. However, it still satisfies the reciprocal property of Green's function such as

$$G_{km}(\mathbf{x}, \mathbf{x}') = G_{km}(\mathbf{x}', \mathbf{x}) \tag{6.59}$$

which implies

$$C^0_{ijkl} \frac{\partial^2 G_{km}}{\partial x'_l \partial x'_j}(\mathbf{x}, \mathbf{x}') + \delta_{im}\delta(\mathbf{x} - \mathbf{x}') = 0, \mathbf{x}' \in D \tag{6.60}$$

Substituting Equation 6.56 into Equation 6.55 and using the above properties of Equation 6.56 Green's function, one can easily confirm that Equation 6.56 is the solution. In Equation 6.56, one can further write

$$
\begin{aligned}
&\int_{\partial D} C^0_{mjkl} \frac{\partial G_{ki}}{\partial x'_l} \epsilon^0_{mn} x'_n n_j d\mathbf{x}' \\
&= \int_D C^0_{mjkl} \frac{\partial G_{ki}}{\partial x'_l} \epsilon^0_{mn} \delta_{jn} d\mathbf{x}' + \int_D C^0_{mjkl} \frac{\partial G_{ki}}{\partial x'_l \partial x'_j} \epsilon^0_{mn} x'_n d\mathbf{x}' \\
&= \int_{\partial D} C^0_{mjkl} G_{ki} \epsilon^0_{mj} n_l dS - \int_D \delta_{im} \delta\left(\mathbf{x} - \mathbf{x}'\right) \epsilon^0_{mn} x'_n d\mathbf{x}' \\
&= -\epsilon^0_{in} x_n
\end{aligned}
\tag{6.61}
$$

Then Equation 6.56 can be rewritten as

$$u_i(\mathbf{x}) = \epsilon^0_{in} x_n + \int_D G_{mi}(\mathbf{x}, \mathbf{x}') \frac{\partial \tau_{mn}}{\partial x'_n}(\mathbf{x}') d\mathbf{x}' \tag{6.62}$$

Using the divergence theorem and Equation 6.58, the above equation can be rewritten as

$$u_i(\mathbf{x}) = \epsilon^0_{in} x_n - \int_D \frac{\partial G_{mi}}{\partial x'_n} \tau_{mn}(\mathbf{x}') d\mathbf{x}' \tag{6.63}$$

Using Equation 6.56 in the strain–displacement relation provides

$$\epsilon_{ij} = \epsilon^0_{ij} - \int_D \Gamma_{ijmn}(\mathbf{x}, \mathbf{x}') \tau_{mn}(\mathbf{x}') dx \tag{6.64}$$

where

$$\Gamma_{ijmn} = \frac{1}{4} \left(\frac{\partial^2 G_{mi}}{\partial x_j \partial x'_n} + \frac{\partial^2 G_{mj}}{\partial x_i \partial x'_n} + \frac{\partial^2 G_{in}}{\partial x_m \partial x'_j} + \frac{\partial^2 G_{jn}}{\partial x_i \partial x'_m} \right) \tag{6.65}$$

is similar but different from the modified Green's function for the infinite domain.

For an arbitrarily chosen $\hat{\tau}_{ij}(\mathbf{x})$, the displacement for $\mathbf{x} \in \partial D$ can be obtained from Equation 6.62 as

$$\hat{u}_i(\mathbf{x}) = \epsilon^0_{in} x_n \tag{6.66}$$

where $G_{mi}(\mathbf{x}, \mathbf{x}') = G_{mi}(\mathbf{x}', \mathbf{x}) = 0$ for $\mathbf{x} \in \partial D$ is used. Therefore, the displacement is always admissible for any $\hat{\tau}_{ij}(\mathbf{x})$. The corresponding strain field is written as

$$\hat{\epsilon}_{ij}(\mathbf{x}) = \epsilon^0_{ij} - \hat{\epsilon}'_{ij}(\mathbf{x}) \tag{6.67}$$

where

$$\hat{\epsilon}'_{ij}(\mathbf{x}) = \int_D \Gamma_{ijmn}(\mathbf{x}, \mathbf{x}')\hat{\tau}_{mn}(\mathbf{x}')d\mathbf{x}' \tag{6.68}$$

The minimum potential energy theorem in Equation 6.36 provides

$$\Pi(u_i) \leq \frac{1}{2}\int_D \hat{\epsilon}_{ij}(\mathbf{x})C_{ijkl}(\mathbf{x})\hat{\epsilon}_{kl}(\mathbf{x})d\mathbf{x}$$

where the RHS of the above inequality is rewritten as

$$\begin{aligned}
2\Pi(\hat{\tau}_{ij}) &= \int_D \hat{\epsilon}_{ij}(\mathbf{x})C_{ijkl}(\mathbf{x})\hat{\epsilon}_{kl}(\mathbf{x})d\mathbf{x} \\
&= \int_D \hat{\epsilon}_{ij}(\mathbf{x})C^0_{ijkl}\hat{\epsilon}_{kl}(\mathbf{x})d\mathbf{x} + \int_D \hat{\epsilon}_{ij}(\mathbf{x})C^d_{ijkl}(\mathbf{x})\hat{\epsilon}_{kl}(\mathbf{x})d\mathbf{x} \\
&= 2\Pi^0 - 2\int_D \hat{\epsilon}'_{ij}(\mathbf{x})C^0_{ijkl}\epsilon^0_{kl}d\mathbf{x} + \int_D \hat{\epsilon}'_{ij}(\mathbf{x})C^0_{ijkl}\hat{\epsilon}'_{kl}(\mathbf{x})d\mathbf{x} \\
&\quad + \int_D \epsilon^0_{ij}C^d_{ijkl}(\mathbf{x})\epsilon^0_{kl}d\mathbf{x} - 2\int_D \hat{\epsilon}'_{ij}(\mathbf{x})C^d_{ijkl}(\mathbf{x})\epsilon^0_{kl}d\mathbf{x} \\
&\quad + \int_D \hat{\epsilon}'_{ij}(\mathbf{x})C^d_{ijkl}(\mathbf{x})\hat{\epsilon}'_{kl}(\mathbf{x})d\mathbf{x}
\end{aligned} \tag{6.69}$$

where $\Pi^0 = \frac{1}{2}\int_D \epsilon^0_{ij}C^0_{ijkl}\epsilon^0_{kl}d\mathbf{x} = \frac{1}{2}V_D\epsilon^0_{ij}C^0_{ijkl}\epsilon^0_{kl}$. Considering

$$\int_D \frac{\partial^2 G_{mi}(\mathbf{x}, \mathbf{x}')}{\partial x_j \partial x'_n}d\mathbf{x} = \frac{\partial}{\partial x'_n}\int_{\partial D} G_{mi}(\mathbf{x}, \mathbf{x}')n_j dS = 0 \tag{6.70}$$

one can obtain $\int_D \Gamma_{ijmn}(\mathbf{x}, \mathbf{x}')d\mathbf{x} = 0$ and then $\int_D \hat{\epsilon}'_{ij}(\mathbf{x})d\mathbf{x} = 0$. Therefore, the inequality can be rewritten as

$$\begin{aligned}
\Pi(u_i) - \Pi^0 &\leq \frac{1}{2}\int_D \hat{\epsilon}'_{ij}(\mathbf{x})C^0_{ijkl}\hat{\epsilon}'_{kl}(\mathbf{x})d\mathbf{x} + \frac{1}{2}\int_D \epsilon^0_{ij}C^d_{ijkl}(\mathbf{x})\epsilon^0_{kl}d\mathbf{x} \\
&\quad - \int_D \hat{\epsilon}'_{ij}(\mathbf{x})C^d_{ijkl}(\mathbf{x})\epsilon^0_{kl}d\mathbf{x} + \frac{1}{2}\int_D \hat{\epsilon}'_{ij}(\mathbf{x})C^d_{ijkl}(\mathbf{x})\hat{\epsilon}'_{kl}(\mathbf{x})d\mathbf{x}
\end{aligned} \tag{6.71}$$

Using the identity $\int_D \Gamma_{ijmn}(\mathbf{x},\mathbf{x}')C^0_{ijkl}\Gamma_{klpq}(\mathbf{x},\mathbf{x}'')d\mathbf{x} = \Gamma_{mnpq}(\mathbf{x}',\mathbf{x}'')$, which can be proved by

$$
\begin{aligned}
&\int_D \frac{\partial^2 G_{mi}(\mathbf{x},\mathbf{x}')}{\partial x_j \partial x'_n} C^0_{ijkl} \frac{\partial^2 G_{kp}(\mathbf{x},\mathbf{x}'')}{\partial x_l \partial x''_q} d\mathbf{x} \\
&= \frac{\partial^2}{\partial x''_q \partial x'_n} \int_D \frac{\partial G_{mi}(\mathbf{x},\mathbf{x}')}{\partial x_j} C^0_{ijkl} \frac{\partial G_{kp}(\mathbf{x},\mathbf{x}'')}{\partial x_l} d\mathbf{x} \\
&= \frac{\partial^2}{\partial x''_q \partial x'_n} \int_D \left\{ \frac{\partial}{\partial x_j} \left[G_{mi}(\mathbf{x},\mathbf{x}') C^0_{ijkl} \frac{\partial G_{kp}(\mathbf{x},\mathbf{x}'')}{\partial x_l} \right] - G_{mi}(\mathbf{x},\mathbf{x}') C^0_{ijkl} \frac{\partial^2 G_{kp}(\mathbf{x},\mathbf{x}'')}{\partial x_j \partial x_l} \right\} d\mathbf{x} \\
&= -\frac{\partial^2}{\partial x''_q \partial x'_n} \int_D G_{mi}(\mathbf{x},\mathbf{x}') C^0_{ijkl} \frac{\partial^2 G_{kp}(\mathbf{x},\mathbf{x}'')}{\partial x_j \partial x_l} d\mathbf{x} \\
&= \frac{\partial^2}{\partial x''_q \partial x'_n} \int_D G_{mi}(\mathbf{x},\mathbf{x}') \delta_{ip} \delta(\mathbf{x}-\mathbf{x}'') d\mathbf{x} \\
&= \frac{\partial^2}{\partial x''_q \partial x'_n} G_{pm}(\mathbf{x}'',\mathbf{x}').
\end{aligned}
\tag{6.72}
$$

one can write

$$
\begin{aligned}
&\int_D \hat{\epsilon}'_{ij}(\mathbf{x}) C^0_{ijkl} \hat{\epsilon}'_{kl}(\mathbf{x}) d\mathbf{x} \\
&= \int_D \left[\int_D \Gamma_{ijmn}(\mathbf{x},\mathbf{x}') \hat{\tau}_{mn}(\mathbf{x}') d\mathbf{x}' \right] C^0_{ijkl} \left[\int_D \Gamma_{klpq}(\mathbf{x},\mathbf{x}'') \hat{\tau}_{pq}(\mathbf{x}'') d\mathbf{x}'' \right] d\mathbf{x} \\
&= \int_D \left\{ \int_D \Gamma_{ijmn}(\mathbf{x},\mathbf{x}') C^0_{ijkl} \left[\int_D \Gamma_{klpq}(\mathbf{x},\mathbf{x}'') \hat{\tau}_{pq}(\mathbf{x}'') d\mathbf{x}'' \right] d\mathbf{x} \right\} \cdot \hat{\tau}_{mn}(\mathbf{x}') d\mathbf{x}' \\
&= \int_D \left\{ \int_D \left[\int_D \Gamma_{ijmn}(\mathbf{x},\mathbf{x}') C^0_{ijkl} \Gamma_{klpq}(\mathbf{x},\mathbf{x}'') d\mathbf{x} \right] \cdot \hat{\tau}_{pq}(\mathbf{x}'') d\mathbf{x}'' \right\} \cdot \hat{\tau}_{mn}(\mathbf{x}') d\mathbf{x}' \\
&= \int_D \left\{ \int_D \Gamma_{mnpq}(\mathbf{x},\mathbf{x}') \hat{\tau}_{pq}(\mathbf{x}') d\mathbf{x}' \right\} \cdot \hat{\tau}_{mn}(\mathbf{x}) d\mathbf{x}
\end{aligned}
\tag{6.73}
$$

In addition, through a lengthy but straightforward simplification, one can write

$$\int_D \hat{\epsilon}'_{ij}(\mathbf{x})C^d_{ijkl}(\mathbf{x})\hat{\epsilon}'_{kl}(\mathbf{x})d\mathbf{x}$$

$$= \int_D \left[\int_D \Gamma_{ijmn}(\mathbf{x},\mathbf{x}')\hat{\tau}_{mn}(\mathbf{x}')d\mathbf{x}'\right] C^d_{ijkl}(\mathbf{x}) \left[\int_D \Gamma_{klpq}(\mathbf{x},\mathbf{x}'')\hat{\tau}_{pq}(\mathbf{x}'')d\mathbf{x}''\right] d\mathbf{x}$$

$$= \int_D \eta_{ij}(\mathbf{x})C^d_{ijkl}(\mathbf{x})\eta_{kl}(\mathbf{x})d\mathbf{x} - \int_D \hat{\tau}_{ij}(\mathbf{x})\left[C^d_{ijkl}(\mathbf{x})\right]^{-1}\hat{\tau}_{kl}(\mathbf{x})d\mathbf{x}$$

$$- 2\int_D \hat{\tau}_{ij}(\mathbf{x})\left[\int_D \Gamma_{ijmn}(\mathbf{x},\mathbf{x}')\hat{\tau}_{mn}(\mathbf{x}')d\mathbf{x}'\right] d\mathbf{x} - \int_D \epsilon^0_{ij}C^d_{ijkl}(\mathbf{x})\epsilon^0_{kl}d\mathbf{x}$$

$$+ 2\int_D \epsilon^0_{ij}\left[\hat{\tau}_{ij}(\mathbf{x}) + C^d_{ijkl}(\mathbf{x})\int_D \Gamma_{klmn}(\mathbf{x},\mathbf{x}')\hat{\tau}_{mn}(\mathbf{x}')d\mathbf{x}'\right] d\mathbf{x} \tag{6.74}$$

where

$$\eta_{ij}(\mathbf{x}) = \left[C^d_{ijkl}(\mathbf{x})\right]^{-1}\hat{\tau}_{kl}(\mathbf{x}) + \int_D \Gamma_{ijmn}(\mathbf{x},\mathbf{x}')\hat{\tau}_{mn}(\mathbf{x}')d\mathbf{x}' - \epsilon^0_{ij} \tag{6.75}$$

Therefore, the inequality becomes

$$\Pi(u_i) \leq \Pi^0 + \frac{1}{2}\int_D \eta_{ij}(\mathbf{x})C^d_{ijkl}(\mathbf{x})\eta_{kl}(\mathbf{x})d\mathbf{x} - \frac{1}{2}\int_D \hat{\tau}_{ij}(\mathbf{x})\eta_{ij}(\mathbf{x})d\mathbf{x} + \frac{1}{2}\int_D \epsilon^0_{ij}\hat{\tau}_{ij}(\mathbf{x})d\mathbf{x} \tag{6.76}$$

Hashin–Shtrikman's Variational Principle Reads

When $C^d_{ijkl}(\mathbf{x})$ is negative semidefinite, the actual solution minimizes the following functional

$$\Pi^h\left(\hat{\tau}_{ij}\right) = \Pi^0 - \frac{1}{2}\int_D \hat{\tau}_{ij}(\mathbf{x})\eta_{ij}(\mathbf{x})d\mathbf{x} + \frac{1}{2}\int_D \epsilon^0_{ij}\hat{\tau}_{ij}(\mathbf{x})d\mathbf{x} \tag{6.77}$$

and the minimum of Π^h is the strain energy of the heterogeneous material under the displacement boundary condition, that is,

$$min\left\{\Pi^h\left(\hat{\tau}_{ij}\right)\right\} = \frac{1}{2}V_D\epsilon^0_{ij}\left\langle C_{ijkl}\right\rangle_D \epsilon^0_{kl} \tag{6.78}$$

On the other hand, when $C^d_{ijkl}(\mathbf{x})$ is positive semidefinite, the actual solution of the stress polarization $\hat{\tau}_{ij}(\mathbf{x})$ maximizes the above functional in Equation 6.77 and the maximum of Π_h is also the strain energy of the heterogeneous material under the displacement boundary condition.

Proof: When $C^d_{ijkl}(\mathbf{x})$ is negative semidefinite, for an arbitrary $\eta_{ij}(\mathbf{x})$, one can always write $\frac{1}{2}\int_D \eta_{ij}(\mathbf{x})C^d_{ijkl}(\mathbf{x})\eta_{kl}(\mathbf{x})d\mathbf{x} \le 0$. Therefore, Equation 6.76 provides

$$\Pi(u_i) \le \Pi^0 - \frac{1}{2}\int_D \hat{\tau}_{ij}(\mathbf{x})\eta_{ij}(\mathbf{x})d\mathbf{x} + \frac{1}{2}\int_D \epsilon^0_{ij}\hat{\tau}_{ij}(\mathbf{x})d\mathbf{x} = \Pi_h\left(\hat{\tau}_{ij}\right)$$

Because the actual solution

$$\epsilon_{ij}(\mathbf{x}) = \epsilon^0_{ij} - \int_D \Gamma_{ijmn}(\mathbf{x},\mathbf{x}')\tau_{mn}(\mathbf{x}')dx = \left[C^d_{ijkl}(\mathbf{x})\right]^{-1}\tau_{kl}(\mathbf{x})$$

makes $\eta_{ij}(\mathbf{x}) = 0$, one can write

$$\begin{aligned} min\left\{\Pi^h\left(\hat{\tau}_{ij}\right)\right\} &= \Pi^0 + \frac{1}{2}\int_D \epsilon^0_{ij}\tau_{ij}(\mathbf{x})d\mathbf{x} \\ &= \frac{1}{2}\int_D \epsilon^0_{ij}\left(C^0_{ijkl}\epsilon^0_{kl} + \tau_{ij}(\mathbf{x})\right)d\mathbf{x} \\ &= \frac{1}{2}V_D\epsilon^0_{ij}\left\langle\sigma_{ij}\right\rangle_D \\ &= \frac{1}{2}V_D\epsilon^0_{ij}\left\langle C_{ijkl}\right\rangle_D\epsilon^0_{kl} \end{aligned} \tag{6.79}$$

where $\left\langle\epsilon_{ij}\right\rangle_D = \epsilon^0_{ij}$ is used in the above derivation.

When $C^d_{ijkl}(\mathbf{x})$ is positive semidefinite, one can use the complementary energy principle to obtain the second part of Hashin–Shtrikman's variational principle.

For an arbitrarily given $\hat{\tau}_{ij}(\mathbf{x})$, the displacement is statically admissible. The stress field corresponding to the displacement can be written as

$$\hat{\sigma}_{ij}(\mathbf{x}) = C^0_{ijkl}\epsilon^0_{kl} - C^0_{ijkl}\int_D \Gamma_{klmn}(\mathbf{x},\mathbf{x}')\hat{\tau}_{mn}(\mathbf{x}')d\mathbf{x}' + \hat{\tau}_{ij}(\mathbf{x}) \tag{6.80}$$

If $\hat{\tau}_{ij}(\mathbf{x})$ is appropriately chosen to make $\hat{\sigma}_{ij}(\mathbf{x})$ statically admissible, that is, $\hat{\sigma}_{ij,i}(\mathbf{x}) = 0$, the complementary energy corresponding to $\hat{\sigma}_{ij}(\mathbf{x})$ is written as

$$\begin{aligned} \Pi_c(\hat{\sigma}_{ij}) &= \frac{1}{2}\int_D \hat{\sigma}_{ij}(\mathbf{x})C^{-1}_{ijkl}(\mathbf{x})\hat{\sigma}_{kl}(\mathbf{x})d\mathbf{x} - \int_{\partial D} \epsilon^0_{jk}x_k\hat{\sigma}_{ij}(\mathbf{x})n_i d\mathbf{x} \\ &= \frac{1}{2}\int_D \hat{\sigma}_{ij}(\mathbf{x})\left[(C^0_{ijkl})^{-1} + M^d_{ijkl}(\mathbf{x})\right]\hat{\sigma}_{kl}(\mathbf{x})d\mathbf{x} - V_D\epsilon^0_{ij}\left\langle\sigma_{ij}\right\rangle_D \end{aligned} \tag{6.81}$$

where $M^d_{ijkl}(\mathbf{x}) = C^{-1}_{ijkl}(\mathbf{x}) - (C^0_{ijkl})^{-1}$. One can obtain

$$\mathbf{C}^0 : \mathbf{M}^d + \mathbf{C}^d : \left(\mathbf{M}^d + (\mathbf{C}^0)^{-1}\right) = \mathbf{C}^0 : \mathbf{M}^d + \mathbf{C}^d : \mathbf{C}^{-1} = \mathbf{0} \tag{6.82}$$

Following the similar procedure from Equation 6.68 to Equation 6.76, one can also obtain that

$$\Pi_c(\sigma) \leq \Pi_c^0 + \frac{1}{2}\int_D \mathbf{C}^0 : \boldsymbol{\eta}(\mathbf{x}) : \mathbf{M}^d(\mathbf{x}) : \mathbf{C}^0 : \boldsymbol{\eta}(\mathbf{x})d\mathbf{x} + \frac{1}{2}\int_D \hat{\boldsymbol{\tau}}(\mathbf{x}) : \boldsymbol{\eta}(\mathbf{x})d\mathbf{x} - \frac{1}{2}\int_D \boldsymbol{\epsilon}^0 : \hat{\boldsymbol{\tau}}(\mathbf{x})d\mathbf{x} \tag{6.83}$$

Obviously, when $C^d_{ijkl}(\mathbf{x})$ is positive semidefinite, from Equation 6.82, $\mathbf{M}^d$ is negative semidefinite. Therefore, one can write

$$\Pi_c(\sigma) \leq \Pi_c^0 + \frac{1}{2}\int_D \hat{\boldsymbol{\tau}}(\mathbf{x}) : \boldsymbol{\eta}(\mathbf{x})d\mathbf{x} - \frac{1}{2}\int_D \boldsymbol{\epsilon}^0 : \hat{\boldsymbol{\tau}}(\mathbf{x})d\mathbf{x} \tag{6.84}$$

For linear elastic materials, $\Pi_c(\sigma) = -\frac{1}{2}V_D\epsilon^0_{ij}\left\langle C_{ijkl}\right\rangle_D \epsilon^0_{kl}$, $\Pi_c^0 = -\frac{1}{2}V_D\epsilon^0_{ij}C^0_{ijkl}\epsilon^0_{kl} = -\Pi^0$. Therefore, one can write

$$\frac{1}{2}V_D\epsilon^0_{ij}\left\langle C_{ijkl}\right\rangle_D \epsilon^0_{kl} \geq \Pi^0 - \frac{1}{2}\int_D \hat{\boldsymbol{\tau}}(\mathbf{x}) : \boldsymbol{\eta}(\mathbf{x})d\mathbf{x} + \frac{1}{2}\int_D \boldsymbol{\epsilon}^0 : \hat{\boldsymbol{\tau}}(\mathbf{x})d\mathbf{x} = \Pi^h\left(\hat{\tau}_{ij}\right) \tag{6.85}$$

Therefore, the actual solution of the stress polarization $\hat{\tau}_{ij}(\mathbf{x})$ maximizes the functional in Equation 6.77 and the maximum of Π_h is also the strain energy of the heterogeneous material under the displacement boundary condition.

6.5 HASHIN–SHTRIKMAN'S BOUNDS

For a composite material with N material elements, whose stiffness tensors are written as $\mathbf{C}^r$ and volume domain as Ω^r with $r = 1, 2, \ldots, N$, to obtain Hashin–Shtrikman's bounds, one can solve the stationary value of Hashin–Shtrikman's functional:

$$\Pi^h\left(\hat{\tau}\right) = \Pi^0 - \frac{1}{2}\int_D \hat{\tau} : \eta d\mathbf{x} + \frac{1}{2}\int_D \epsilon^0 : \hat{\tau} d\mathbf{x} \tag{6.86}$$

For each material element Ω_r with stiffness C^r, one can assume $\hat{\tau}(x)$ as a piecewise uniform polarization stress field as

$$\hat{\boldsymbol{\tau}}(\mathbf{x})\mid_{\mathbf{x}\in\Omega^r} = \hat{\boldsymbol{\tau}}^r \tag{6.87}$$

Then,

$$\boldsymbol{\eta}(x)\mid_{\mathbf{x}\in\Omega^r} = \mathbf{C}_r^{d^{-1}} : \hat{\boldsymbol{\tau}}^r + \sum_{s=1}^{N}\int_{\Omega_s} \boldsymbol{\Gamma}(\mathbf{x}, \mathbf{x}')d\mathbf{x}' : \hat{\boldsymbol{\tau}}^s - \boldsymbol{\epsilon}^0 \tag{6.88}$$

thus,

$$\int_D \hat{\boldsymbol{\tau}} : \boldsymbol{\eta} d\mathbf{x}$$

$$= \sum_{r=1}^{N} \int_{\Omega_r} \hat{\boldsymbol{\tau}}^r : \mathbf{C}_r^{d^{-1}} : \hat{\boldsymbol{\tau}}^r d\mathbf{x} + \sum_{r=1}^{N}\sum_{s=1}^{N} \int_{\Omega_r}\int_{\Omega_s} \hat{\boldsymbol{\tau}}^r : \boldsymbol{\Gamma}(\mathbf{x},\mathbf{x}')d\mathbf{x}' : \hat{\boldsymbol{\tau}}^s d\mathbf{x} - \sum_{r=1}^{N} \int_{\Omega_r} \hat{\boldsymbol{\tau}}^r : \boldsymbol{\epsilon}^0 d\mathbf{x}$$

$$= V_D \sum_r \phi_r \hat{\boldsymbol{\tau}}^r : \mathbf{C}_r^{d^{-1}} : \hat{\boldsymbol{\tau}}^r + V_D^2 \sum_r \sum_s \phi_r \phi_s \hat{\boldsymbol{\tau}}^r : \mathbf{P}^{rs} : \hat{\boldsymbol{\tau}}^s - V_D \sum_r \phi_r \hat{\boldsymbol{\tau}}^r : \boldsymbol{\epsilon}^0 \tag{6.89}$$

where

$$P^{rs} = \frac{1}{\Omega_r \Omega_s} \int_{\Omega_r}\int_{\Omega_s} \Gamma(\mathbf{x},\mathbf{x}')d\mathbf{x}' d\mathbf{x} \tag{6.90}$$

For a random distributed composite with the spherical particles, Willis [2] provided an approximate solution to the integral of the above modified Green's function as

$$\begin{aligned} P_{rs} &= \frac{1}{\Omega_r \Omega_s} \int_{\Omega_r}\int_{\Omega_s} \Gamma(\mathbf{x},\mathbf{x}')d\mathbf{x}' d\mathbf{x} \\ &= \frac{1}{\Omega_r \Omega_s} \int_{\Omega_r}\int_{\Omega_s} \Gamma^{\infty}(\mathbf{x},\mathbf{x}')d\mathbf{x}' d\mathbf{x} \\ &= \frac{1}{\Omega_s} \left(\delta_{rs} - \phi_s\right) D^{\Omega} \end{aligned} \tag{6.91}$$

where

$$D^{\Omega} = \int_{\Omega} \Gamma^{\infty}(\mathbf{x},\mathbf{x}')d\mathbf{x}' \quad x \in \Omega - \text{spherical domain} \tag{6.92}$$

Therefore, Hashin–Shtrikman's functional can be rewritten as

$$\begin{aligned} \Pi^h\left(\hat{\tau}\right) = \Pi^0 - \frac{1}{2} V_D \sum_r \phi_r \hat{\boldsymbol{\tau}}^r : \mathbf{C}_r^{d^{-1}} : \hat{\boldsymbol{\tau}}^r \\ - \frac{1}{2} V_D^2 \sum_r \sum_s \phi_r \phi_s \hat{\boldsymbol{\tau}}^r : \mathbf{P}^{rs} : \hat{\boldsymbol{\tau}}^s + V_D \sum_r \phi_r \hat{\boldsymbol{\tau}}^r : \boldsymbol{\epsilon}^0 \end{aligned} \tag{6.93}$$

To solve for the stationary value of $\Pi^h\left(\hat{\tau}\right)$, $\hat{\tau}$ should be chosen to satisfy the following equations:

$$\frac{\partial \Pi^h\left(\hat{\tau}\right)}{\partial \hat{\tau}^r} = 0 \quad r = 1, 2, \ldots, N. \tag{6.94}$$

Then, one can obtain N linear equations for N unknown $\hat{\tau}^r$ $(r = 1, 2, \ldots, N)$ as follows:

$$\mathbf{C}_r^{d^{-1}} : \hat{\tau}^r + V_D \sum_{s=1}^{N} \phi_s \mathbf{P}^{rs} : \hat{\tau}^s = \epsilon^0 \quad r = 1, 2, \ldots, N \tag{6.95}$$

The solution of the linear equation system can be written as

$$\hat{\tau}^r = \mathbf{R}^r : \epsilon^0 \tag{6.96}$$

Under the assumptions of piecewise uniform polarization stress distribution, one can obtain that the stress $\hat{\tau}^r$ in Equation 6.96 makes the extreme value of Hashin–Shtrikman's functional $\Pi^h(\hat{\tau})$ as

$$\Pi^h(\hat{\tau}) = \frac{1}{2} V_D \epsilon^0 : C^0 : \epsilon^0 + \frac{1}{2} V_D \sum_r \phi_r \epsilon^0 : \mathbf{R}^r : \epsilon^0 \tag{6.97}$$

For an isotropic composite material, the stiffness of each material element can be written with $\mathbf{C}^r$ in terms of two material constants, say μ^r and k^r as the shear and bulk moduli, respectively. To obtain Hashin–Shtrikman's bounds, one can consider the following two bounds:

6.5.1 The Lower Bound

Choose the reference material stiffness $\mathbf{C}^{min}$ with

$$\begin{cases} \mu^{min} = min\{\mu^r\} \\ k^{min} = min\{k^r\} \end{cases} \quad (r = 1, 2, \ldots, N) \tag{6.98}$$

Then, solve for $\mathbf{P}^{rs}$ and $\mathbf{R}^r$ through the above procedure from Equation 6.86 to Equation 6.96, which is written as $\mathbf{P}^{min}$ and $\mathbf{R}^{min}$ for convenience later. Based on Hashin–Shtrikman's variational principle, because $\mathbf{C}^d(\mathbf{x})$ is positive semidefinite over the domain D, the actual solution will maximize $\Pi^h(\hat{\tau})$, which is the actual strain energy. Therefore, the actual strain energy is even larger than or equal to that in Equation 6.97. Therefore, we can write

$$\frac{1}{2} V_D \epsilon^0 : \langle \mathbf{C} \rangle_D : \epsilon^0 \geq I^h(\hat{\tau}) = \frac{1}{2} V_D \epsilon^0 : \mathbf{C}^0 : \epsilon^0 + \frac{1}{2} V_D \epsilon^0 : \sum_{r=1}^{N} \phi_r \mathbf{R}^{r,min} : \epsilon^0 \tag{6.99}$$

Considering ϵ^0 is an arbitrary symmetric tensor, one can obtain

$$\langle \mathbf{C} \rangle_D \geq \mathbf{C}^{min} + \sum_{r=1}^{N} \phi_r \mathbf{R}^{r,min} \tag{6.100}$$

6.5.2 The Upper Bound

Choose the reference material stiffness $\mathbf{C}^{max}$ with

$$\begin{cases} \mu^0 = max\{\mu^r\} \\ k^0 = max\{k^r\} \end{cases} \quad (r = 1, 2, \ldots, N) \tag{6.101}$$

One can also solve for $\mathbf{P}^{rs}$ and $\mathbf{R}^r$ through the same procedure written for $\mathbf{P}^{max}$ and $\mathbf{R}^{max}$. Because $\mathbf{C}^d(\mathbf{x})$ is negative semidefinite over the domain D, one can similiarly obtain

$$\langle \mathbf{C} \rangle_D \leq \mathbf{C}^{max} + \sum_{r=1}^{N} \phi_r \mathbf{R}^{r,max} \tag{6.102}$$

Therefore, we can obtain the two bounds of the effective stiffness of a composite with N material phases as

$$\mathbf{C}^{min} + \sum_{r=1}^{N} \phi_r \mathbf{R}^{r,min} \leq \langle \mathbf{C} \rangle_D \leq \mathbf{C}^{max} + \sum_{r=1}^{N} \phi_r \mathbf{R}^{r,max} \tag{6.103}$$

Notice that $\mathbf{C}^{min}$ or $\mathbf{C}^{max}$ do not have to be the stiffness of an actual material phase. As long as it can make $\mathbf{C}^d(\mathbf{x})$ positive or negative semidefinite, the above process can be valid. However, the wider the range between $\mathbf{C}^{min}$ and $\mathbf{C}^{max}$, the larger the range between the two bounds, which will provide less valuable prediction.

Example: A two-phase composite material has the material constants of each phase satisfying $k_1 > k_0$ and $\mu_1 > \mu_0$. The volume fraction of each phase satisfies $\phi_0 + \phi_1 = 1$. The effective material constants will satisfy the following equations based on Hashin–Shtrikman's bounds.

$$k_0 + \frac{\phi_1}{1/(k_1 - k_0) + 3\phi_0/'[/[(3k_0 + 4\mu_0)} \leq \langle k \rangle_D \leq k_1 + \frac{\phi_0}{1/(k_0 - k_1) + 3\phi_1/(3k_1 + 4\mu_1)} \tag{6.104}$$

$$\mu_0 + \frac{\phi_1}{\frac{1}{\mu_1 - \mu_0} + \frac{6\phi_0(k_0 + 2\mu_0)}{5\mu_0(3k_0 + 4\mu_0)}} \leq \langle \mu \rangle_D \leq \mu_1 + \frac{\phi_0}{\frac{1}{\mu_0 - \mu_1} + \frac{6\phi_1(k_1 + 2\mu_1)}{5\mu_1(3k_1 + 4\mu_1)}} \tag{6.105}$$

6.6 EXERCISES

1. Consider a polymer composite material with two material phases: a polymer matrix $E_0 = 20$ GPa, $v_0 = 0.4$, and a ceramic reinforcement $E_1 = 50$ GPa, $v_1 = 0.3$. Assume that the reinforcement has a spherical shape. Formulate and illustrate the effective shear modulus and bulk modulus of the composite versus volume fraction ϕ using Hill's bounds and Hashin–Shtrikman's bounds.

2. Show that for a finite domain D subjected to a displacement boundary condition $u_i^0 = \epsilon_{ij}^0 x_j$, the solution to equation $C_{ijkl}^0 u_{k,li} + f_j = 0$ can be written in terms of Green's function as

$$u_i(\mathbf{x}) = \int_D G_{mi}(\mathbf{x}, \mathbf{x}') f_m(\mathbf{x}') d\mathbf{x}' - \int_{\partial D} C_{mjkl}^0 \frac{\partial G_{ki}}{\partial x_l'} \epsilon_{mn}^0 x_n' n_j dS(\mathbf{x}')$$

where Green's function for the above boundary value problem satisfies $C_{ijkl}^0 G_{km,lj}(\mathbf{x}, \mathbf{x}') + \delta_{im}\delta(\mathbf{x} - \mathbf{x}') = 0, \mathbf{x}' \in D$ and $G_{km}(\mathbf{x}, \mathbf{x}') = 0, \mathbf{x}' \in \partial D$.

3. Consider an RVE denoted by D subjected to a displacement boundary condition $u_i = \epsilon_{ik}^0 x_k$. The stiffness in the RVE is denoted by $C_{ijkl}(\mathbf{x})$. With a reference stiffness C_{ijkl}^0 and an arbitrary stress polarization $\hat{\tau}_{ij}(\mathbf{x})$, show the detail steps that

$$\Pi(u_i) \leq \Pi^0 + \frac{1}{2} \int_D \eta_{ij}(\mathbf{x}) C_{ijkl}^d(\mathbf{x}) \eta_{kl}(\mathbf{x}) d\mathbf{x} - \frac{1}{2} \int_D \hat{\tau}_{ij}(\mathbf{x}) \eta_{ij}(\mathbf{x}) d\mathbf{x} + \frac{1}{2} \int_D \epsilon_{ij}^0 \hat{\tau}_{ij}(\mathbf{x}) d\mathbf{x}$$

where $C_{ijkl}^d(\mathbf{x}) = C_{ijkl}(\mathbf{x}) - C_{ijkl}^0$, $\Pi^0 = \frac{1}{2} V_D \epsilon_{ij}^0 C_{ijkl}^0 \epsilon_{kl}^0$, and $\eta_{ij}(\mathbf{x}) = \left[C_{ijkl}^d(\mathbf{x})\right]^{-1} \hat{\tau}_{kl}(\mathbf{x}) + \int_D \Gamma_{ijmn}(\mathbf{x}, \mathbf{x}') \hat{\tau}_{mn}(\mathbf{x}') d\mathbf{x}' - \epsilon_{ij}^0$.

4. Under the same conditions in Question 2, derive

$$\Pi_c(\sigma) \leq \Pi_c^0 + \frac{1}{2} \int_D \mathbf{C}^0 : \eta(\mathbf{x}) : \mathbf{M}^d(\mathbf{x}) : \mathbf{C}^0 : \eta(\mathbf{x}) d\mathbf{x} + \frac{1}{2} \int_D \hat{\tau}(\mathbf{x}) : \eta(\mathbf{x}) d\mathbf{x}$$

$$- \frac{1}{2} \int_D \epsilon^0 : \hat{\tau}(\mathbf{x}) d\mathbf{x}$$

where $M_{ijkl}^d(\mathbf{x}) = C_{ijkl}^{-1}(\mathbf{x}) - (C_{ijkl}^0)^{-1}$.

CHAPTER 7

Homogenization for Effective Elasticity Based on the Vectorial Methods

IN A COMPOSITE containing a matrix and multiple dispersed phases, the effective elastic behavior can be represented by the relation between the average stress and strain tensors over an RVE. Given a stress or displacement load on the boundary of the RVE, when the size of the RVE is large enough, the average stress and strain may converge to unique values, respectively, which can be used to derive the elastic moduli. Because these types of methods directly correlate the change of tensors or vectors and thus obtain the homogenized properties, we classify them as the vectorial methods.

7.1 EFFECTIVE MATERIAL BEHAVIOR AND MATERIAL PHASES

Consider a mutiple-phase composite containing a matrix and N types of dispersed phases, as shown in Figure 7.1. An RVE is constructed to reflect the volume fraction of each material phase and the corresponding microstructure. If each phase occupies a volume fraction ϕ_r, in the RVE, one can write

$$\sum_{r=0}^{N} \phi_r = 1; \quad r = 0, 1, 2, 3, \ldots, N \tag{7.1}$$

The interface between material phases are considered to be perfectly bonded so that the stress and displacement continuities are satisfied. Therefore, the average stress and the average strain can be written as

$$\langle \sigma \rangle_D = \sigma^0 = \sum_{r=0}^{N} \phi_r \langle \sigma \rangle_{\Omega_r} \tag{7.2}$$

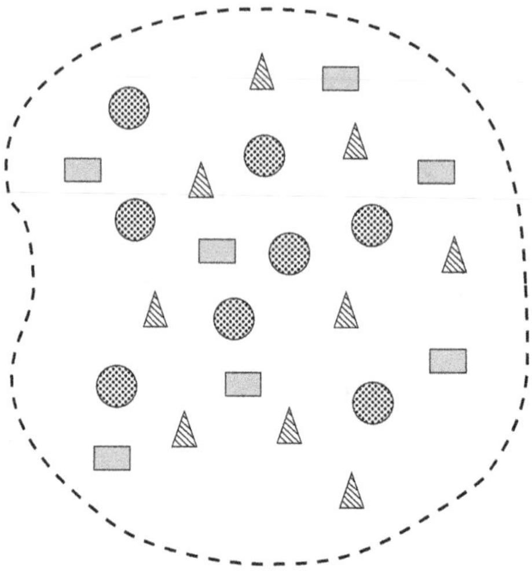

FIGURE 7.1 Mutiple-phase composite containing a matrix and N types of dispersed phases.

and

$$\langle \boldsymbol{\epsilon} \rangle_D = \sum_{r=0}^{N} \phi_r \langle \boldsymbol{\epsilon} \rangle_{\Omega_r} \tag{7.3}$$

where Ω_0 is a matrix. The stiffness of each material phase r is uniform and is written as $\mathbf{C}^r$. The average stress and strain for each material phase satisfies

$$\langle \boldsymbol{\sigma} \rangle_{\Omega_r} = \mathbf{C}^r : \langle \boldsymbol{\epsilon} \rangle_{\Omega_r} \tag{7.4}$$

If we can find the relation between $\langle \boldsymbol{\epsilon} \rangle_{\Omega_r}$ and $\boldsymbol{\sigma}^0$, such as

$$\langle \boldsymbol{\epsilon} \rangle_{\Omega_r} = \mathbf{T}^r : \boldsymbol{\sigma}^0 \tag{7.5}$$

where $\mathbf{T}^r$ is a fourth-rank tensor. Then, one can obtain

$$\langle \boldsymbol{\epsilon} \rangle_D = \left(\sum_{r=0}^{N} \phi_r \mathbf{T}^r \right) : \boldsymbol{\sigma}^0 \tag{7.6}$$

According to Equation 7.6,

$$\langle \boldsymbol{\sigma} \rangle_D = \boldsymbol{\sigma}^0 = \left(\sum_{r=0}^{N} \phi_r \mathbf{T}^r \right)^{-1} : \langle \boldsymbol{\epsilon} \rangle_D \tag{7.7}$$

Thus, the effective stiffness can be written as

$$\langle \mathbf{C} \rangle_D = \left(\sum_{r=0}^{N} \phi_r \mathbf{T}^r \right)^{-1} \tag{7.8}$$

The tensor $\mathbf{T}^r$ actually depends on the microstructure and the mechanical properties of the material phases. It may change case by case. However, to predict the effective material behavior, some assumptions or simplifications are commonly used, so that Eshelby's solution of one inhomogeneity embedded in a matrix can be used. In the following, we use a two-phase composite as an example to demonstrate the popular micromechanics models to predict the effective stiffness.

Consider a two-phase composite under a uniform stress with particle volume fraction ϕ and stiffness $\mathbf{C}^0$ for the matrix phase and $\mathbf{C}^1$ for the particle phase, respectively. One can write the average stress and strain as

$$\langle \sigma \rangle_D = \phi \mathbf{C}^1 : \langle \epsilon \rangle_\Omega + (1 - \phi)\mathbf{C}^0 : \langle \epsilon \rangle_M = \sigma^0 \tag{7.9}$$

$$\langle \epsilon \rangle_D = \phi \, \langle \epsilon \rangle_\Omega + (1 - \phi) \, \langle \epsilon \rangle_M \tag{7.10}$$

To find the relation between $\langle \sigma \rangle_D$ and $\langle \epsilon \rangle_D$, one more equation is needed, because there are three unknowns including $\langle \epsilon \rangle_\Omega$, $\langle \epsilon \rangle_M$, and $\langle \epsilon \rangle_D$, or $\langle \sigma \rangle_\Omega$, $\langle \sigma \rangle_M$, and $\langle \sigma \rangle_D$. However, there are only two equations existing. Once one more equation is provided, one can solve $\mathbf{T}^r$ and then obtain $\langle \mathbf{C} \rangle_D$. However, it is unnecessary to explicitly derive $\mathbf{T}^r$.

7.2 MICROMECHANICS-BASED MODELS FOR TWO-PHASE COMPOSITES

7.2.1 The Voigt Model

Assume the strains on both phases are the same:

$$\langle \epsilon \rangle_\Omega = \langle \epsilon \rangle_M = \bar{\epsilon} \tag{7.11}$$

$$\langle \sigma \rangle_D = \sigma^0 = \left[\phi \mathbf{C}^1 + (1 - \phi)\mathbf{C}^0 \right] : \bar{\epsilon} \tag{7.12}$$

$$\langle \epsilon \rangle_D = [\phi \bar{\epsilon} + (1 - \phi)\bar{\epsilon}] = \bar{\epsilon} \tag{7.13}$$

According to Equation 7.13, Equation 7.12 becomes

$$\langle \sigma \rangle_D = \sigma^0 = \left[\phi \mathbf{C}^1 + (1 - \phi)\mathbf{C}^0 \right] : \langle \epsilon \rangle_D \tag{7.14}$$

Then, the effective stiffness could be obtained as

$$\langle \mathbf{C} \rangle_D = \phi \mathbf{C}^1 + (1 - \phi)\mathbf{C}^0 \tag{7.15}$$

When both phases are isotropic materials, one can write it in terms of the bulk modulus and shear modulus as [48]:

$$\langle k \rangle_D = \phi k^1 + (1 - \phi)k^0 = k^0 + \phi \left(k^1 - k^0 \right) \tag{7.16}$$

and

$$\langle\mu\rangle_D = \phi\mu^1 + (1-\phi)\mu^0 = \mu^0 + \phi\left(\mu^1 - \mu^0\right) \tag{7.17}$$

from which we can obtain all other isotropic elastic moduli.

7.2.2 The Reuss Model

Assume the stresses on both phases are the same:

$$\langle\sigma\rangle_\Omega = \langle\sigma\rangle_M = \bar{\sigma} \tag{7.18}$$

and

$$\mathbf{C}^1 : \langle\epsilon\rangle_\Omega = \mathbf{C}^0 : \langle\epsilon\rangle_M \tag{7.19}$$

The average stress and strain are

$$\langle\sigma\rangle_D = \bar{\sigma} \tag{7.20}$$

and

$$\begin{aligned}\langle\epsilon\rangle_D &= \phi\,\langle\epsilon\rangle_\Omega + (1-\phi)\,\langle\epsilon\rangle_M \\ &= \phi\left(\mathbf{C}^1\right)^{-1} : \bar{\sigma} + (1-\phi)\left(\mathbf{C}^0\right)^{-1} : \bar{\sigma} \\ &= \left[\phi\left(\mathbf{C}^1\right)^{-1} + (1-\phi)\left(\mathbf{C}^0\right)^{-1}\right] : \bar{\sigma}.\end{aligned} \tag{7.21}$$

Thus, one can obtain the relation between the average stress and strain as

$$\langle\sigma\rangle_D = \left[\phi\left(\mathbf{C}^1\right)^{-1} + (1-\phi)\left(\mathbf{C}^0\right)^{-1}\right]^{-1} : \langle\epsilon\rangle_D \tag{7.22}$$

Then, the effective stiffness could be obtained as the denominators in the above equation may reduce to zero

$$\langle\mathbf{C}\rangle_D = \left[\phi\left(\mathbf{C}^1\right)^{-1} + (1-\phi)\left(\mathbf{C}^0\right)^{-1}\right]^{-1} \tag{7.23}$$

When both phases are isotropic materials, one can write it in terms of the bulk modulus and shear modulus as [49]:

$$\langle k\rangle_D = \left[\phi\left(k^1\right)^{-1} + (1-\phi)\left(k^0\right)^{-1}\right]^{-1} = \frac{k^0k^1}{k^1 - \phi\left(k^1 - k^0\right)} \tag{7.24}$$

and

$$\langle\mu\rangle_D = \left[\phi\left(\mu^1\right)^{-1} + (1-\phi)\left(\mu^0\right)^{-1}\right]^{-1} = \frac{\mu^0\mu^1}{\mu^1 - \phi\left(\mu^1 - \mu^0\right)} \tag{7.25}$$

Actually, the Voigt model and the Reuss model are the same as Hill's upper and lower bounds.

7.2.3 The Dilute Model

The above two models simply assumed that the stress or strain is uniform over two material phases. However, for a particulate composite, the local stress and strain fields will be significantly disturbed by the material difference based on Eshelby's solution for one inhomogeneity embedded in the infinite domain. The dilute model directly uses Eshelby's solution to obtain the relation between particle's average strain and the applied stress as follows:

For a dilute distribution of particles in a uniform matrix, because the distance between particles is much larger than the radius of the particles, the particle interactions can be disregarded. The particle's averaged strain can be directly obtained from the local solution for one inhomogeneity embedded in the infinite domain as

$$\begin{aligned}
\langle \boldsymbol{\epsilon} \rangle_\Omega &= \boldsymbol{\epsilon}^0 + \boldsymbol{\epsilon}' = \boldsymbol{\epsilon}^0 - \mathbf{D}^{\Omega} : \mathbf{C}^0 : \boldsymbol{\epsilon}^* \\
&= \left[\mathbf{I} - \mathbf{D}^{\Omega} : \mathbf{C}^0 : (\mathbf{C}^0)^{-1} : \left(\mathbf{D}^{\Omega} - \Delta \mathbf{C}^{-1} \right)^{-1} \right] : \boldsymbol{\epsilon}^0 \\
&= -\Delta \mathbf{C}^{-1} : \left(\mathbf{D}^{\Omega} - \Delta \mathbf{C}^{-1} \right)^{-1} : \left(\mathbf{C}^0 \right)^{-1} : \boldsymbol{\sigma}^0 \\
&= \left(\mathbf{I} - \mathbf{D}^{\Omega} : \Delta \mathbf{C} \right)^{-1} : \left(\mathbf{C}^0 \right)^{-1} : \boldsymbol{\sigma}^0
\end{aligned} \tag{7.26}$$

and

$$\langle \boldsymbol{\epsilon} \rangle_M = \frac{1}{1-\phi} \left(\mathbf{C}^0 \right)^{-1} : \left(\boldsymbol{\sigma}^0 - \phi \mathbf{C}^1 : \langle \boldsymbol{\epsilon} \rangle_\Omega \right) \tag{7.27}$$

Therefore, one can write

$$\begin{aligned}
\langle \boldsymbol{\epsilon} \rangle_D &= \phi \langle \boldsymbol{\epsilon} \rangle_\Omega + \left(\mathbf{C}^0 \right)^{-1} \boldsymbol{\sigma}^0 - \phi \left(\mathbf{C}^0 \right)^{-1} : \mathbf{C}^1 : \langle \boldsymbol{\epsilon} \rangle_\Omega \\
&= \left(\mathbf{C}^0 \right)^{-1} : \boldsymbol{\sigma}^0 - \phi \left[\left(\mathbf{C}^0 \right)^{-1} : \mathbf{C}^0 - \left(\mathbf{C}^0 \right)^{-1} : \mathbf{C}^1 \right] : \langle \boldsymbol{\epsilon} \rangle_\Omega \\
&= \left(\mathbf{C}^0 \right)^{-1} : \boldsymbol{\sigma}^0 + \phi \left(\mathbf{C}^0 \right)^{-1} : \left(\mathbf{D}^{\Omega} - \Delta \mathbf{C}^{-1} \right)^{-1} : \left(\mathbf{C}^0 \right)^{-1} : \boldsymbol{\sigma}^0.
\end{aligned} \tag{7.28}$$

Then, the effective stiffness could be written as

$$\begin{aligned}
\langle \mathbf{C} \rangle_D &= \left[\left(\mathbf{C}^0 \right)^{-1} + \phi \left(\mathbf{C}^0 \right)^{-1} : \left(\mathbf{D}^{\Omega} - \Delta \mathbf{C}^{-1} \right)^{-1} : \left(\mathbf{C}^0 \right)^{-1} \right]^{-1} \\
&= \mathbf{C}^0 : \left[\mathbf{I} + \phi \left(\mathbf{C}^0 \right)^{-1} : \left(\mathbf{D}^{\Omega} - \Delta \mathbf{C}^{-1} \right)^{-1} \right]^{-1}
\end{aligned} \tag{7.29}$$

When both phases are isotropic materials and the particles are spherical, one can write it in terms of the bulk modulus and shear modulus as

$$\langle k\rangle_D = k^0 + \frac{\phi\left(k^1 - k^0\right)\left(3k^0 + 4\mu^0\right)}{3k^1 + 4\mu^0} \tag{7.30}$$

$$\langle \mu\rangle_D = \mu^0 + \frac{5\phi\mu^0\left(\mu^1 - \mu^0\right)\left(3k^0 + 4\mu^0\right)}{3k^0\left(3\mu^0 + 2\mu^1\right) + 4\mu^0\left(2\mu^0 + 3\mu^1\right)} \tag{7.31}$$

Because one particle in the matrix represents the microstructure, if the particle is not spherical, the effective material behavior in a different direction will be different, which causes an anisotropic material behavior. In the dilute model, the effect of the particle's volume and interaction on the particle's average strain is not considered. Therefore, this model is only valid for composites with small volume fractions of particles. When volume fraction increases to a certain value, the results can be divergent.

7.2.4 The Mori–Tanaka Model

To consider the effect of volume fraction on the particle's average strain, Mori and Tanaka [15] proposed complex manipulations of the field variables. Yin and Sun [43] proposed a more straightforward method to derive this model. In this method, a particle is embedded into the matrix with a uniform strain the same as the matrix's averaged strain, and the particle's averaged strain is calculated from the solution for one particle embedded in the infinite matrix in the equation

$$\langle \epsilon\rangle_\Omega = \left(\mathbf{I} - \mathbf{D}^\Omega : \Delta\mathbf{C}\right)^{-1} : \langle \epsilon\rangle_M \tag{7.32}$$

and

$$\langle \epsilon\rangle_D = \left[\phi\left(\mathbf{I} - \mathbf{D}^\Omega : \Delta\mathbf{C}\right)^{-1} + (1-\phi)\mathbf{I}\right] : \langle \epsilon\rangle_M \tag{7.33}$$

According to Equation 7.33, one can write

$$\langle \epsilon\rangle_M = \left[\phi\left(\mathbf{I} - \mathbf{D}^\Omega : \Delta\mathbf{C}\right)^{-1} + (1-\phi)\mathbf{I}\right]^{-1} : \langle \epsilon\rangle_D \tag{7.34}$$

Substituting Equation 7.34 into Equation 7.32 yields

$$\langle \epsilon\rangle_\Omega = \left[\phi\mathbf{I} + (1-\phi)\left(\mathbf{I} - \mathbf{D}^\Omega : \Delta\mathbf{C}\right)\right]^{-1} : \langle \epsilon\rangle_D \tag{7.35}$$

Finally, one can obtain

$$\langle \sigma\rangle_D = \phi\mathbf{C}^1 : \langle \epsilon\rangle_\Omega + (1-\phi)\mathbf{C}^0 : \langle \epsilon\rangle_M = \left\{\mathbf{C}^0 + \phi\left[\Delta\mathbf{C}^{-1} - (1-\phi)\mathbf{D}^\Omega\right]^{-1}\right\} : \langle \epsilon\rangle_D \tag{7.36}$$

The effective stiffness could be written as

$$\langle \mathbf{C} \rangle_D = \mathbf{C}^0 + \phi \left[\Delta\mathbf{C}^{-1} - (1-\phi)\mathbf{D}^{\Omega} \right]^{-1} \tag{7.37}$$

When both phases are isotropic materials and the particles are spherical, one can write it in terms of the bulk modulus and shear modulus as

$$\langle k \rangle_D = k^0 + \frac{\phi \left(k^1 - k^0\right)\left(3k^0 + 4\mu^0\right)}{3k^0 + 4\mu^0 + 3(1-\phi)\left(k^1 - k^0\right)} \tag{7.38}$$

$$\langle \mu \rangle_D = \mu^0 + \frac{5\phi\mu^0\left(\mu^1 - \mu^0\right)\left(3k^0 + 4\mu^0\right)}{5\mu^0\left(3k^0 + 4\mu^0\right) + 6(1-\phi)\left(\mu^1 - \mu^0\right)\left(k^0 + 2\mu^0\right)} \tag{7.39}$$

7.2.5 The Self-Consistent Model

In the self-consistent model [19,50], to solve the average strain in a material phase, one can choose a material point on that phase and isolate it as an infinitesimal volume element. Then, the rest of the material is homogenized as a uniform material, whose mechanical property is still identical to the composite itself. The particle averaged strain is derived from the solution for one particle embedded in the infinite domain which has the same material properties as the composite as follows:

$$\langle \epsilon \rangle_{\Omega} = -\left(\mathbf{C}^1 - \bar{\mathbf{C}}\right)^{-1} : \left(\mathbf{D}^{\Omega} - \Delta\mathbf{C}^{-1}\right)^{-1} : \langle \epsilon \rangle_D \tag{7.40}$$

where $\Delta\mathbf{C}$ and $\mathbf{D}^{\Omega}$ are based on the composite's effective stiffness $\langle \mathbf{C} \rangle_D$, for simplicity written as $\bar{\mathbf{C}}$ here, as follows:

$$\Delta\mathbf{C} = \mathbf{C}^1 - \bar{\mathbf{C}} \tag{7.41}$$

and

$$\mathbf{D}^{\Omega} = \mathbf{D}^{\Omega}\left(\bar{\mathbf{C}}\right) \tag{7.42}$$

Therefore, one can obtain,

$$\langle \epsilon \rangle_M = \frac{1}{1-\phi}\left[\mathbf{I} - \phi\left(\mathbf{I} - \mathbf{D}^{\Omega} : \Delta\mathbf{C}\right)^{-1}\right] : \langle \epsilon \rangle_D \tag{7.43}$$

$$\langle \sigma \rangle_D = \left\{\mathbf{C}^0 + \phi\left(\mathbf{C}^1 - \mathbf{C}^0\right) : \left[\mathbf{I} - \mathbf{D}^{\Omega} : \left(\mathbf{C}^1 - \bar{\mathbf{C}}\right)\right]^{-1}\right\} : \langle \epsilon \rangle_D \tag{7.44}$$

The effective stiffness could be obtained as

$$\langle \mathbf{C} \rangle_D = \bar{\mathbf{C}} = \mathbf{C}^0 + \phi\left(\mathbf{C}^1 - \mathbf{C}^0\right) : \left[\mathbf{I} - \mathbf{D}^{\Omega} : \left(\mathbf{C}^1 - \bar{\mathbf{C}}\right)\right]^{-1} \tag{7.45}$$

When both phases are isotropic materials and the particles are spherical, one can write it in terms of the bulk modulus and shear modulus as

$$\langle k\rangle_D = k^0 + \frac{\phi\left(k^1 - k^0\right)\left(3\langle k\rangle_D + 4\langle\mu\rangle_D\right)}{3k^1 + 4\langle\mu\rangle_D} \tag{7.46}$$

$$\langle\mu\rangle_D = \mu^0 + \frac{5\phi\langle\mu\rangle_D\left(\mu^1 - \mu^0\right)\left(3\langle k\rangle_D + 4\langle\mu\rangle_D\right)}{\langle k\rangle_D\left(3\langle\mu\rangle_D + 2\mu^1\right) + 4\langle\mu\rangle_D\left(2\langle\mu\rangle_D + 3\mu^1\right)} \tag{7.47}$$

Unlike the dilute model and Mori–Tanaka's model, one cannot get the explicit expressions of the composite's moduli directly with the self-consistent model. The simultaneous equations of Equations 7.46 and 7.47 have to be solved. This is because in the self-consistent model, an inhomogeneity is assumed to be embedded into a matrix with the effective moduli of $\langle k\rangle_D$ and $\langle k\rangle_D$, which are still needed to determine when used.

7.2.6 The Differential Scheme

When the volume fraction of particles is high, the difference between the modeling prediction and the actual results may increase. Especially, for a composite containing identical shapes of spherical particles, it is subjected to the maximum volume fraction at 74%. However, the volume fraction in many composite materials may reach 80%–90% due to the gradation of particles. The differential scheme was proposed with the observation that a composite with a finite volume fraction of particles can be fabricated through the following process:

(1) Start with the pure matrix with stiffness $\mathbf{C}^0$ and volume V_0, and add a small volume of particle V_1 into the matrix for a volume fraction $\phi_1 = \frac{V_1}{V_0+V_1}$. The particle's average strain can be obtained through any of the above methods, say the dilute model, or the Mori–Tanaka model, or even the self-consistent model, as

$$\langle\epsilon\rangle_{V_1} = \mathbf{A}_1 : \langle\epsilon\rangle_D \tag{7.48}$$

where $\mathbf{A}_1$ is the strain concentration tensor. The average strain is

$$\langle\epsilon\rangle_D = \phi_1\mathbf{A}_1 : \langle\epsilon\rangle_D + (1 - \phi_1)\langle\epsilon\rangle_M \tag{7.49}$$

In consequence, one can obtain

$$\langle\epsilon\rangle_M = \frac{1}{1 - \phi_1}\left(\mathbf{I} - \phi_1\mathbf{A}_1\right)\langle\epsilon\rangle_D \tag{7.50}$$

and

$$\langle\sigma\rangle_D = \phi_1\mathbf{C}^1\langle\epsilon\rangle_{V_1} + (1 - \phi_1)\mathbf{C}^0 : \langle\epsilon\rangle_M = \left(\mathbf{C}^0 + \phi_1\Delta\mathbf{C}(\phi_1) : \mathbf{A}_1\right) : \langle\epsilon\rangle_D \tag{7.51}$$

Therefore, we obtain

$$\bar{\mathbf{C}}(\phi_1) = \mathbf{C}^0 + \phi_1 \Delta\mathbf{C} : \mathbf{A}_1 \tag{7.52}$$

where

$$\Delta\mathbf{C}(\phi_1) = \mathbf{C}^1 - \mathbf{C}^0 \tag{7.53}$$

(2) Treat the new composite as a matrix, add another volume of particle, and update the volume fraction and effectiveness. Step by step, one can cumulate the overall volume fraction of the particles to a finite number less than 100%. From step i to step $i+1$, the volume of the particle increases from V_i to V_{i+1} with an increment ΔV_i, so that the volume fractions at the two steps are written as

$$\phi_{i+1} = \frac{V_i + \Delta V_i}{V_0 + V_i + \Delta V_i}; \quad \phi_i = \frac{V_i}{V_0 + V_i} \tag{7.54}$$

The volume fraction increment is written as

$$\Delta\phi_i = \phi_{i+1} - \phi_i = (1 - \phi_i)\frac{\Delta V_i}{V_0 + V_i + \Delta V_i} \tag{7.55}$$

At step i, the effective stiffness is written as $\bar{\mathbf{C}}(\phi_i)$. With the incremental volume of particles, similarly to Equation 7.52, one can write

$$\bar{\mathbf{C}}(\phi_{i+1}) = \bar{\mathbf{C}}(\phi_i) + \frac{\Delta V_i,}{V_0 + V_i + \Delta V_i}\Delta\mathbf{C} : \mathbf{A}_i \tag{7.56}$$

where

$$\Delta\mathbf{C}(\phi_{i+1}) = \mathbf{C}^1 - \bar{\mathbf{C}}(\phi_i) \tag{7.57}$$

and $\mathbf{A}_i$ is obtained by considering one particle with $\mathbf{C}^1$ embedded in the matrix with $\bar{\mathbf{C}}(\phi_i)$. Then, Equation 7.56 can be written as

$$\bar{\mathbf{C}}(\phi_{i+1}) - \bar{\mathbf{C}}(\phi_i) = \frac{\Delta\phi_i}{1 - \phi_i}\left[\mathbf{C}^1 - \bar{\mathbf{C}}(\phi_i)\right] : \mathbf{A}_i \tag{7.58}$$

(3) Considering the increment ΔV_i as an infinitesimal value, the above equation can be written as

$$\frac{d\mathbf{C}(\#)}{d\phi} = \frac{1}{1 - \phi}\left[\mathbf{C}^1 - \mathbf{C}(\phi)\right] : \mathbf{A}(\mathbf{C}(\#)) \tag{7.59}$$

where $\mathbf{A}(\mathbf{C}(\phi))$ means the fourth-rank tensor relating the particle's average strain to the overall average strain for one particle with $\mathbf{C}^1$ embedded in the matrix with $\mathbf{C}(\phi)$. To solve for the above equation, one needs to determine the method to solve $\mathbf{A}$. An initial condition is also needed, which can be obtained by the first step. When $\phi = 0$, one may write $\mathbf{C}(\phi) = \mathbf{C}^0$.

When both phases are isotropic materials and the particles are spherical, one can write it in terms of the bulk modulus and shear modulus as

$$\frac{d\langle k\rangle_D}{d\phi} = \frac{\left(\langle k\rangle_D - k^1\right)\left(3\langle k\rangle_D + 4\langle\mu\rangle_D\right)}{(1-\phi)\left(3k^1 + 4\langle\mu\rangle_D\right)} \tag{7.60}$$

$$\frac{d\langle\mu\rangle_D}{d\phi} = \frac{\left(\langle\mu\rangle_D - \mu^1\right)\left(3\langle k\rangle_D + 4\langle\mu\rangle_D\right)}{(1-\phi)\left[3\langle k\rangle_D\left(3k^1 + 4\langle\mu\rangle_D\right) + 4\langle\mu\rangle_D\left(2\langle\mu\rangle_D + 3\mu^1\right)\right]} \tag{7.61}$$

with initial condition

$$\phi = 0: \quad \langle k\rangle_D = k^0, \ \langle\mu\rangle_D = \mu^0 \tag{7.62}$$

7.3 EXERCISES

1. Use one particle in a finite domain as an RVE of a composite. Apply a test load on the boundary and derive the average stress and strain by using the image stress concept. Evaluate the effective stiffness of this RVE and compare with the Mori–Tanaka model.

2. Consider a porous material containing a polymer material with $E = 2$ GPa and $v = 0.4$ and spherical air voids. Predict Young's modulus and Poisson's ratio of this material changing with the volume fraction of the air voids from 0% to 50% using the Voigt model, the Reuss model, the dilute model, the Mori–Tanaka model, the self-consistent model, and Hashin–Shtrikman's bounds, respectively. Illustrate your results through figures.

3. Consider a metal particulate polymer matrix material with $E_0 = 2$ GPa and $v_0 = 0.4$ for the polymer and $E_1 = 200$ GPa and $v_1 = 0.3$. Predict the shear modulus and bulk modulus of the composite material changing with the volume fraction of the metal particle from 0% to 90% using the differential scheme.

CHAPTER 8

Homogenization for Effective Elasticity Based on the Perturbation Method

8.1 INTRODUCTION

Homogenization theory based on the perturbation methods uses a periodic structure to represent the microstructure and sets up the governing equations at both the microscopic and macroscopic scales. Using ergodicity, one can use the perturbation method to study general material settings such as randomly dispersed composites. Through homogenization, the material features at microscale can be considered and the homogenized governing partial differential equation in the macroscale can be obtained. In general, the homogenization theory based on the perturbation method should provide consistent results with other homogenization methods if the microstructure of the RVE can be well-defined.

Consider a linear elastic heterogeneous material under an external load. In the macroscopic scale, denoted by D, the effective material properties can be obtained by certain weighted averages of the microstructure at the microscopic scale, denoted by d. To use homogenization, the length scales satisfy

$$D \gg d$$

or the ratio of the two size scales satisfies

$$\epsilon = d/D \ll 1$$

Here, we set up two coordinates: $\mathbf{x}$ is a global coordinate at the macroscale, where the elastic fields change slowly; and $\mathbf{y}$ is a local coordinate, shown in Figure 8.1, which are related to each other through the scale ratio as

$$\mathbf{y} = \mathbf{x}/\epsilon$$

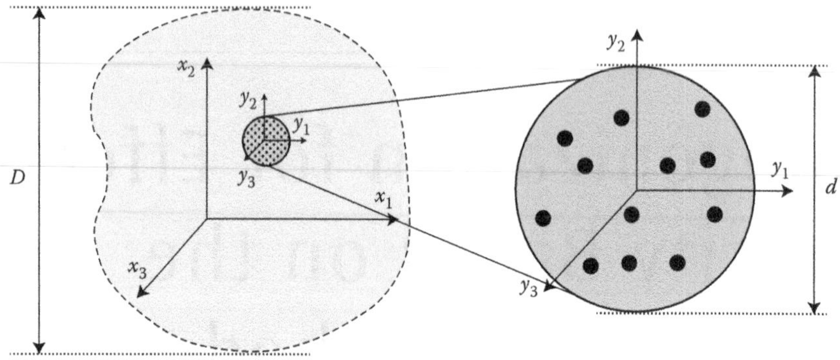

FIGURE 8.1 Different scale coordinates.

As schematically illustrated in Figure 8.2, when the heterogeneous material is subjected to an applied load, the elastic field may continuously change at the macroscale slowly. However, at the microscale, the local elastic field may fluctuate rapidly due to the material change in the heterogeneous microstructure. Considering the periodicity of the microstructure, the fluctuation of local elastic field in the $\mathbf{y}$ coordinate will follow a certain pattern but its average still follows the trend in the $\mathbf{x}$ coordinate. Therefore, the local field can be written in terms of both $\mathbf{x}$ and $\mathbf{y}$. The perturbation of the local field and material properties will be periodic, say $f(\mathbf{x}, \mathbf{y}) = f(\mathbf{x}, \mathbf{y} + d\boldsymbol{\delta})$ with f representing a general local field function.

The homogenization procedure based on the perturbation method generally involves the following steps:

1. Set up the the averaged governing equations, perform the perturbation expansion, and decouple the governing equations in the two scales.
2. Solve the effective properties through characterization of the microstructure and mechanical analysis of the local fields.
3. Solve the homogenized equations of the governing equation for material behavior at the macroscopic scale with a certain boundary and initial conditions.
4. Solve the local fields by using the global material behavior.

Following we will demonstrate the concept by a one-dimensional case and then extend the method to general periodic and random composites.

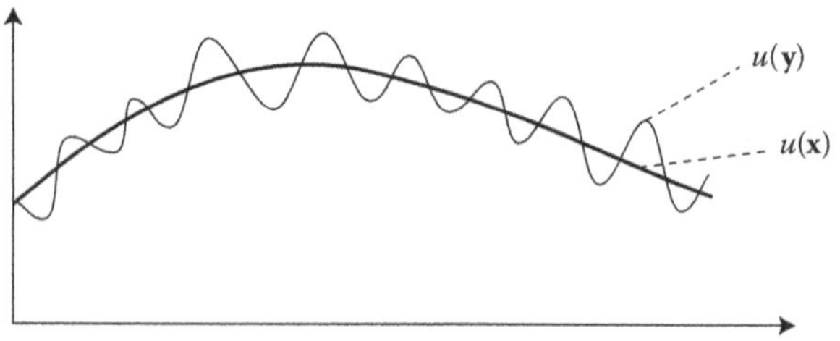

FIGURE 8.2 Perturbation of the displacement field.

8.2 ONE-DIMENSIONAL ASYMPTOTIC HOMOGENIZATION

Consider the equilibrium of a one-dimensional bar

$$\frac{d}{dx}\left[E\frac{du}{\partial x}\right]=0, \quad 0<x<D \tag{8.1}$$

where E is the elastic modulus and u the displacement in the x direction. Here, the bar exhibits periodic heterogenous microstructure when it is observed at microscale with a characteristic length of a unit cell d, which is the period. Therefore, E is a periodic function in the microscale. The displacement and stress field in the microscale of d satisfy the continuity conditions. In the macroscale of D, the displacement and stress also satisfy the global boundary conditions.

To obtain the response at the macroscale instead of the detailed behavior at the microscale, we can use the asymptotic homogenization as follows:

First, we can set up the global and local coordinates x and y through the scaling factor $\epsilon = d/D \ll 1$, that is, $x = \epsilon y$. The displacement u may be written in terms of two coordinates as $u = u(x, y)$. Starting at the microscale, the governing equation can be rewritten as

$$\frac{d}{dx}\left[E^{\epsilon}(y)\frac{du^{\epsilon}}{\partial x}\right]=0, \quad 0<x<D \tag{8.2}$$

where the elastic modulus $E^{\epsilon}(y)$ is a periodic function of y. Expand the displacement in an asymptotic form,

$$u^{\epsilon}=u^0(x,y)+\epsilon u^1(x,y)+\epsilon^2u^2(x,y)+\cdots \tag{8.3}$$

where u^i represent the displacement variation in the ith order of the scaling factor ϵ. Using the chain rule of $y = x/\epsilon$, $\frac{d}{dx}=\frac{\partial}{\partial x}+\frac{1}{\epsilon}\frac{\partial}{\partial y}$, Equation 8.2 can be rewritten as

$$\left(\frac{\partial}{\partial x}+\frac{1}{\epsilon}\frac{\partial}{\partial y}\right)\left[E(y)\left(\frac{\partial}{\partial x}+\frac{1}{\epsilon}\frac{\partial}{\partial y}\right)\left(u_k^0+\epsilon u_k^1+\epsilon^2u_k^2+\cdots\right)\right]=0 \tag{8.4}$$

After reorganization, the above equation can be rewritten as

$$\left(\epsilon^{-2}L^0+\epsilon^{-1}L^1+L^2\right)\left(u^0+\epsilon u^1+\epsilon^2u^2+\cdots\right)=0 \tag{8.5}$$

where

$$L^0=\frac{\partial}{\partial y}E(y)\frac{\partial}{\partial y} \tag{8.6}$$

$$L^1=E(y)\frac{\partial}{\partial x}\frac{\partial}{\partial y}+\frac{\partial}{\partial y}E(y)\frac{\partial}{\partial x} \tag{8.7}$$

$$L^2=E(y)\frac{\partial^2}{\partial x\partial x} \tag{8.8}$$

Then Equation 8.5 is expanded in terms of ϵ

$$\epsilon^{-2}L^0u^0 + \epsilon^{-1}\left(L^1u^0 + L^0u^1\right) + \left(L^2u^0 + L^1u^1 + L^0u^2\right) + O(\epsilon) = 0 \tag{8.9}$$

when $\epsilon \to 0$, the effective properties of the unit cell can exactly be obtained as the system volume approaches the infinity. Therefore, to make the above equation satisfied, one can write the following three equations at different orders of ϵ:

$$L^0u^0 = 0 \tag{8.10}$$

$$L^1u^0 + L^0u^1 = 0 \tag{8.11}$$

$$L^2u^0 + L^1u^1 + L^0u^2 = 0 \tag{8.12}$$

From the first equation, we can write

$$\frac{\partial}{\partial y}u^0(x,y) = \frac{f^0(x)}{E(y)} \tag{8.13}$$

where $f^0(x)$ is an integral constant on y. It can be further written as

$$u^0(x,y) = f^0(x)\int_{y_0}^{y}\frac{dy}{E(y)} + g^0(x) \tag{8.14}$$

where $g^0(x)$ is another integral constant on y. Because $u^0(x,y) = u^0(x, y+d)$, one can write

$$f^0(x)\int_{y_0}^{y_0+d}\frac{1}{E(y)}dy = 0 \tag{8.15}$$

or $f^0(x) = 0$. Therefore,

$$u^0(x,y) = u^0(x)$$

is the displacement in the global scale.

The second equation can be simplified as

$$\frac{\partial}{\partial y}E(y)\frac{\partial}{\partial y}u^1(x,y) = -\frac{\partial}{\partial y}E(y)\frac{\partial u^0}{\partial x} \tag{8.16}$$

We can integrate the above equation to obtain

$$u^1(x,y) = \int_{y_0}^{y}\left[-\frac{\partial u^0}{\partial x} + \frac{f^1(x)}{E(y)}\right]dy + g^1(x) \tag{8.17}$$

where $f^1(x)$ and $g^1(x)$ are integral constants again on y. Considering the y-periodicity of $u^1(x, y)$ with a period d, we can obtain

$$\int_{y_0}^{y_0+d} \left[-\frac{\partial u^0}{\partial x} + \frac{f^1(x)}{E(y)} \right] dy = 0$$

so that

$$f^1(x) = \frac{\frac{\partial u^0}{\partial x} d}{\int_{y_0}^{y_0+d} \frac{dy}{E(y)}} \tag{8.18}$$

Therefore, $u^1(x, y)$ can be obtained as

$$u^1(x, y) = -\frac{\partial u^0}{\partial x} y + \frac{\partial u^0}{\partial x} \frac{\int_{y_0}^{y} \frac{1}{E(y)} dy}{\frac{1}{d} \int_{y_0}^{y_0+d} \frac{dy}{E(y)}} + g^1(x) \tag{8.19}$$

Because the integral of displacement $u^\epsilon(x, y)$ over the period $y \in (y_0, y_0 + d)$ should be convergent to $du^0(x)$, one can obtain

$$\int_{y_0}^{y_0+d} u^1(x, y) dy = 0 \tag{8.20}$$

so that $g^1(x)$ can be determined.

The third equilibrium equation at the order of ϵ^0, say Equation 8.12, can be rewritten as

$$\frac{\partial}{\partial y} E(y) \left[\frac{\partial u^1}{\partial x} + \frac{\partial u^2}{\partial y} \right] + E(y) \left[\frac{\partial^2 u^0}{\partial x^2} + \frac{\partial^2 u^1}{\partial x \partial y} \right] = 0 \tag{8.21}$$

which stands for the actual equilibrium in the local coordinate. Using Equation 8.19, one can obtain

$$\frac{\partial^2 u^0}{\partial x^2} + \frac{\partial^2 u^1}{\partial x \partial y} = \frac{1}{E(y) \left[\frac{1}{d} \int_{y_0}^{y_0+d} \frac{dy}{E(y)} \right]} \frac{\partial^2 u^0}{\partial x^2} \tag{8.22}$$

Therefore, substituting the above equation into Equation 8.21 yields

$$\frac{\partial}{\partial y}E(y)\left[\frac{\partial u^1}{\partial x}+\frac{\partial u^2}{\partial y}\right]+\frac{1}{\frac{1}{d}\int\limits_{y_0}^{y_0+d}\frac{dy}{E(y)}}\frac{\partial^2 u^0}{\partial x^2}=0 \tag{8.23}$$

Then, one can write

$$\frac{\partial u^1}{\partial x}+\frac{\partial u^2}{\partial y}=-\frac{1}{\frac{1}{d}\int\limits_{y_0}^{y_0+d}\frac{dy}{E(y)}}\frac{\partial^2 u^0}{\partial x^2}\frac{y}{E(y)}+\frac{f^2(x)}{E(y)}$$

or

$$\frac{\partial u^2}{\partial y}=-\frac{1}{\frac{1}{d}\int\limits_{y_0}^{y_0+d}\frac{dy}{E(y)}}\frac{\partial^2 u^0}{\partial x^2}\frac{y}{E(y)}+\frac{f^2(x)}{E(y)}+\frac{\partial^2 u^0}{\partial x^2}y-\frac{\partial^2 u^0}{\partial x^2}\frac{\int\limits_{y_0}^{y}\frac{1}{E(y)}dy}{\frac{1}{d}\int\limits_{y_0}^{y_0+d}\frac{dy}{E(y)}}-\frac{dg^1(x)}{dx} \tag{8.24}$$

Considering the periodicity of $\frac{\partial u^2(x,y)}{\partial y}$ over the period $y \in (y_0, y_0+d)$, one can obtain

$$\frac{1}{\frac{1}{d}\int\limits_{y_0}^{y_0+d}\frac{dy}{E(y)}}\frac{\partial^2 u^0}{\partial x^2}=0 \tag{8.25}$$

and $u^2(x,y)$ can be explicitly obtained in terms of u^0 by integrating Equation 8.24 and considering $\int_{y_0}^{y_0+d} u^2(x,y)dy=0$. Notice that Equation 8.25 can be rewritten as

$$\frac{\partial}{\partial x}\left[\frac{1}{\frac{1}{d}\int\limits_{y_0}^{y_0+d}\frac{dy}{E(y)}}\frac{\partial u^0}{\partial x}\right]=0 \tag{8.26}$$

Comparing with Equation 8.2, one can write the effective elastic modulus for the homogenized differential equation as

$$\langle E\rangle = \frac{1}{\frac{1}{d}\int_{y_0}^{y_0+d} \frac{dy}{E(y)}} \tag{8.27}$$

which is consistent with the Reuss model or the lower bound. Using this equilibrium equation to solve the global boundary value problem for u^0, one can straightforwardly derive the local fields of u^1 and u^2 based on the local material properties in the unit cell and u^0.

8.3 HOMOGENIZATION OF A PERIODIC COMPOSITE

Similarly, the asymptotic homogenization method can be extended to 3D cases. Consider a composite with periodically distributed microstructure. Then, the local coordinate $\mathbf{y}$ can be set up over a unit cell Ω, which sufficiently represents the microstructure as shown in Figure 8.3.

The governing equation at the macroscale or equilibrium equation can be written as

$$\frac{\partial}{\partial x_i}\left[C_{ijkl}\frac{\partial u_k}{\partial x_l}\right] = -f_j(\mathbf{x}) \tag{8.28}$$

where the body force f_j is provided at the macroscale, so it only depends on the global coordinate $\mathbf{x}$. In the two-scale context, the above equation can be written as

$$\frac{\partial}{\partial x_i}\left[C^{\epsilon}_{ijkl}\frac{\partial u^{\epsilon}_k}{\partial x_l}\right] = -f_j(\mathbf{x}) \tag{8.29}$$

where the superscripts ϵ of the stiffness C^{ϵ}_{ijkl} and the displacement u^{ϵ}_k indicate that at the macroscale they change slowly but at the microscale they depend on the scale ratio ϵ. On the boundary, one can write stress and displacement boundary conditions as

$$\begin{cases} n_i C^{\epsilon}_{ijkl}(x)u^{\epsilon}_{k,l} = t^0_j; & \mathbf{x} \in \partial D^t \\ u^{\epsilon}_j = u^0_j; & \mathbf{x} \in \partial D^u \end{cases} \tag{8.30}$$

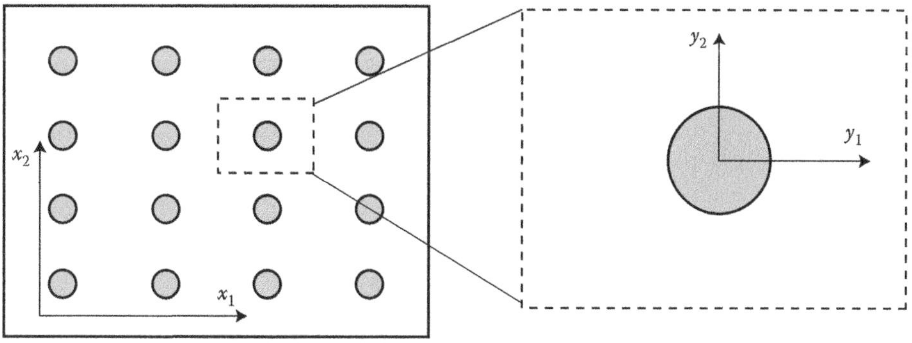

FIGURE 8.3 Periodic composite.

where ∂D^t and ∂D^u are complementary, that is, $\partial D^t + \partial D^u = \partial D$. At the local coordinate $\mathbf{y}$, the displacement can be written as

$$u_i^\epsilon = u_i(\mathbf{x}, \mathbf{y}) = u_i\left(\mathbf{x}, \frac{\mathbf{x}}{\epsilon}\right) \tag{8.31}$$

and the stiffness tensor

$$C_{ijkl}^\epsilon(\mathbf{x}) = C_{ijkl}(\mathbf{y}) \tag{8.32}$$

Using the chain rule, one can write the derivative of u_i on x_j as

$$\frac{\partial u_i^\epsilon}{\partial x_j} = \frac{\partial u_i(\mathbf{x}, \mathbf{y})}{\partial x_j} = \frac{\partial u_i(\mathbf{x}, \mathbf{y})}{\partial x_j} + \frac{1}{\epsilon}\frac{\partial u_i(\mathbf{x}, \mathbf{y})}{\partial y_j} \tag{8.33}$$

The perturbation expansion of u_i^ϵ can be written in an asymptotic form as

$$u_i^\epsilon = u_i^0(\mathbf{x}) + \epsilon u_i^1(\mathbf{x}, \mathbf{y}) + \epsilon^2 u_i^2(\mathbf{x}, \mathbf{y}) + \cdots \tag{8.34}$$

Notice that here we directly take the first term $u_i^0(\mathbf{x})$ based on the one-dimensional case. The proof can be found in References [51,52].

Equation 8.28 can be written as

$$\left(\frac{\partial}{\partial x_i} + \frac{1}{\epsilon}\frac{\partial}{\partial y_i}\right)\left[C_{ijkl}(\mathbf{y})\left(\frac{\partial}{\partial x_l} + \frac{1}{\epsilon}\frac{\partial}{\partial y_l}\right)\left(u_k^0 + \epsilon u_k^1 + \epsilon^2 u_k^2 + \cdots\right)\right] = -f_j \tag{8.35}$$

After reorganization, the above equation yields

$$\left(\epsilon^{-2} A_{jk}^0 + \epsilon^{-1} A_{jk}^1 + A_{jk}^2\right)\left(u_k^0 + \epsilon u_k^1 + \epsilon^2 u_k^2\right) = -f_j \tag{8.36}$$

where

$$A_{jk}^0 = \frac{\partial}{\partial y_i} C_{ijkl}(\mathbf{y}) \frac{\partial}{\partial y_l} \tag{8.37}$$

$$A_{jk}^1 = C_{ijkl}(\mathbf{y}) \frac{\partial}{\partial x_i}\frac{\partial}{\partial y_l} + \frac{\partial}{\partial y_i} C_{ijkl}(\mathbf{y}) \frac{\partial}{\partial x_l} \tag{8.38}$$

$$A_{jk}^2 = C_{ijkl}(\mathbf{y}) \frac{\partial^2}{\partial x_i \partial x_l} \tag{8.39}$$

Then, equilibrium Equation 8.36 is expanded in terms of ϵ:

$$\epsilon^{-2} A_{jk}^0 u_k^0 + \epsilon^{-1}\left(A_{jk}^1 u_k^0 + A_{jk}^0 u_k^1\right) + \left(A_{jk}^2 u_k^0 + A_{jk}^1 u_k^1 + A_{jk}^0 u_k^2\right) + O(\epsilon) = -f_j \tag{8.40}$$

when $\epsilon \rightarrow 0$, the effective properties of the unit cell can exactly be obtained as the system volume approaches the infinity. Therefore, to make the above equation satisfied, one can write

$$A^0_{jk}u^0_k = 0 \tag{8.41}$$

$$A^1_{jk}u^0_k + A^0_{jk}u^1_k = 0 \tag{8.42}$$

$$A^2_{jk}u^0_k + A^1_{jk}u^1_k + A^0_{jk}u^2_k = -f_j \tag{8.43}$$

Notice that the body force f_i is given in order ϵ^0 as it only depends on $\mathbf{x}$. Because u^0_k depends on $\mathbf{x}$ only, the first equilibrium equation $A^0_{jk}u^0_k = 0$ is automatically satisfied. The second equilibrium equation can be rewritten as

$$\frac{\partial}{\partial y_i}C_{ijkl}(\mathbf{y})\frac{\partial}{\partial y_l}u^1_k(\mathbf{x},\mathbf{y}) + \frac{\partial}{\partial y_i}C_{ijkl}(\mathbf{y})\frac{\partial}{\partial x_l}u^0_k = 0 \tag{8.44}$$

The above equation can be solved by numerical methods or Green's function technique as $C_{ijkl}(\mathbf{y})$ is provided by the microstructure and $\frac{\partial}{\partial x_l}u^0_k$ is independent from $\mathbf{y}$. In the local coordinate of the unit cell, periodic stress and displacement boundary conditions are given; the local field can be solved in a general form as

$$u^1_m(\mathbf{x},\mathbf{y}) = \int_Y G_{mj}(\mathbf{x},\mathbf{y})\left[\frac{\partial}{\partial y_i}C_{ijkl}(\mathbf{y})\frac{\partial}{\partial x_l}u^0_k\right]d\mathbf{y} \tag{8.45}$$

where G_{mj} is an influence function to consider the source of $\frac{\partial}{\partial x_l}u^0_k$, which can be determined by the equivalent inclusion method as we elaborated before or directly calculated by numerical methods, such as the finite element or boundary element methods. Notice that the average of displacement $u^\epsilon_i(\mathbf{x},\mathbf{y})$ in the unit cell Ω should be consistent with the global displacement $u_i(\mathbf{x})$ or $u^0_i(\mathbf{x})$ without considering any rigid-body motion, that is,

$$\int_\Omega u^1_m(\mathbf{x},\mathbf{y})d\mathbf{y} = \mathbf{0} \tag{8.46}$$

As a special case, one can approximately assume

$$u^1_m(\mathbf{x},\mathbf{y}) = \mathbf{0} \text{ for } \mathbf{y} \in \partial\Omega \tag{8.47}$$

which reduces the periodic boundary of the unit cell to a fixed boundary. Based on the continuity of displacement field, Equation 8.46 will be automatically satisfied. The solution can be obtained using Green's function similar to Section 6.4. However, this rarely happens in the actual materials. Therefore, real periodic boundary conditions should be enforced to higher accuracy. Using $\mathbf{u}^0$ and $\mathbf{u}^1$, one can derive the strain and stress field in the local coordinate of the unit cell as

$$\epsilon_{ij}(\mathbf{x},\mathbf{y}) = \epsilon_{ij}^0 + \frac{1}{2}(u_{i,j'}^1 + u_{j,i'}^1) \tag{8.48}$$

and

$$\sigma_{ij}(\mathbf{x},\mathbf{y}) = C_{ijkl}(\mathbf{y})\epsilon_{kl}(\mathbf{x},\mathbf{y}) \tag{8.49}$$

where $\epsilon_{ij}^0 = \frac{1}{2}(u_{i,j}^0 + u_{j,i}^0)$. Here, the derivative, i is on x_i whereas, i' on y_i. Because $\mathbf{u}^1$ is essentially written in terms of $\mathbf{u}^0$, Equation 8.48 can be rewritten as

$$\epsilon_{ij}(\mathbf{x},\mathbf{y}) = E_{ijkl}(\mathbf{y})\epsilon_{kl}^0(\mathbf{x}) \tag{8.50}$$

and then

$$\sigma_{ij}(\mathbf{x},\mathbf{y}) = C_{ijmn}(\mathbf{y})E_{mnkl}(\mathbf{y})\epsilon_{kl}^0(\mathbf{x}) \tag{8.51}$$

where $E_{ijkl}(\mathbf{y})$ is an influence function for strain field. Due to the displacement continuity and periodic boundary condition

$$\int\limits_{\Omega} \left(u_{i,j'}^1 + u_{j,i'}^1\right) d\mathbf{y} = \int\limits_{\partial\Omega} \left(u_i^1 n_j + u_j^1 n_i\right) d\mathbf{y} = \mathbf{0}$$

one can obtain

$$\int\limits_{\Omega} \epsilon_{ij}(\mathbf{x},\mathbf{y})d\mathbf{y} = \int\limits_{\Omega} \epsilon_{ij}^0(\mathbf{x})d\mathbf{y} \tag{8.52}$$

Now we can consider the equilibrium equation of ϵ^0. Equation 8.43 can be rewritten as

$$C_{ijkl}(\mathbf{y})\frac{\partial^2 u_k^0}{\partial x_i \partial x_l} + C_{ijkl}(\mathbf{y})\frac{\partial}{\partial x_i}\frac{\partial u_k^1}{\partial y_l} + \frac{\partial}{\partial y_i}C_{ijkl}(\mathbf{y})\frac{\partial u_k^1}{\partial x_l} + \frac{\partial}{\partial y_i}C_{ijkl}(\mathbf{y})\frac{\partial u_k^2}{\partial y_l} = -f_j \tag{8.53}$$

Taking the volume integral of the above equation over the domain $\mathbf{y} \in \Omega$, one can obtain

$$\int\limits_{\Omega} \frac{\partial \sigma_{ij}(\mathbf{x},\mathbf{y})}{\partial x_i} d\mathbf{y} + \int\limits_{\partial\Omega} C_{ijkl}(\mathbf{y})\frac{\partial u_k^1}{\partial x_l} n_i d\mathbf{y} + \int\limits_{\Omega} \frac{\partial}{\partial y_i}C_{ijkl}(\mathbf{y})\frac{\partial u_k^2}{\partial y_l} d\mathbf{y} = -\int\limits_{\Omega} f_j d\mathbf{y} \tag{8.54}$$

Here, the periodic boundary condition yields $\int_{\partial\Omega} C_{ijkl}(\mathbf{y})\frac{\partial u_k^1}{\partial x_l} n_i d\mathbf{y} = 0$ and equilibrium on stress caused by u_k^2 randers $\int_{\Omega} \frac{\partial}{\partial y_i} C_{ijkl}(\mathbf{y})\frac{\partial u_k^2}{\partial y_l} d\mathbf{y} = 0$. Therefore, taking volume average of

Equation 8.54 can provide the equilibrium equation in terms of the global coordinate **x** as

$$\frac{\partial}{\partial x_i}\left[\int_{\Omega} \sigma_{ij}(\mathbf{x},\mathbf{y})d\mathbf{y}\right] = -f_j(\mathbf{x}) \tag{8.55}$$

Substituting Equation 8.51 into the above equation yields

$$\frac{\partial}{\partial x_i}\left[\int_{\Omega} C_{ijmn}(\mathbf{y})E_{mnkl}(\mathbf{y})d\mathbf{y}u^0_{k,l}(\mathbf{x})\right] = -f_j(\mathbf{x}) \tag{8.56}$$

where Equation 8.52 is used. Comparing Equation 8.56 with Equation 8.28, one can obtain the effective elastic modulus for the homogenized differential equation as

$$\langle C_{ijkl}\rangle = \int_{\Omega} C_{ijmn}(\mathbf{y})E_{mnkl}(\mathbf{y})d\mathbf{y} \tag{8.57}$$

Therefore, once the microstructure is given, if the strain influence function $E_{mnkl}(\mathbf{y})$ is provided, the effective elastic modulus can be solved. Then by solving the global boundary value problem, one can obtain $u_i^0(\mathbf{x})$. Then the local field can be solved.

8.4 EXCERCISES

1. Use a regular perturbation expansion to solve the following differential equations with corresponding boundary conditions:

   a.

$$\begin{cases} u_{,xx} + u + \epsilon u^2 = 0, & 0 < x < \infty. \\ u(0) = 1 & u_{,x}(0) = 0. \end{cases}$$

   b.

$$\begin{cases} \epsilon u_{,xx} + u_{,x} + u = 0, & 0 < x < 1. \\ u(0) = a & u_{,x}(0) = b. \end{cases}$$

   c.

$$\begin{cases} u_{,xx} + u_{,yy} = \epsilon u, & 0 < x < 1, 0 < y < 1. \\ u(0,y) = u(1,y) = 0; & u_{,x}(x,0) = 0, u_{,x}(x,1) = x. \end{cases}$$

2. Given a 1D column standing on a rigid foundation bearing a load of 1 kN on the top. The length is 1 m. The self-gravity of the column is 1 kN/m. Set up the global coordinate from the bottom to the top. The material's Young's modulus varies in the local coordinate periodically as $E(y) = 70 + 2\sin\left(\frac{y}{d}\right)$ GPa, where $d = 1$ mm. Determine the effective Young's modulus and the local displacement distribution of $u^1(x,y)$ for $x = 0, 0.5$, and 1.

3. A 2D cantilever beam is subjected to a distributed force of w, where the height of the beam h is much less than the length L ($h \ll L$). The beam is made of composite materials containing a simple lattice distribution of spherical particles in a matrix. The stiffness of particle and matrix phases are $\mathbf{C}^1$ and $\mathbf{C}^0$, respectively. The particle diameter d is much less than h. The volume fraction of particles ϕ is very small too. Construct the strain influence function $\mathbf{E}$ by Green's function method. Determine the effective stiffness. Calculate the local strain distribution at the four corner points of the beam.

CHAPTER 9

Defects in Materials: Void, Microcrack, Dislocation, and Damage

ALTHOUGH COMPOSITE MATERIALS are commonly designed to maximize the advantages of different material phases for synergic benefits. For example, in fiber-reinforced polymer composites, the strength of polymer materials can be reinforced with fibers whereas fibers can be protected by polymer matrix. However, manufacturing multiphase materials is prone to defects, such as air voids, microcracks or discontinuities, and interfacial debondings. When the material is subjected to fatigue, creep, and plastic loads, cracking, dislocation, and damage may be induced at the microscale, which may lead to yielding, softening, cracking, and failure of the material at the macroscale. The defects can be treated as a disturbing source or a new material phase in the micromechanics.

9.1 VOIDS

Stress concentration due to an air void in a solid has been well studied analytically and experimentally. Some stress components in the neighborhood of an air void may be much higher than a uniform far-field stress due to the inhomogeneity. We can use Eshelby's solution to study the stress field considering the inhomogeneity with a zero stiffness tensor as illustrated in Figure 9.1.

In Figure 9.1, an ellipsoidal void with stiffness $\mathbf{C}^1 = 0$ is embedded in a matrix with stiffness $\mathbf{C}^0$. The mismatch stiffess can be written as

$$\triangle\mathbf{C} = \mathbf{C}^1 - \mathbf{C}^0 = -\mathbf{C}^0 \tag{9.1}$$

and the strain field can be written as

$$\epsilon = \epsilon^0 + \epsilon' = \epsilon^0 - \mathbf{D}(\mathbf{x}) : \mathbf{C}^0 : \epsilon^* \tag{9.2}$$

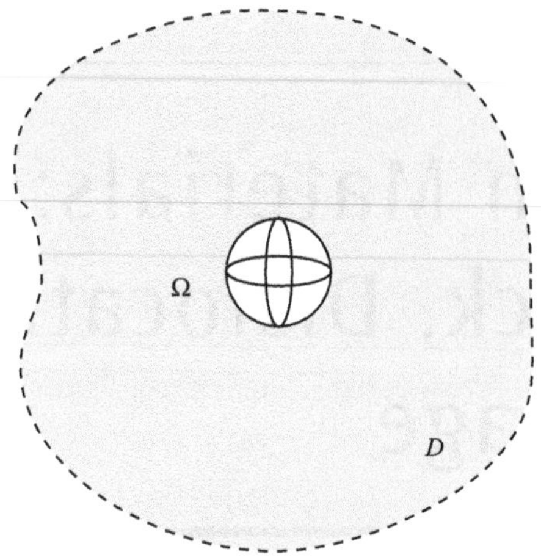

FIGURE 9.1 Void in the matrix.

Using the equivalent inclusion condition yields

$$\mathbf{C}^0 : \left(\epsilon^0 - \mathbf{S} : \epsilon^* - \epsilon^*\right) = \mathbf{C}^1 : \epsilon = \mathbf{0} \tag{9.3}$$

where $\mathbf{S} = \mathbf{D}^{\Omega} : \mathbf{C}^0$. Considering Equation 9.2, one can see that the strain in the air void $\epsilon = \epsilon^0 - \mathbf{S} : \epsilon^*$ should be equal to the eigenstrain of the equivalent inclusion. Then, one can obtain

$$\epsilon^* = (I + \mathbf{S})^{-1} : \epsilon^0 \tag{9.4}$$

Substituting the above equation into Equation 9.2 yields

$$\epsilon = \left[\mathbf{I} - \mathbf{D}(\mathbf{x}) : \mathbf{C}^0 : (\mathbf{I} + \mathbf{S})^{-1}\right] : \epsilon^0 \tag{9.5}$$

For a point on the particle $x \in \Omega$, one can write $\mathbf{D}(\mathbf{x}) = \mathbf{D}^{\Omega}$ and then we can confirm

$$\epsilon = \epsilon^* \tag{9.6}$$

However, because the strain will be discontinuous a cross the interface of Ω, the stress distribution along the void surface cannot be calculated directly from the eigenstrain. Instead, Equation 9.5 should be used in the constitutive law for stress calculation. In the equivalent inclusion, the stress can be treated as the superposition of the uniform stress $\sigma^0 = \mathbf{C}^0 : \epsilon^0$ and the one caused by disturbed strain and eigenstrain

$$\sigma^{\Omega} = \mathbf{C}^0 : \left(-\mathbf{S} : \epsilon^* - \epsilon^*\right) = -\sigma^0. \tag{9.7}$$

From Equation 5.53, the potential energy for a solid containing a void subjected to a uniform stress σ^0_{ij} can be written as

$$\Pi = -\frac{1}{2}\int_D \sigma^0_{ij} u^0_{i,j} dx - \frac{1}{2}\int_{\Omega} \sigma^0_{ij} \epsilon^*_{ij} dx \tag{9.8}$$

Comparing it with the reference material system without any void, one can obtain the potential energy reduced by the void as

$$\Delta\Pi = -\frac{1}{2}V_{\Omega}\sigma^0_{ij}\epsilon^*_{ij} \tag{9.9}$$

9.2 MICROCRACKS

9.2.1 Penny-Shape Crack

When one semi-axis of the ellipsoidal void approaches zero, that is, $a_3 = c \to 0$, and the other two are finite, that is, $a_1 = a_2 = a$, as illustrated in Figure 9.2.

When a uniform far-field stress σ^0, which causes a far-field strain ϵ^0, is applied, using Equation 9.4, the eigenstrain can be written as

$$\epsilon^* = (\mathbf{I} + \mathbf{S})^{-1} : (\mathbf{C}^0)^{-1} : \sigma^0 \tag{9.10}$$

Each nonzero component can be obtained as [3]

$$S_{1111} = S_{2222} = \frac{\pi(13 - 8v)}{32(1 - v)}\frac{c}{a} \tag{9.11}$$

$$S_{3333} = 1 - \frac{\pi(1 - 2v)}{4(1 - v)}\frac{c}{a} \tag{9.12}$$

$$S_{1122} = S_{2211} = \frac{\pi\,(8v - 1)}{32(1 - v)}\frac{c}{a} \tag{9.13}$$

$$S_{1133} = S_{2233} = \frac{\pi\,(2v - 1)}{8(1 - v)}\frac{c}{a} \tag{9.14}$$

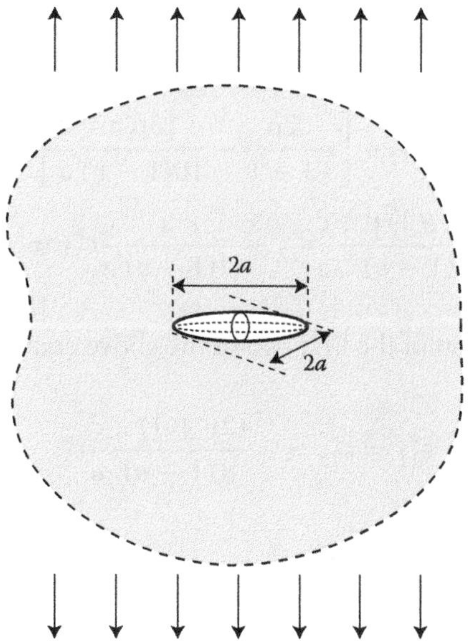

FIGURE 9.2 Penny-shaped crack in the matrix.

$$S_{3311} = S_{3322} = \frac{v}{1-v}\left(1 - \frac{\pi\,(4v+1)}{8}\frac{c}{a}\right) \tag{9.15}$$

$$S_{1212} = \frac{\pi\,(7-8v)}{32(1-v)}\frac{c}{a} \tag{9.16}$$

$$S_{1313} = S_{2323} = \frac{1}{2}\left[1 + \frac{\pi\,(v-2)}{4(1-v)}\frac{c}{a}\right] \tag{9.17}$$

and the other components can be obtained

$$S_{ijkl} = S_{jikl} = S_{ijlk} \tag{9.18}$$

Notice that the major symmetry cannot be satisfied, that is,

$$S_{ijkl} \neq S_{klij} \tag{9.19}$$

for example, $S_{1133} \neq S_{3311}$. Using Eshelby's tensor, one can write the eigenstrain, strain, and stress fields.

For an isotropic solid with a penny-shape crack (void), neither the stiffness tensor nor the Eshelby's tensor mix shear and normal strain components, which means a tensile or shear stress will only produce tensile or shear strain, respectively, in the given coordinates. One can seperately consider the two cases. When a uniform tensile stress σ^0_{33} is applied in the far-field, it may cause the penny-shape crack propagate as a Mode-I crack.

To calculate the eigenstrain or strain of the "crack inclusion," Equation 9.7 can be rewritten as

$$\left[-\frac{2\mu}{1-v} + \frac{13\mu\pi}{16(1-v)}\frac{c}{a}\right]\epsilon^*_{11} + \left[-\frac{2\mu v}{1-v} + \frac{(16v-1)\,\mu\pi}{16(1-v)}\frac{c}{a}\right]\epsilon^*_{22} - \frac{(2v+1)\,\mu\pi}{4(1-v)}\frac{c}{a}\epsilon^*_{33} = 0$$

$$\left[-\frac{2\mu v}{1-v} + \frac{(16v-1)\,\mu\pi}{16(1-v)}\frac{c}{a}\right]\epsilon^*_{11} + \left[-\frac{2\mu}{1-v} + \frac{13\mu\pi}{16(1-v)}\frac{c}{a}\right]\epsilon^*_{22} - \frac{(2v+1)\,\mu\pi}{4(1-v)}\frac{c}{a}\epsilon^*_{33} = 0$$

$$-\frac{(2v+1)\,\mu\pi}{4(1-v)}\frac{c}{a}\epsilon^*_{11} - \frac{(2v+1)\,\mu\pi}{4(1-v)}\frac{c}{a}\epsilon^*_{22} - \frac{\mu\pi}{2(1-v)}\frac{c}{a}\epsilon^*_{33} = -\sigma^0_{33}$$

As $c \to 0$, the combination of the first two of the above equations yields

$$\epsilon^*_{11} = \epsilon^*_{22} = -\frac{(2v+1)\,\pi}{8\,(1+v)}\frac{c}{a}\epsilon^*_{33}$$

Using the last equation, one can write

$$\epsilon^*_{33} = \frac{2(1-v)a}{\mu\pi}\frac{\sigma^0_{33}}{c} \tag{9.20}$$

Then, we can obtain

$$\epsilon_{11}^{*} = \epsilon_{22}^{*} = -\frac{(1+2v)(1-v)}{4(1+v)\mu}\sigma_{33}^{0} \tag{9.21}$$

Notice that ϵ_{33}^{*} approaches infinite as $c \to 0$ and the shear stress components are zero. The potential energy caused by the crack is written as

$$\Delta\Pi = -\frac{1}{2}\int_{\Omega} \sigma_{ij}^{0}\epsilon_{ij}^{*}dx = -\frac{1}{2}\int_{\Omega} \sigma_{33}^{0}\epsilon_{33}^{*}dx = -\frac{(1-v)a\left(\sigma_{33}^{0}\right)^{2}}{\mu\pi}\frac{4}{c}\frac{4}{3}\pi a^{2}c = -\frac{4(1-v)}{3\mu\pi}\left(\sigma_{33}^{0}\right)^{2}a^{3} \tag{9.22}$$

The driving force for crack growth by uniaxial tension can be written as

$$f^{I} = -\frac{\partial(\Pi)}{\partial a} = \frac{4(1-v)}{\mu\pi}\left(\sigma_{33}^{0}\right)^{2}a^{2} \tag{9.23}$$

Notice that $\Pi = \Pi^{0} + \Delta\Pi$ with Π^{0} being a constant with a, so it can be disregarded.

Considering the crack growth process of the penny-shape crack, the Gibbs free energy of this system can be written in terms of potential energy and surface energy as

$$G = \Pi + A\gamma \tag{9.24}$$

where A is the surface area of the penny-shaped crack, that is, $A = 2\pi a^{2}$ and γ is the surface energy per unit area. The Griffith fracture criterion [53] assumes that the variation of Gibbs free energy is zero when the load increases to make the crack grow, that is,

$$\frac{\partial G}{\partial a} = \frac{\partial\left(-\frac{4(1-v)}{3\mu\pi}\left(\sigma_{33}^{0}\right)^{2}a^{3} + 2\pi\gamma a^{2}\right)}{\partial a} = 0 \tag{9.25}$$

Therefore, when the fracture driving force from the Mode-I loading satisfies

$$f^{I} = \frac{4(1-v)}{\mu}\left(\sigma_{33}^{0}\right)^{2}a^{2} = 4\pi a\gamma \tag{9.26}$$

the crack will propagate; otherwise, it will remain inactive. The critical condition implies: (a) given a penny crack, when the applied stress reaches

$$\sigma_{33}^{cr} = \sqrt{\frac{\pi\mu\gamma}{(1-v)a}} \tag{9.27}$$

the crack will grow; and (b) given a load σ_{33}^{0}, if the penny crack radius is less than

$$a^{cr} = \frac{\pi\mu\gamma}{(1-v)\left(\sigma_{33}^{0}\right)^{2}} \tag{9.28}$$

the crack will remain inactive. Notice that given a penny-shaped crack, if the load σ_{33}^0 is fixed, once the crack starts to grow, the crack size a increases and the critical stress σ_{33}^{cr} for the current crack decreases, so that the crack becomes unstable.

Similarly, we can also investigate the simple shear loading situation. If a shear stress σ_{13}^0 is applied in the far-field of a penny crack, the eigenstrain can be solved in the similiar fashion. One can find that all components of ϵ_{ij}^* are zero except that

$$\epsilon_{13}^* = \frac{2(1-\upsilon)a}{\mu\pi(2-\upsilon)}\frac{\sigma_{13}^0}{c} \tag{9.29}$$

Using the Griffith fracture criterion, one can obtain the fracture driving force from the Mode-II loading as

$$f^{II} = \frac{8(1-\upsilon)}{\mu(2-\upsilon)}\left(\sigma_{33}^0\right)^2 a^2 = 4\pi a\gamma \tag{9.30}$$

Then, the critical stress for a Mode-II crack is

$$\sigma_{13}^{cr} = \sqrt{\frac{\pi\mu(2-\upsilon)\gamma}{2(1-\upsilon)a}} \tag{9.31}$$

Notice that Griffith's theory is applicable to elastic problem, and thus provides excellent agreement with experimental data for brittle materials such as glass. However, for ductile materials, such as metals or polymers, the plastic deformation may play a dominant role so that the Gibbs free energy should be modified to include the plastic energy dissipation around the fracture tip and higher critical stress may be resulted in.

9.2.2 Slit-Like Crack

In a plane strain problem, a slit-like crack with length $2a$ in Figure 9.3 can also be obtained by an ellipsoidal void taking the limit $a_2 = b \rightarrow 0$ and $a_3 = c \rightarrow \infty$ whereas $a_1 = a$. Then,

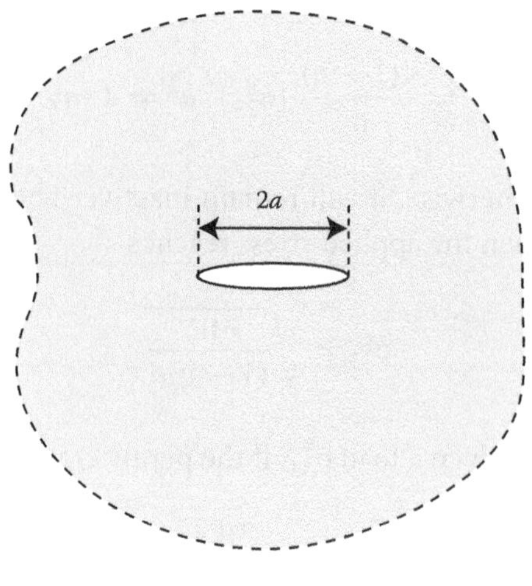

FIGURE 9.3 Slit-like crack.

we can still use Eshelby's equivalent inclusion method to conduct the stress and fracture analysis. Eshelby's tensor for isotropic material in $x_1 - x_2$ plane can be written as

$$S_{1111} = \frac{1}{2(1-v)}\left[\frac{b^2+2ab}{(a+b)^2} + (1-2v)\frac{b}{a+b}\right] \tag{9.32}$$

$$S_{2222} = \frac{1}{2(1-v)}\left[\frac{a^2+2ab}{(a+b)^2} + (1-2v)\frac{a}{a+b}\right] \tag{9.33}$$

$$S_{1122} = \frac{1}{2(1-v)}\left[\frac{b^2}{(a+b)^2} - (1-2v)\frac{b}{a+b}\right] \tag{9.34}$$

$$S_{2211} = \frac{1}{2(1-v)}\left[\frac{a^2}{(a+b)^2} - (1-2v)\frac{a}{a+b}\right] \tag{9.35}$$

Similarly to a penny-shaped crack, under the Mode-I loading condition, a tensile stress σ_{22}^0 is applied and the corresponding eigenstrain can be obtained as

$$\epsilon_{11}^* = -\frac{(1-\nu)}{2\mu}\sigma_{22}^0 \tag{9.36}$$

and

$$\epsilon_{22}^* = \frac{(1-\nu)a}{\mu b}\sigma_{22}^0 \tag{9.37}$$

Here, ϵ_{22}^* approaches infinite as $b \to 0$.

The Gibbs free energy of this system can be similarly written in terms of potential energy and surface energy as

$$\begin{aligned} G &= \Pi + A\gamma \\ &= -\frac{1}{2}\int_D \sigma_{ij}^0 u_{ij}^0 dx - \frac{1}{2}\int_\Omega \sigma_{ij}^0 \epsilon_{ij}^* dx + 4\gamma ac \\ &= -\frac{1}{2}\int_D \sigma_{ij}^0 u_{ij}^0 dx - \frac{1}{2}\frac{(1-\nu)a}{\mu b}\left(\sigma_{22}^0\right)^2 \pi abc + 4\gamma ac \\ &= -\frac{1}{2}\int_D \sigma_{ij}^0 u_{ij}^0 dx - \frac{(1-\nu)\pi}{2\mu}\left(\sigma_{22}^0\right)^2 a^2 c + 4\gamma ac. \end{aligned} \tag{9.38}$$

The total driving force of crack growth by the tension σ_{22}^0 can be written as

$$f^I = -\frac{\partial \Pi}{\partial a} = \frac{(1-\nu)\pi}{\mu}\left(\sigma_{22}^0\right)^2 ac \tag{9.39}$$

For a unit length in x_3 direction, the driving force can be written as

$$\frac{f^I}{c} = \frac{(1-\nu)\pi}{\mu}\left(\sigma_{22}^0\right)^2 a \tag{9.40}$$

The critical condition of $\frac{\partial G}{\partial a} = 0$ implies: (a) given a slit-like crack with length $2a$, when the applied stress reach

$$\sigma_{22}^{cr} = \sqrt{\frac{4\mu\gamma}{(1-v)\pi a}} \tag{9.41}$$

the crack will grow; and (b) given a load σ_{22}^0, if the penny crack radius is less than

$$a^{cr} = \frac{4\mu\gamma}{(1-v)\pi\left(\sigma_{22}^0\right)^2} \tag{9.42}$$

the crack will stay inactive.

Similarly, if a shear stress σ_{12}^0 is applied, given a slit-like crack, the critical stress for a Mode-II crack is

$$\sigma_{12}^{cr} = \sqrt{\frac{4\mu\gamma}{(1-v)\pi a}} \tag{9.43}$$

The above analysis can be straightforwardly extended to more complex cases such as flat ellipsoidal cracking.

9.2.3 Flat Ellipsoidal Crack

For a general flat ellipsoidal crack, without loss of any generality, we assume a crack with $a > b$ and $c \to 0$. The above two cases will be the two extreme cases with $a = b$ or $a \to \infty$. When a solid with such a crack is subjected to a far-field uniaxial tension σ_{33}^0, one can obtain

$$\epsilon_{33}^* = \frac{(1-v)}{\mu E(k)}\frac{b}{c}\sigma_{33}^0 \tag{9.44}$$

where $E(k)$ is the elliptic integral

$$E(k) = \int_0^{\pi/2} \sqrt{1-k^2 sin^2\phi d\phi} \tag{9.45}$$

and

$$k = \sqrt{1-b^2/a^2} \tag{9.46}$$

Notice that ϵ_{33}^{*} approaches infinite as $c \to 0$. The potential energy caused by the crack is written as

$$\Delta\Pi = -\frac{1}{2}\int_{\Omega} \sigma_{33}^{0}\epsilon_{33}^{*}dx = -\frac{2}{3}\pi abc\frac{(1-\upsilon)}{\mu E(k)}\frac{b}{c}\left(\sigma_{33}^{0}\right)^{2} = -\frac{2\pi(1-\upsilon)}{3\mu}\frac{a^{2}b}{E(k)}\left(\sigma_{33}^{0}\right)^{2} \tag{9.47}$$

The driving force for crack growth by uniaxial tension in two axial directions can be written as

$$f_{a}^{I} = -\frac{\partial(\Pi)}{\partial a} = \frac{2\pi(1-\upsilon)}{3\mu}\left(\sigma_{33}^{0}\right)^{2}\left(\frac{2ab}{E(k)} - \frac{E'(k)}{E^{2}(k)}\frac{b^{3}}{ak}\right) \tag{9.48}$$

$$f_{b}^{I} = -\frac{\partial(\Pi)}{\partial b} = \frac{2\pi(1-\upsilon)}{3\mu}\left(\sigma_{33}^{0}\right)^{2}\left(\frac{a^{2}}{E(k)} + \frac{E'(k)}{E^{2}(k)}\frac{b^{2}}{k}\right) \tag{9.49}$$

where

$$E'(k) = \frac{E(k) - F(k)}{k} \tag{9.50}$$

in which

$$F(k) = \int_{0}^{\pi/2} \frac{1}{\sqrt{1 - k^{2}sin^{2}\phi}}d\phi \tag{9.51}$$

The Griffith fracture criterion [53] provides that when the fracture driving force along the two axes caused by the Mode-I loading satisfies

$$f_{a}^{I} = 2\pi b\gamma \tag{9.52}$$

$$f_{b}^{I} = 2\pi a\gamma \tag{9.53}$$

the crack will propagate in x_1 or x_2 direction, respectively; otherwise, it will remain inactive. As the loading increases, when one of the above two equations satisfies, the crack will start to grow. Actually, the second equation will play the controlling role, which means that the crack grows from the minor axis of a flat ellipsoidal crack for Mode-I loading condition. When the applied stress reaches

$$\sigma_{33}^{cr} = \sqrt{\frac{3\mu a\gamma}{(1-\upsilon)\left(\frac{a^{2}}{E(k)} + \frac{E'(k)}{E^{2}(k)}\frac{b^{2}}{k}\right)}} \tag{9.54}$$

the crack will start to grow in the minor axis and release the stress. It becomes a penny-shaped crack, it will follow the case in Section 9.2.1. Similarly, we can also investigate the simple shear loading situation. Notice that because the load is not axial symmetric, given a shear load, say σ_{12}^{0} or σ_{13}^{0}, the crack propagation may follow different patterns [3], which can propagate in either axis depending on the ratio of a/b.

9.2.4 Crack Opening Displacement, Stress Intensity Factor, and J-Integral

For a slit-like crack under a Mode-I loading condition, the displacement distribution along the crack surface forms the crack opening displacement. Here, plane strain condition is considered. As shown in Figure 9.4, define $d(x_1)$ to be the opening displacement between two crack surface points as a function of x_1 caused by the Mode-I load. Therefore, $d(\pm a) = 0$. The slit-like crack is given as

$$\frac{x_1^2}{a^2} + \frac{x_2^2}{b^2} = 1 \tag{9.55}$$

where $b \to 0$. Because the actual strain in the equivalent inclusion of the above ellipse domain is equal to the eigenstrain, which is uniform, one can obtain the crack opening displacement as

$$d(x_1) = 2\epsilon_{22}^* y(x_1) \tag{9.56}$$

where $y(x_1)$ is the upper surface point of the ellipse depending on x_1. Using Equation 9.37, one can rewrite the above equation as

$$d(x_1) = 2\sigma_{22}^0 \frac{(1-\nu)}{\mu} \sqrt{a^2 - x_1^2} \tag{9.57}$$

which is the crack opening displacement. When $b \to 0$, one can still see that the deformed slit crack will form an ellipse with the maximum opening of $2\sigma_{22}^0 \frac{(1-\nu)}{\mu} a$ at the x_2 axis.

If the same traction σ_{22}^0 is applied on the crack surface, the crack opening displacement should be recovered and the total potential energy will be the same as a solid under uniaxal

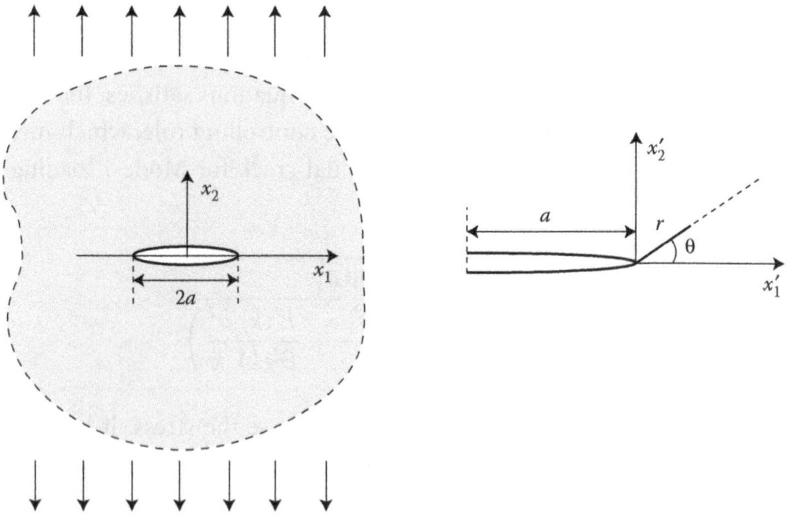

FIGURE 9.4 Solid with cracks under remote stress.

loading without a crack. Therefore, the potential energy change by the crack is equal to the work done by the traction σ_{22}^0 on the crack opening displacement done by itself, that is,

$$\Delta\Pi = -\frac{1}{2}\int_{-a}^{a} d(x_1)\sigma_{22}^0 dx_1 \tag{9.58}$$

which is counted at the unit thickness in the x_3 direction. Using Equation 9.57 in the above integral, one can write

$$\Delta\Pi = -\frac{(1-\nu)\pi}{2\mu}\left(\sigma_{22}^0\right)^2 a^2 \tag{9.59}$$

which is consistent with the one we derived before.

Based on the linear elastic fracture mechanics (LEFM) theory, the stress field in the neighborhood of the crack tip of a silt-like crack undergoing uniform far-field stress σ_{22}^0 (Figure 9.4) can be written in terms of the polar coordinates r and θ

$$\sigma_{11} = \frac{K_I}{\sqrt{2\pi r}}\cos\left(\frac{\theta}{2}\right)\left[1-\sin\left(\frac{\theta}{2}\right)\sin\left(\frac{3\theta}{2}\right)\right] \tag{9.60}$$

$$\sigma_{22} = \frac{K_I}{\sqrt{2\pi r}}\cos\left(\frac{\theta}{2}\right)\left[1+\sin\left(\frac{\theta}{2}\right)\sin\left(\frac{3\theta}{2}\right)\right] \tag{9.61}$$

$$\sigma_{12} = \frac{K_I}{\sqrt{2\pi r}}\cos\left(\frac{\theta}{2}\right)\sin\left(\frac{\theta}{2}\right)\cos\left(\frac{3\theta}{2}\right) \tag{9.62}$$

and the displacements near the crack tip ($r \ll a$) are

$$u_1 = \frac{K_I}{2\mu}\sqrt{\frac{r}{2\pi}}\cos\left(\frac{\theta}{2}\right)\left[\kappa - 1 + 2\sin^2\left(\frac{\theta}{2}\right)\right] \tag{9.63}$$

$$u_2 = \frac{K_I}{2\mu}\sqrt{\frac{r}{2\pi}} sin\left(\frac{\theta}{2}\right)\left[\kappa + 1 - 2cos^2\left(\frac{\theta}{2}\right)\right] \tag{9.64}$$

where

$$K_I = \lim_{r\to 0}\sigma_{22}(r,0)\sqrt{2\pi r} \tag{9.65}$$

is called the stress intensity factor of the Mode-I crack and $r = \sqrt{(x_1')^2 + (x_2')^2} = \sqrt{(x_1 - a)^2 + (x_2)^2}$, $\theta = \arctan\left(\frac{x_2'}{x_1'}\right) = \arctan\left(\frac{x_2}{x_1 - a}\right)$.

Notice that when $r \to 0$, the stresses (9.60,9.61,9.62) approach infinity, which means that the existence of a crack might cause stress singularity at the tip of the crack. Notice that the solution is restricted to those problems without considering plastic deformations.

Actually, fracture mechanics solution indicates that the stress field near a crack tip varies with $\frac{1}{\sqrt{r}}$, regardless of the modes of the crack including Mode-I, Mode-II, and Mode III as illustrated in the following Figure 9.5. Now we continue to use the Mode-I as an example. The stress intensity factor in Equation 9.65 defines the amplitude of the crack-tip singularity, which means stresses near the crack tip increase in proportion to K_I. Moreover, the stress intensity factor completely defines the crack tip conditions; if K_I is known, the stress, strain, and displacement fields can be expressed explicitly. Therefore, the stress intensity factor is considered as a characteristic parameter of a crack, one of the most important concepts in fracture mechanics.

Using the eigenstrain of Equations 9.36 and 9.37, one can write the stress field as

$$\sigma(\mathbf{x}) = \sigma^0 - \mathbf{D}(\mathbf{x}) : \mathbf{C}^0 : \epsilon^* \tag{9.66}$$

After a lengthy but straightforward derivation of $\mathbf{D}(\mathbf{x})$, one can eventually obtain the stress at $(a^+, 0)$ as

$$\sigma_{22}(r, 0) = \sigma_{22}^0 \sqrt{\frac{a}{2r}} \tag{9.67}$$

where $r = x_1 - a$. Substituting Equation 9.65 with Equation 9.67, one can obtain, for a slit-like crack,

$$K_I = \sigma_{22}^0 \sqrt{\pi a} \tag{9.68}$$

When the material behavior is intrinsically nonlinear, concepts from LEFM lose their meaning. For such cases, new theories and concepts to analyze the behavior of cracks should be introduced. Two typical theories are crack tip opening displacement (CTOD) theory and J-integral theory.

The above solutions based on LEFM theory is still valid if the nonlinear material deformation is confined to a small area near the crack tip. Wells [54] suggest a CTOD parameter to measure the toughness of materials.

Considering Equation 9.64, notice $\theta = \pi$ along the x axis, then the crack opening displacement is

$$\delta = 2u_2 = \frac{(\kappa + 1)}{\mu} K_I \sqrt{\frac{a - x_1}{2\pi}} \tag{9.69}$$

where $\kappa = 3 - 4\nu$ for plane strain problems and $\kappa = \frac{3-\nu}{1+\nu}$ for plane stress problem. It has a slight difference from the COD in Equation 9.57.

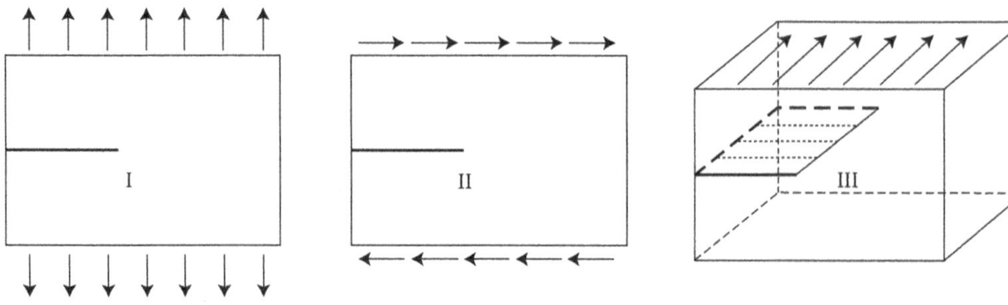

FIGURE 9.5 Solid with cracks under remote stress.

It is obvious that the opening at the crack tip (CTOD) is zero if $x_1 = a$. Here, the crack size a is an effective one, when plasticity at the crack tip is taken into account.

Consider a crack with a small plastic zone, as illustrated in Figure 9.6. Because of the existence of the plastic area, the size of the crack will be considered a little longer, called the effective length of the crack. Thus, we can estimate the CTOD by solving for the displacement at the physical crack tip, assuming an effective crack length of $a + r_y$, and the Irwin plastic zone correction [55] is

$$a_{eff} = a + r_y = a + \frac{1}{2\pi}\left(\frac{K_I}{\sigma_y}\right)^2 \tag{9.70}$$

Then Irwin plastic zone is calculated as

$$\delta = 2u_y = \frac{4}{\pi}\frac{K_1^2}{\sigma_y E} \tag{9.71}$$

for plane stress and

$$\delta = 2u_y = \frac{1}{\sqrt{3}}\frac{4\left(1-\nu^2\right)}{\pi}\frac{K_1^2}{\sigma_y E} \tag{9.72}$$

for plane strain. CTOD is a measure for the deformation at the crack tip, which can be used to predict the crack growth criterion, by comparing with a critical value.

Notice that in many cases, the size of the nonlinear deformation area is so large, maybe even larger than the size of the crack, that it is impossible to characterize the fracture behavior only using LEFM.

J-integral was introduced by Rice in fracture mechanics in 1968 [57], which has achieved a great success in describing a crack's behavior.

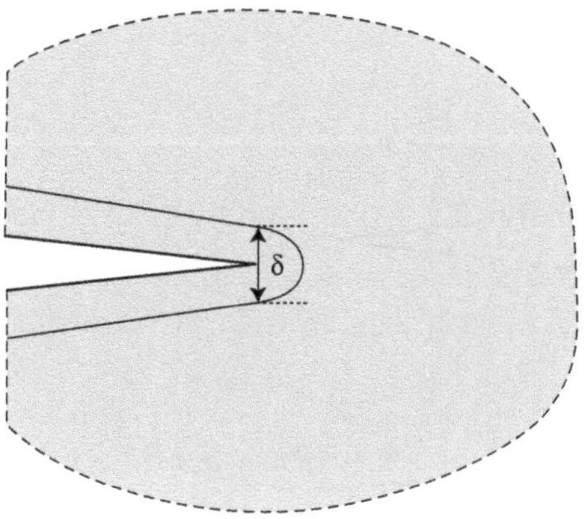

FIGURE 9.6 Crack tip opening displacement.

Consider an arbitrary counterclockwise path Γ around the tip of a crack, as in Figure 9.7. The J-integral is defined as

$$J = \int_{\Gamma} \left(w dy - t_i \frac{\partial u_i}{\partial x} ds \right) \tag{9.73}$$

where w, t_i, u_i represent the strain energy density, components of the traction vector, and displacement vector components, respectively.

$$w = \int_{0}^{\varepsilon_{ij}} \sigma_{ij} d\varepsilon_{ij} \tag{9.74}$$

The value of the J-integral is a path-independent integral around the crack tip. Refer to Figure 9.8.

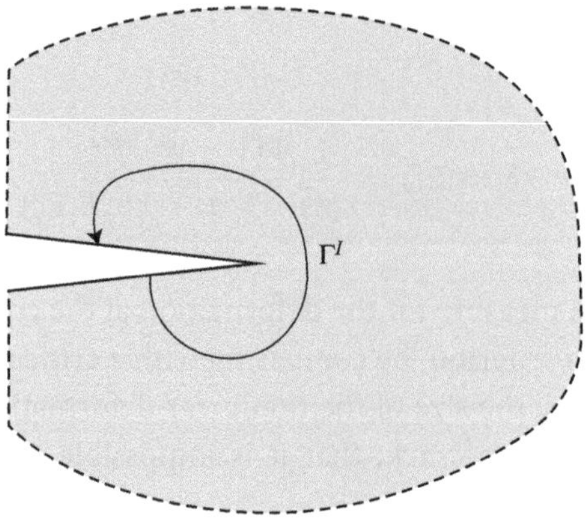

FIGURE 9.7 J-integral.

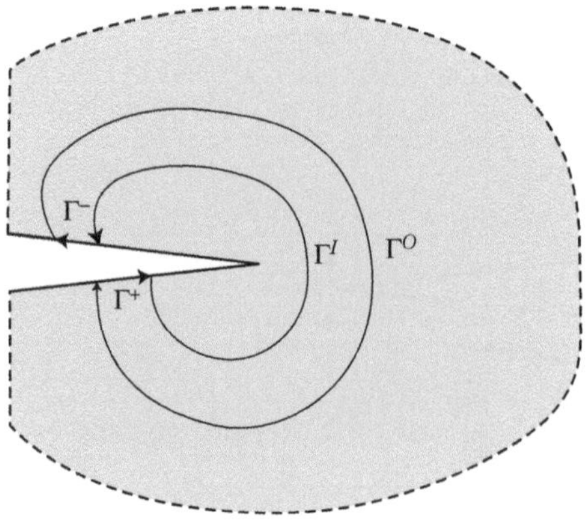

FIGURE 9.8 Path independence of J-integral.

Proof

$$J_k = \int_{\Gamma} \left(w dx_k - t_i \frac{\partial u_i}{\partial x_k} ds \right) = \int_{\Gamma} \left(w\delta_{jk} - \sigma_{ij} u_{i,k} \right) n_j ds$$
$$= \int_{\Omega} \left(\frac{\partial w}{\partial \varepsilon_{mn}} \frac{\partial \varepsilon_{mn}}{\partial x_j} - \sigma_{ij,j} u_{i,k} - \sigma_{ij} u_{i,kj} \right) d\Omega$$
$$= \int_{\Gamma^I + \Gamma^O + \Gamma^+ + \Gamma^-} \left(w dx_k - t_i \frac{\partial u_i}{\partial x_k} ds \right) \tag{9.75}$$

Notice

$$\sigma_{mn} = \frac{\partial w}{\partial \varepsilon_{mn}}$$
$$\varepsilon_{mn} = \frac{1}{2} \left(u_{m,n} + u_{n,m} \right)$$
$$\sigma_{ij,j} = 0$$

Equation 9.75 is simplified as

$$\int_{\Gamma^I + \Gamma^O + \Gamma^+ + \Gamma^-} \left(w dx_k - t_i \frac{\partial u_i}{\partial x_k} ds \right) = 0 \tag{9.76}$$

Notice that on Γ^+ and Γ^-, $dx_1 = 0$, $t_i = 0$, then we have

$$\int_{\Gamma^I + \Gamma^O} \left(w dx_k - t_i \frac{\partial u_i}{\partial x_k} ds \right) = 0 \tag{9.77}$$

Proved.

It is straightforward to verify that for linear elastic problems, J-integral is equivalent to the energy release rate G, and it can also be expressed by the stress intensity factor,

$$J_1 = \frac{(\kappa + 1)(\nu + 1)}{4E} K_I^2 = G = \frac{1}{E} K_I^2 = G \quad \text{plane stress} \tag{9.78}$$

$$J_1 = \frac{(\kappa + 1)(\nu + 1)}{4E} K_I^2 = G = \frac{1 - \nu^2}{E} K_I^2 \quad \text{plane strain} \tag{9.79}$$

J-integral can replace the energy release rate in LEFM and is related to the stress intensity factor, which means it can be used as the growth criterion of a crack too.

$$J = J_C \tag{9.80}$$

In this equation, critical value J_C can be determined through experiments. The integral is easily obtained because the path around the crack tip can be chosen arbitrarily.

9.3 DISLOCATION

Dislocation is a crystallographic defect associated with the crystalline lattice structure, which may cause the plastic deformation and low strength of the material. The dislocation theory can be traced back to Volterra [56] and Taylor [60,61] among others. This section will introduce the fundamental solution caused by a dislocation.

9.3.1 Introduction

Dislocations, in forms of line defects in crystal lattice, have been used to interpret the low strength and plastic deformation of crystal materials. In Figure 9.9, a periodic crystal structure is subjected to a shear stress, which should be written as a periodic function. A simple model can be written as

$$\tau = \frac{Gb}{2\pi a} \sin \frac{2\pi x}{b} \tag{9.81}$$

where G is the shear modulus or the modulus of rigidity. For small deformation, it can be rewritten as

$$\tau = G\frac{x}{a} = G\gamma \tag{9.82}$$

The critical shear stress (strength) can be written as

$$\tau_{cr} = \frac{Gb}{2\pi a} \tag{9.83}$$

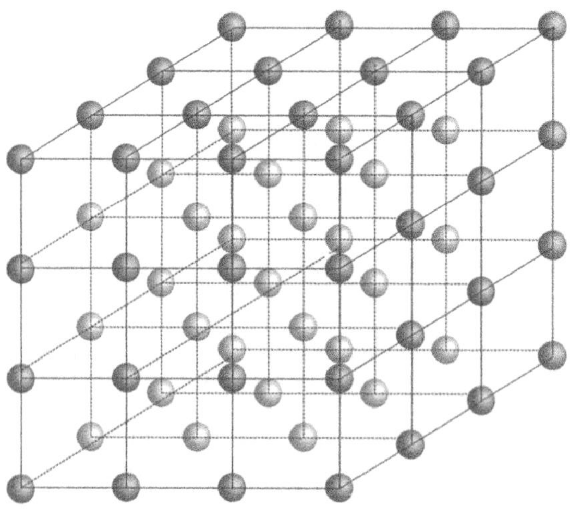

FIGURE 9.9 Crystal.

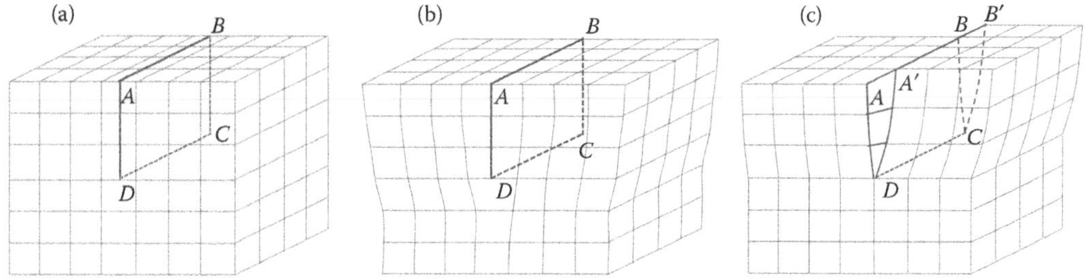

FIGURE 9.10 Dislocation types: (a) cubic lattice of atoms; (b) edge dislocation; and (c) screw dislocation.

Because b and a are at the same scale, the theoretical shear strength should be at the same scale at the shear modulus, that is, $\tau_{cr} \propto G$. However, the actual shear strength of crystal materials is generally observed at 10^{-4} to $10^{-8}G$. Orowan [58], Polanyi [59], and Taylor [60,61] independently interpreted this significant difference by the presence of dislocations.

There are two basic geometric types. In Figure 9.10a shows a simple cubic lattice of atoms with bonding forces balanced and thus the crystal structure is stable. Now imagine all the bonds on the surface *ABCD* are broken, an extra half-plane of atoms are inserted in the slot, and the simple cubic lattice is distorted as shown in Figure 9.10b. Above the line *DC*, the lattice has been displaced with one atom spacing; whereas below the line *DC*, the lattice keeps the similiar configuration. The disturbance of the interatomic bonds decrease with increasing distance to the line *DC*. The line *DC* is called an **edge dislocation**. On the other side, if the crystal at one side has a relative displacement in the ABCD plane referred to the other side, as shown in Figure 9.10c. The line *DC* is called a **screw dislocation**.

9.3.2 Burgers Vector and Burgers Circuit

A dislocation can be quantified through the Burgers vector using the Burgers circuit. A Burgers circuit is a closed atom-to-atom loop in a crystal lattice enclosing dislocations. For example, Figure 9.11 illustrates a **Burgers circuit** containing a dislocation, that is, *ABCDE*, where *A* and *E* are overlapped forming a closed loop. The positive direction is along the clockwise direction in a 2D space or along the right-handed screw direction in a 3D space. In the corresponding dislocation-free configuration, the same atom-to-atom sequence can be made but the ending point *E* will not be closed with *A*, which can be indicated by a vector $\overrightarrow{EA}$, namely a **Burgers vector**, which is normal to the dislocation line. Similiarly, a Burgers circuit can be drawn containing a screw dislocation. The Burgers vector will be parallel to the line of a screw dislocation.

9.3.3 Continuum Model for Dislocation

The concept of dislocation can be extended to a continuum body. A dislocation can be considered as the boundary of a slip plane. In Figure 9.10a, a bounded slip plane is embedded within the material. The dislocation line L has a directional vector **v**, which moves along the boundary of the slip plane. A Burgers circuit can be constructed as follows: for any

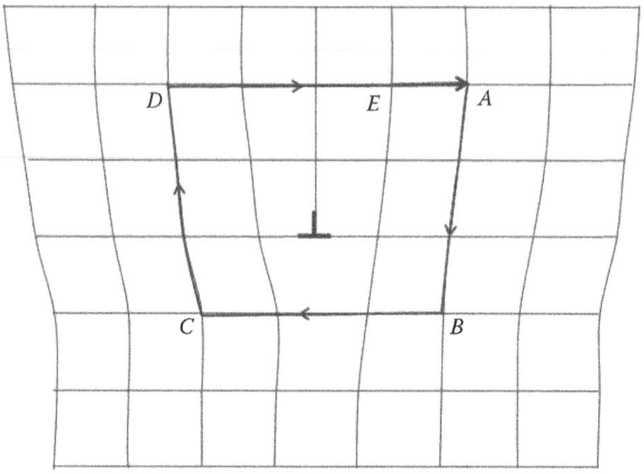

FIGURE 9.11 Burgers circuit and Burgers vector.

boundary point $\mathbf{x}$, we can obtain the direction vector $\mathbf{v}(\mathbf{x})$, draw a loop around the vector in the right-handed screw direction, which starts at the slip plane, denoted as the lower plane S^+, and ends at the slip plane, denoted as the upper plane S^-. Referred to the material without dislocation, if the slip displacement is denoted as a vector of $\mathbf{b}$, then the Burgers vector is written as $\mathbf{b}$, which indicates the lower surface S^+ slips by $\mathbf{b}$ relative to the upper plane S^-.

In Figure 9.10b, the slip of the surfaces is illustrated. Although the spacing between the two surfaces S^+ and S^- approaches zero, here we define the space as h and the out normal direction as $\mathbf{n}^+$ and $\mathbf{n}^-$. In Figure 9.10c, using a point A at the upper plane as reference, along the direction of $\mathbf{n}^+$, the displacement of another point $\mathbf{x}$ between the two surfaces is assumed to proportionally increase up to $\mathbf{b}$, that is,

$$u_i = b_i \frac{t}{h} \tag{9.84}$$

where t indicates the distance from the point to the origin, that is,

$$t = \left|\mathbf{x} - \mathbf{x}^A\right| = \sqrt{\left(x_1 - x_1^A\right)^2 + \left(x_2 - x_2^A\right)^2 + \left(x_3 - x_3^A\right)^2}$$

One can write

$$u_{i,j} = \frac{\partial u_i}{\partial t}\frac{\partial t}{\partial x_j} = -n_j \frac{b_i}{h} \tag{9.85}$$

where $n_j = \frac{x_j - x_j^A}{|\mathbf{x} - \mathbf{x}^A|}$. Therefore, the strain can be written as

$$\epsilon_{ij}^* = -\frac{n_i b_j + n_j b_i}{2h} \tag{9.86}$$

For $h \to 0$, one can write

$$\epsilon_{ij}^* = -\frac{n_i b_j + n_j b_i}{2}\delta(\mathbf{s} - \mathbf{x}) \tag{9.87}$$

where

$$\int_{\Omega} \delta(\mathbf{s} - \mathbf{x})\, d\mathbf{x} = \int_{s} \delta(\mathbf{x} - \mathbf{x}')\, d\mathbf{x} = \begin{cases} 0 & \mathbf{x} \notin s. \\ \infty & \mathbf{x} \in s. \end{cases} \tag{9.88}$$

where

$$\int_{\infty} \delta(\mathbf{x} - \mathbf{x}')\, d\mathbf{x}' = 1, \text{ for } \mathbf{x} \in s \tag{9.89}$$

Therefore, the strain caused by the dislocation can be considered as a singular eigenstrain in Equation 9.87. Using Green's function technique, the displacement field in the soild caused by the source of dislocation can be written as

$$\begin{aligned} u_i(\mathbf{x}) &= -\int_{\infty} C_{jlmn}\epsilon_{mn}^* G_{ij,l}(\mathbf{x}, \mathbf{x}')\, d\mathbf{x}' \\ &= \int_{\infty} C_{jlmn}\frac{n_n b_m + n_m b_n}{2}\delta(\mathbf{s} - \mathbf{x})\, G_{ij,l}(\mathbf{x}, \mathbf{x}')\, d\mathbf{x}' \\ &= \int_{\infty} C_{jlmn} n_m b_n G_{ij,l}(\mathbf{x}, \mathbf{x}')\, d\mathbf{x}'. \end{aligned} \tag{9.90}$$

from which we can obtain the stress and strain fields.

9.4 DAMAGE

Although the J-Integral and stress intensity factor have played important roles in fracture mechanics in recent years, they are difficult to use in describing the cases of time-dependent behavior, large strain plasticity, and for composites, the delamination phenomena. Fracture mechanic mainly focuses on macro-scale cracks and their extensions. But in many engineering cases, the initiation of a macroscopic crack is more essential. Before a macroscopic crack extends, a lot of micro-scale cavities or cracks, called micro-defects, will appear under the external loadings due to the breakage of atom bonds. The continuum damage mechanics [62,63], builds the bridge between microscopic cavities and microscopic cracks, which describes the damage progression arising from internal defects. It can be regarded as a meso-scale process. This theory describes the damage and fracture process from the initiation of microscopic cavities or cracks to the final fracture of materials caused by the macroscopic cracks. As it is named, the theory is in the framework of continuum mechanics.

It is impossible to discuss the evolution details of all the micro-scale cavities or cracks. In order to study the effect of micro-scale defects, a homogenized continuum meso-scale RVE is selected to represent the mechanical behavior of a damaged material. The RVE must be large enough to contain many defects and small enough to be considered as a material point of a continuum. Once the RVE is chosen, the damage at any point $\mathbf{x}$ of a material can be described by a state variable $D(\mathbf{x})$, that is, damage variable.

Consider the damage of a cylindrical bar with cross-section A under uniaxis tensile force F, as shown in Figure 9.12. Because of the existence of the damage, the actual load-carrying area is the effective area $\tilde{A}$, while $A_D = A - \tilde{A}$ is the area of the micro-scale cavities and cracks on the cross-section.

In CDM, the damage parameter is defined as

$$D = \frac{A_D}{A} = \frac{A - \tilde{A}}{A} \tag{9.91}$$

For special cases, $D = 0$ and $D = 1$ represent the initial undamaged state and final fractured state, respectively,

Then, the effective stress $\tilde{\sigma}$ can be written as

$$\tilde{\sigma} = \frac{F}{\tilde{A}} = \frac{\sigma}{1 - D} \tag{9.92}$$

From Equations 9.91 and 9.92, we can postulate that the damaged material Figure 9.12a with cross-section A under external force F is mechanically equivalent to a fictitious undamaged material Figure 9.12b, subject to the identical force F, which has a cross-sectional area $\tilde{A}$. The fictitious undamaged configuration shown in Figure 9.12b gives an important notion for the damage modeling.

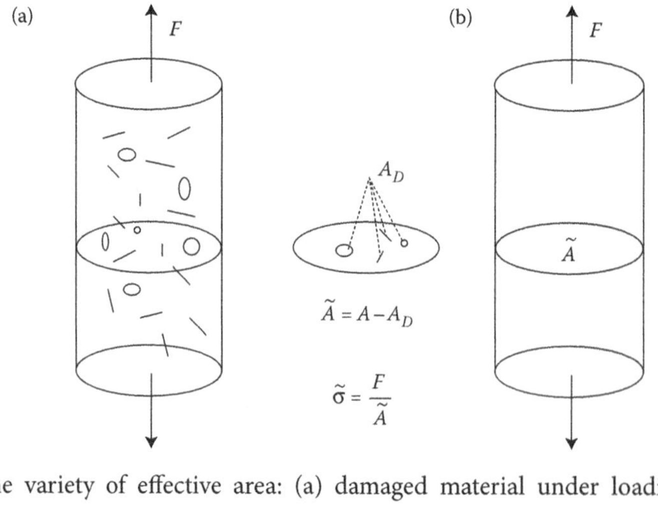

FIGURE 9.12 The variety of effective area: (a) damaged material under loading; (b) fictitious undamaged state.

For composite materials, the debonding of the interface between the matrix and the particles is one of the most important defects as Figure 9.13.

Now we consider the damage by using the elastic moduli. Because Figures 9.12a and 9.12b are equivalent states, the elastic strain under stress $\tilde{\sigma}$ in case (a) should be equal to that in case (b) under σ, then we have

$$\tilde{\sigma} = E\varepsilon \tag{9.93}$$

$$\sigma = E^{\mathrm{d}}\varepsilon \tag{9.94}$$

From Equations 9.93 and 9.94, one has

$$\tilde{\sigma} = \frac{E}{E^{\mathrm{d}}}\sigma \tag{9.95}$$

Comparing with Equation 9.92, we have

$$1 - D = \frac{E^{\mathrm{d}}}{E}, \quad \text{or} \quad E^{\mathrm{d}} = (1 - D)E \tag{9.96}$$

Equation 9.96 shows that the material becomes more compliant with the existence of defects.

Now consider the defects of composite material. Composites might be degraded by a number of mechanisms, such as static overload, impact, overheating, creep, and fatigue. Among all the types of the defect of composite, delaminations, bond failures, and cracks are the major failure ways, which are all related with interfacial debonding.

Since debonding is popular in composite, the evaluation of the interfacial property and the overall property of composite with debonding is one of the focuses in researching.

Let us consider a simple model of interfacial debonding between a particle and the matrix, as shown in Figure 9.13. In a particle-reinforcement material, consider the one typical particle embedded in the infinite matrix. Under applied external loading, the particle starts to

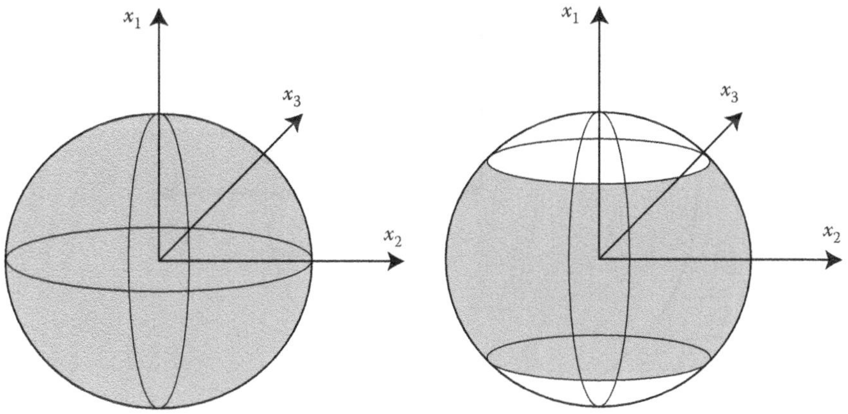

FIGURE 9.13 Interfacial debonding of composites.

debond, as shown in the figure, at the x_1 direction. With the debonding interface, shown as white-colored area, the local stress field inside the particles changes apparently, which would affect the overall property of the composite. The stress on the debonded surface area is released, the effective stiffness of the particles reduces as a result of the increase of debonding area.

In order to use Eshelby's theory for perfectly bonded particle, the equivalent stiffness method is used [64–66]. Partially debonded isotropic particles are replaced by fictitious orthotropic yet perfectly bonded particles, that is, the original stiffness matrix $\mathbf{C^1}$ of the particle is replaced by a new one $\mathbf{C^d}$, which indicates the released stiffness caused by the debonding.

With this concept of fictitious particles, we can rewrite the eigenstrain Equation 3.50 as

$$\epsilon^* = \left(\mathbf{C}^{\mathrm{d}}\right)^{-1} : \left(\mathbf{D}^{\Omega} - \Delta\mathbf{C}^{-1}\right)^{-1} : \epsilon^{\mathbf{0}} \tag{9.97}$$

Then, the corresponding stress field of the new fictitious inhomogeneity which takes the place of the debonding one is

$$\sigma = \sigma^{\mathbf{0}} + \mathbf{C^0} : (\mathbf{S} - I) : \epsilon^* \tag{9.98}$$

where $\mathbf{C^d}$ is an equivalent elastic stiffness tensor of the debonded particle , and $\Delta\mathbf{C} = \mathbf{C^d} - \mathbf{C^0}$, ϵ^0 is the strain tensor due to the remote stress, $\mathbf{D}^{\Omega}$ is defined in Equation 3.15, $\mathbf{S} = \mathbf{D}^{\Omega} : \mathbf{C^0}$ is Eshelby's tensor.

Then the following process is to determine the property of fictitious particle of the damage model.

Consider a particle in a local Cartesian coordinate system, and coincide with the principle directions of the local stress, as shown in Figure 9.14. For simplicity, the averaged stress of

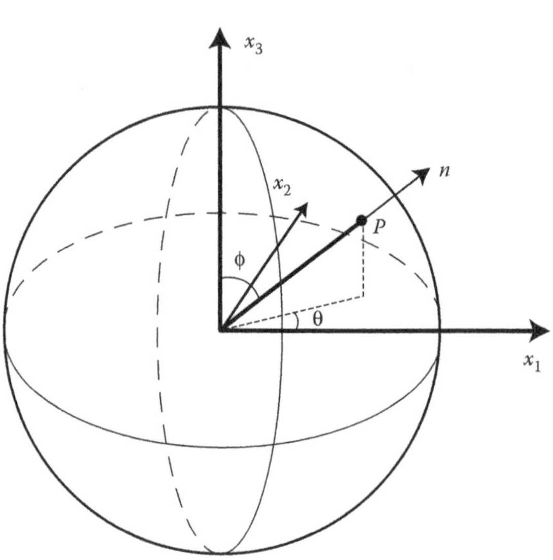

FIGURE 9.14 Local coordinate.

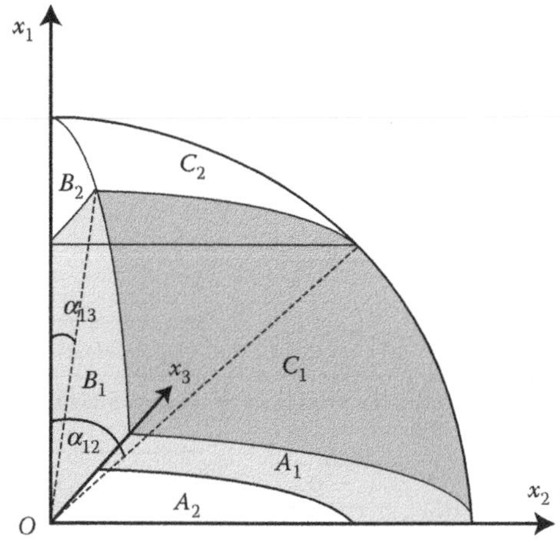

FIGURE 9.15 Projections of the debonded area.

particles is assumed to represent the interfacial stress so that the interfacial normal stress is expressed as normal

$$\sigma^n = \mathbf{n} \cdot \bar{\sigma} \cdot \mathbf{n} \tag{9.99}$$

where $\mathbf{n} = (\sin\phi\cos\theta, \sin\phi\sin\theta, \cos\phi)$ is the unit outward normal vector at any point along the interface, where θ and ϕ are two Eularian angles shown in Figure 9.14. By comparing the interfacial normal stress with the interfacial debonding strength σ_{cri} at any surface point, we can solve the debonding area on the surface of the particle as a function of debonding angles to the directions of the three principal stresses $(\sigma_1, \sigma_2, \sigma_3)$ in the particles.

Initially, the loading is small so that all the principal stresses are less than the interfacial strength. Thus, all the particles are fully bonded and no debonding process is activated. With further increase in the far-field loading, the interfacial debonding may occur in the interfacial area with the interfacial normal stress larger than the interfacial strength. The interfacial debonding will spread in the following three categories. In each category, for one-eighth of a particle with structural symmetry, as shown in Figure 9.15, the debonding area is described by two debonding angles α_{12} and α_{13}, which are defined in terms of the principal stresses and the interfacial strength. Three damage parameters D_i $(i = 1, 2, 3)$, projections of the debonding area in three principal directions normalized by the total projected area, are derived to evaluate the loss of the particles tensile load-transfer capacity for each category as follows.

9.4.1 Category 1 $\sigma_1 > \sigma_{cri} > \sigma_2 > \sigma_3$

Only the first principal stress is greater than the interfacial strength σ_{cri}. The interfacial debonding initiates from the first principal direction, which is corresponding to the white area on the surface of the particle in Figure 9.15, and propagates toward the other two principal directions. Figure 9.15 shows one-eighth of the debonded particle. And the debonding area can be described by two debonding angles, α_{12} and α_{13}. Correspondingly, the projections of the debonding area can be derived. As shown in the figure, A_2, B_2, and C_2 denote the projections of the debonded area onto the three midplanes of $x_2 - x_3, x_3 - x_1$, and $x_1 - x_2$, respectively; and A_1, B_1, and C_1 denote the corresponding projections of the undebonded area. Therefore, $A_1 + A_2 = B_1 + B_2 = C_1 + C_2 = \pi a^2$. The three damage parameters are expressed as

$$D_1 = \frac{A_2}{\pi a^2/4} = \frac{\sigma_1 - \sigma_{cri}}{\sqrt{(\sigma_1 - \sigma_2)(\sigma_1 - \sigma_3)}} \tag{9.100}$$

$$D_2 = \frac{B_2}{\pi a^2/4} = \frac{2}{\pi}\left(\arcsin\sqrt{\frac{\sigma_1 - \sigma_{cri}}{\sigma_1 - \sigma_3}} - \sqrt{\frac{(\sigma_1 - \sigma_{cri})(\sigma_{cri} - \sigma_2)}{(\sigma_1 - \sigma_3)(\sigma_1 - \sigma_2)}}\right) \tag{9.101}$$

$$D_3 = \frac{C_2}{\pi a^2/4} = \frac{2}{\pi}\left(\arcsin\sqrt{\frac{\sigma_1 - \sigma_{cri}}{\sigma_1 - \sigma_2}} - \sqrt{\frac{(\sigma_1 - \sigma_{cri})(\sigma_{cri} - \sigma_3)}{(\sigma_1 - \sigma_2)(\sigma_1 - \sigma_3)}}\right) \tag{9.102}$$

9.4.2 Category 2 $\sigma_1 > \sigma_2 > \sigma_{cri} > \sigma_3$

Two principal stresses are greater than the interfacial strength σ_{cri}. The interfacial debonding propagates around the particle, which is corresponding to the gray area on the surface of the particle in Figure 9.15, but still bond to the matrix along two ends in the other principal direction. In Figure 9.15, A_1, B_1, and C_1 denote the projections of the debonded area onto the three midplanes of $x_2 - x_3, x_3 - x_1$, and $x_1 - x_2$, respectively. As a result, the three damage parameters are obtained as

$$D_1 = \frac{B_1}{\pi a^2/4} = \frac{2}{\pi}\left(\arcsin\sqrt{\frac{\sigma_2 - \sigma_{cri}}{\sigma_2 - \sigma_3}} - \sqrt{\frac{(\sigma_1 - \sigma_{cri})(\sigma_{cri} - \sigma_3)}{(\sigma_1 - \sigma_3)(\sigma_2 - \sigma_3)}}\right) \tag{9.103}$$

$$D_2 = \frac{C_1}{\pi a^2/4} = \frac{2}{\pi}\left(\arcsin\sqrt{\frac{\sigma_1 - \sigma_{cri}}{\sigma_1 - \sigma_3}} - \sqrt{\frac{\sigma_2 - \sigma_{cri}}{\sigma_1 - \sigma_3}}\sqrt{\frac{\sigma_{cri} - \sigma_3}{\sigma_2 - \sigma_3}}\right) \tag{9.104}$$

$$D_3 = \frac{A_1}{\pi a^2/4} = 1 - \frac{\sigma_{cri} - \sigma_3}{\sqrt{(\sigma_1 - \sigma_3)(\sigma_2 - \sigma_3)}} \tag{9.105}$$

9.4.3 Category 3 $\sigma_1 > \sigma_2 > \sigma_3 > \sigma_{cri}$

In this case, all principal stresses exceed the interfacial strength. Thus, the entire interface is debonded as a void and the particle cannot transfer any tensile loading. The damage parameters are written as

$$D_1 = D_2 = D_3 = 1 \tag{9.106}$$

With the increase of the debonding, the overall composite becomes more compliant. To simulate the stiffness softening, the damaged particle is replaced by the fully bonded one, but with reduced stiffness in certain directions corresponding to the damage parameters.

The above equations give damage parameters D_i of the category $k = 1, 2, 3$. For more general case, we regard the undamaged case as category 0, and the corresponding parameters are all zero.

$$D_1 = D_2 = D_3 = 0 \tag{9.107}$$

By using the above damage parameters, equivalent orthotropic elastic compliance matrix for category k ($k = 0, 1, 2, 3$) can be formulated as

$$S = \begin{pmatrix} \frac{1}{E\xi_{11}} & -\frac{\nu}{E\xi_{12}} & -\frac{\nu}{E\xi_{13}} & 0 & 0 & 0 \\ -\frac{\nu}{E\xi_{21}} & \frac{1}{E\xi_{22}} & -\frac{\nu}{E\xi_{23}} & 0 & 0 & 0 \\ -\frac{\nu}{E\xi_{31}} & -\frac{\nu}{E\xi_{32}} & \frac{1}{E\xi_{33}} & 0 & 0 & 0 \\ 0 & 0 & 0 & \frac{1+\nu}{E\xi_{23}} & 0 & 0 \\ 0 & 0 & 0 & 0 & \frac{1+\nu}{E\xi_{13}} & 0 \\ 0 & 0 & 0 & 0 & 0 & \frac{1+\nu}{E\xi_{12}} \end{pmatrix} \quad (9.108)$$

where

$$\xi_{IJ} = (1 - D_I)(1 - D_k), \quad I, J = 1, 2, 3 \quad (9.109)$$

Thus, the equivalent stiffness tensor of the damaged particle is the inversion of compliance tensor, and it has the following form

$$C^d_{ijkl} = \lambda^d_{IK}\delta_{ij}\delta_{kl} + \lambda^d_{IJ}\left(\delta_{ik}\delta_{jl} + \delta_{il}\delta_{jk}\right) \quad (9.110)$$

where

$$\lambda^d_{IK} = \lambda(1 - D_I)(1 - D_k) \quad (9.111)$$

$$\mu^d_{IK} = \mu(1 - D_I)(1 - D_k) \quad (9.112)$$

It is noted that Mura's tensorial indicial notation is followed in Equations 9.111 and 9.112; that is, uppercase indices have the same representation as the corresponding lowercase ones but are not summed.

For spherical particles, using Mori–Tanaka's model, the effective elastic moduli of composites is written directly as

$$\langle \mathbf{C} \rangle_D = \mathbf{C}^0 + \phi\left[\left(\mathbf{C}^d - \mathbf{C}^0\right)^{-1} - (1 - \phi)\,\mathbf{D}^{\Omega}\right]^{-1} \quad (9.113)$$

where the components of $\mathbf{C}^d$ is has the form as Equation 9.110.

9.5 EXERCISES

1. Calculate the stress concentration factors for a solid ($E = 200$ GPa and $v = 0.3$) under a uniaxial tension with a spherical void and a circular hole, respectively. The axis of the circular hole is perpendicular to the loading direction.

2. Similarly to Section 9.2.1 for uniaxial tension of σ_{33}^0, derive the eigenstrain for a penny-shaped crack under a simple shear of σ_{13}^0 is

$$\epsilon_{13}^* = \frac{2(1-v)}{\mu\pi(2-v)}\frac{a}{c}\sigma_{13}^0, \text{ otherwise } \epsilon_{ij}^* = 0 \text{ with } i,j = 1,2,3$$

3. Similarly to Section 9.2.1, derive the eigenstrain for a slit-like crack under a uniaxial tension of σ_{22}^0 is

$$\epsilon_{11}^* = -\frac{(1-\nu)}{2\mu}\sigma_{22}^0 \text{ and } \epsilon_{22}^* = \frac{(1-\nu)a}{\mu b}\sigma_{22}^0 \text{ otherwise } \epsilon_{ij}^* = 0 \text{ with } i,j = 1,2,3$$

4. Prove the critical stress for the Mode-II slit-like crack is

$$\sigma_{12}^{cr} = \sqrt{\frac{4\gamma\mu}{\pi(1-\nu)a}}$$

5. For a flat ellipsoidal crack with $a > b$ and $c \to 0$ under a shear load σ_{13}^0, derive the critical stress in each axis based on Griffith's criterion and discuss the crack propagation direction based on the ratio of a/b.

CHAPTER 10

Boundary Effects on Particulate Composites

When a composite is uniformly fabricated, in a statistical sense at a larger length scale, the composite can be treated as a statistically homogeneous material, and the stress and strain at a material point can be evaluated by the averages of stress and strain on an RVE. Here an RVE in a continuum body is a material volume that statistically represents the neighborhood of a material point. The microstructure can be periodic, random, or even functionally graded materials. From the relation between averaged stress and strain, we can derive an effective mechanical constitutive law of the RVE. Whether an RVE can provide an accurate prediction of effective material behavior has been an interesting problem. Drugan [34] used ensemble average to obtain the effective elastic properties and found a maximum error of 5% when the minimum Green's function can be obtained in lower dimensions too. An RVE size is no less than the twice of the particle size for any volume fraction. Their formulation has been derived using the Green's function in the infinite domain, which has been used by Eshelby [5]. However, because there commonly exist particles close to the boundary of a composite, Eshelby's assumption for one particle embedded in an infinite matrix cannot be exactly satisfied, and thus the boundary effect has not been included in the modeling [3]. The recent development of advanced materials commonly requires fabrication of materials in thin layers or films. Moreover, the characterization of materials can be conducted at micron or nanometer levels. The miniaturization of material testing and fabrication attracts significant attention about the boundary effect on particulate composites. This chapter will first introduce fundamental solutions for a concentrated force in a semi-infinite domain and then use it in the equivalent inclusion method for investigation of boundary effects on a semi-infinite domain containing one, two or many particles. Then, an algorithm for virtual experiments of a composite sample is introduced.

10.1 FUNDAMENTAL SOLUTION FOR SEMI-INFINITE DOMAINS

The fundamental solution for a semi-infinite domain with a concentrated force can be traced back to Boussinesq's solution [67], in which a concentrated force on the surface of the domain is considered. Mindlin [68] provided the solution for a concentrated force in the interior of the semi-infinite domain with traction-free boundary. Using a similar procedure, Rongved [69] obtained the solution for a concentrated force in the interior of the half space with fixed boundary. He also provided a solution for two joint semi-infinite solids with an interior force [70]. Yu [71] derived the elastic fields of two joint semi-infinite solids in which frictionless and fully bonded interfaces are analyzed separately. In their derivation, Galerkin vectors [68] due to infinitesimal inclusion are first presented, with which the displacement is derived. Then the displacement caused by a finite inclusion is obtained by direct integration over the inclusion Ω. Fully bonded and frictionless interfaces are considered separately and different kinds of eigenstrains are analyzed. Introducing the imaging displacement, Walpole [72] derived the general solutions for the two joined half-space with bonded/smooth interfaces. Green's function of the two fully bonded semi-infinite domains including a concentrated force will be introduced below. Assuming the second half domain is rigid or of zero stiffness, one can obtain the fundamental solution for a semi-infinite domain with fixed or traction free boundaries, respectively [68,70].

Consider an infinite two-phase elastic solid in Figure 10.1, which consists of two different semi-infinite, homogeneous, and isotropic, halves, with a fully bonded interface. The Cartesian coordinate has been set up with the origin on the interface of $x_1 - x_2$ plane at $x_3 = 0$. A concentrated force $\mathbf{F} = (F_1, F_2, F_3)$ is applied in the interior of the upper half at $\mathbf{x}' = (x_1', x_2', x_3')$ with $x_3' > 0$. The stiffness tensors for the two phases of materials are denoted by $\mathbf{C}^0$ and $\bar{\mathbf{C}}^0$ for $x_3 \geq 0$ and $x_3 < 0$, respectively. Specifically, the shear modulus and Poisson's ratio are written as μ^0 and υ^0 for $x_3 \geq 0$ and $\bar{\mu}^0$ and $\bar{\upsilon}^0$ for $x_3 < 0$, respectively.

The image point of $\mathbf{x}'$ is written as $\bar{\mathbf{x}}'$ with the coordinate $(x_1', x_2', -x_3')$. The corresponding image force is $\overline{\mathbf{F}} = (F_1, F_2, -F_3)$. Notice that the image of a vector, say $\mathbf{F}$ and $\mathbf{x}'$, can be

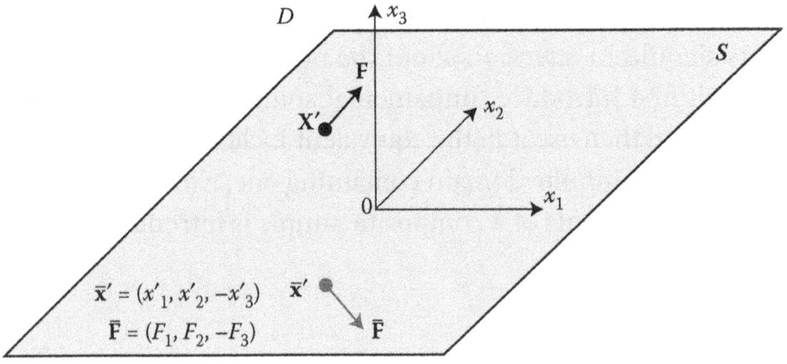

FIGURE 10.1 Two joined semi-infinite domains ($x_3 \geq 0$) containing a concentrated force $\mathbf{F}$ at $\mathbf{x}'$ with the image force $\overline{\mathbf{F}}$ at $\bar{\mathbf{x}}'$.

written by a mirror projection as

$$(\bullet) = \mathbf{Q} \cdot (\bar{\bullet}) \text{ with } \mathbf{Q} = \begin{bmatrix} 1 & 0 & 0 \\ 0 & 1 & 0 \\ 0 & 0 & -1 \end{bmatrix} \tag{10.1}$$

where $(\bullet)$ and $(\bar{\bullet})$ mean any vector in the original and mirror projected coordinate, such as $\bar{F}_i = Q_{ji}F_j$ and $F_i = Q_{ij}\bar{F}_j$.

Based on Walpole's formulation [73], the displacement at $\mathbf{x}$ in the infinite two-phase solid can be rewritten in terms of Green's function as

$$u_i(\mathbf{x}) = G_{ij}(\mathbf{x}, \mathbf{x}')F_j(\mathbf{x}') \tag{10.2}$$

where the form of $G_{ij}(\mathbf{x}, \mathbf{x}')$ depends on the location of $\mathbf{x}$. For $x_3 \geq 0$,

$$\begin{aligned} 4\pi\mu^0 G_{ij}(\mathbf{x}, \mathbf{x}') = & \left[\phi\delta_{ij} - \frac{1}{4(1-v^0)}\psi_{,ij}\right] \\ & + A\bar{\phi}\delta_{ij} - DQ_{im}Q_{jn}\bar{\psi}_{,mn} \\ & - Cx_3\left[Q_{jm}\bar{\psi}_{,im3} + 4(1-v^0)\delta_{j3}\bar{\phi}_{,i} + 2(1-2v^0)\delta_{i3}Q_{jm}\bar{\phi}_{,m} - Q_{jm}x_3\bar{\phi}_{,im}\right] \\ & + B\left(\delta_{i3}\delta_{jk} - \delta_{ik}\delta_{j3}\right)\left[\ln\left(x_3 + x_3' + \bar{\psi}\right)\right]_{,k} \\ & - (G+B)Q_{jm}\left[\left(x_3 + x_3'\right)\ln\left(x_3 + x_3' + \bar{\psi}\right) - \bar{\psi}\right]_{,im} \end{aligned} \tag{10.3}$$

and for $x_3 < 0$,

$$\begin{aligned} 4\pi\mu^0 G_{ij}(\mathbf{x}, \mathbf{x}') = & A\phi\delta_{ij} - D\psi_{,ij} - x_3 F\left[\ln\left(-x_3 + x_3' + \psi\right)\right]_{,ij} \\ & + B\left(\delta_{i3}\delta_{jk} - \delta_{ik}\delta_{j3}\right)\left[\ln\left(-x_3 + x_3' + \psi\right)\right]_{,k} \\ & - (G+B)Q_{im}\left[\left(-x_3 + x_3'\right)\ln\left(-x_3 + x_3' + \psi\right) - \psi\right]_{,jm} \end{aligned} \tag{10.4}$$

where

$$A = \frac{\mu^0 - \bar{\mu}^0}{\mu^0 + \bar{\mu}^0}$$

$$B = \frac{2\mu^0(1-2v^0)\left(\mu^0 - \bar{\mu}^0\right)}{\left(\mu^0 + \bar{\mu}^0\right)\left[\mu^0 + \bar{\mu}^0(3-4v^0)\right]}$$

$$C = \frac{\mu^0 - \bar{\mu}^0}{2(1-v^0)\left(\mu^0 + (3-4v^0)\bar{\mu}^0\right)}$$

$$D = \frac{(3-4v^0)}{2}C$$

$$F = \frac{2\mu^0\left[\mu^0\left(1-2\bar{v}^0\right)-\bar{\mu}^0(1-2v^0)\right]}{\left[\mu^0+\bar{\mu}^0(3-4v^0)\right]\left[\bar{\mu}^0+\mu^0\left(3-4\bar{v}^0\right)\right]}$$

$$G = \frac{\mu^0\left[\bar{\mu}^0\left(1-2\bar{v}^0\right)\left(3-4v^0\right)-\mu^0(1-2v^0)\left(3-4\bar{v}^0\right)\right]}{\left[\mu^0+\bar{\mu}^0(3-4v^0)\right]\left[\bar{\mu}^0+\mu^0\left(3-4\bar{v}^0\right)\right]}$$

and $\psi = \left|\mathbf{x}-\mathbf{x}'\right|$, $\phi = 1/\left|\mathbf{x}-\mathbf{x}'\right|$, $\bar{\psi} = \left|\mathbf{x}-\bar{\mathbf{x}}'\right|$, $\bar{\phi} = 1/\left|\mathbf{x}-\bar{\mathbf{x}}'\right|$.

When $\bar{\mu}^0 = 0$, the above Green's function is reduced to Mindlin's solution [68]:

$$\begin{aligned}
&4\pi\mu^0 G_{ij}(\mathbf{x},\mathbf{x}') \\
&= \left[\phi\delta_{ij} - \frac{1}{4(1-v^0)}\psi_{,ij}\right] + \bar{\phi}\delta_{ij} - \frac{(3-4v^0)}{4(1-v^0)}Q_{im}Q_{jn}\bar{\psi}_{,mn} \\
&\quad - \frac{1}{2(1-v^0)}x_3\left[Q_{jm}\bar{\psi}_{,im3} + 4(1-v^0)\delta_{j3}\bar{\phi}_{,i} + 2(1-2v^0)\delta_{i3}Q_{jm}\bar{\phi}_{,m} - Q_{jm}x_3\bar{\phi}_{,im}\right] \\
&\quad + 2(1-2v^0)\left(\delta_{i3}\delta_{jk}-\delta_{ik}\delta_{j3}\right)\left[\ln\left(x_3+x_3'+\bar{\psi}\right)\right]_{,k} \\
&\quad - (1-2v^0)Q_{jm}\left[\left(x_3+x_3'\right)\ln\left(x_3+x_3'+\bar{\psi}\right)-\bar{\psi}\right]_{,im}
\end{aligned} \tag{10.5}$$

Notice that the above mathematical form of Mindlin's solution is more concise than the existing form solution although they are consistent with each other [68,70–72]. For $\bar{\mu}^0 \to \infty$, the above Green's function is reduced to Rongved's solution [70,74] as:

$$\begin{aligned}
G_{ij}\left(\mathbf{x},\mathbf{x}'\right) &= \frac{1}{4\pi\mu^0}\delta_{ij}\left(\phi-\bar{\phi}\right) - \frac{1}{16\pi\mu^0(1-v^0)}\left(\psi_{,ij}-Q_{im}Q_{jn}\bar{\psi}_{,mn}\right) \\
&\quad + \frac{x_3}{8\pi\mu^0(1-v^0)(3-4v^0)} \\
&\quad \times\left[Q_{jm}\bar{\psi}_{,im3} + 4(1-v^0)\delta_{j3}\bar{\phi}_{,i} + 2(1-2v^0)\delta_{i3}Q_{jm}\bar{\phi}_{,m} - Q_{jm}x_3\bar{\phi}_{,im}\right]
\end{aligned} \tag{10.6}$$

which is applicable to the problem for a concentrated force F_i in the interior point $\mathbf{x}'$ of a semi-infinite domain with a fixed boundary.

10.2 EQUIVALENT INCLUSION METHOD FOR ONE PARTICLE IN A SEMI-INFINITE DOMAIN

Figure 10.2 illustrates a semi-infinite domain D ($x_3 \geq 0$) containing a subdomain Ω centered at $\mathbf{x}^c$ with a surface S ($x_3 = 0$). When an inhomogeneity is subjected to a body force or a far-field stress, the mismatch between inhomogeneity and the matrix can be simulated by an appropriately chosen eigenstrain on the subdomain so that the elastic field can be obtained through solving the inclusion problem [46,74]. The inclusion problems for a semi-infinite domain or two half-infinite domains have been studied [71,72,75,76]. However, this section will show the inhomogeneity problem for the semi-infinite domain subjected to a uniform far-field stress. The formulation can be extended to a many-particle system in a semi-infinite

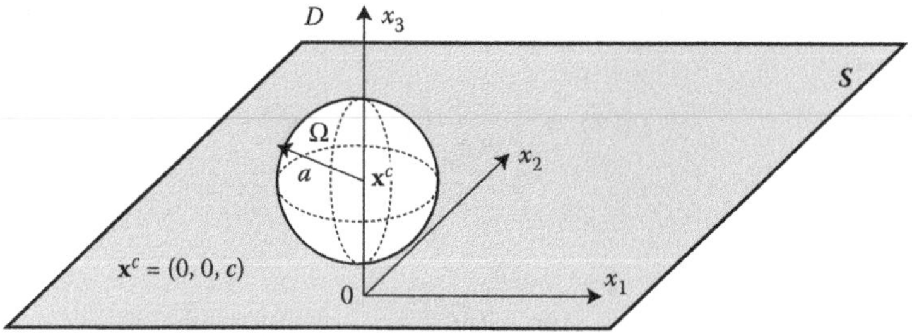

FIGURE 10.2 A semi-infinite domain D ($x_3 \geq 0$) containing a subdomain Ω with a surface S.

solid, which is demonstrated by the cases of two spheres. The local field will depend on the boundary conditions on the surface, which can be in displacement [41] or stress boundary conditions [42].

Below we will use the latter one as an example, as shown in Figure 10.2.

Below we will investigate one inhomogeneity embedded in a semi-infinite solid located at different depths with the surface S subjected to uniform traction to be compatible with the far-field stress. When the inhomogeneity is far from the boundary, the solution should recover Eshelby's existing solution and one can use a uniform eigenstrain to represent the material mismatch. When particle is close to the surface, the eigenstrain should be in a polynomial form. We will first demonstrate the solution to the inclusion problem. Then, we can use the stress equivalent condition to determine the eigenstrain coefficients and determine the local field.

The distribution of the eigenstrain $\boldsymbol{\epsilon}^*(\mathbf{x})$ is typically continuous over Ω, which can be written in terms of the Taylor expansion of $\mathbf{x} - \mathbf{x}^c$ with the reference point at the particle's center, such as

$$\epsilon_{ij}^*(\mathbf{x}) = \epsilon_{ij}^0 + \epsilon_{ijk}^1 \left(x_k - x_k^c\right) + \epsilon_{ijkl}^2 \left(x_k - x_k^c\right)\left(x_l - x_l^c\right) + \cdots, \quad \mathbf{x} \in \Omega \tag{10.7}$$

The constitutive law over the semi-infinite domain can be rewritten as

$$\sigma_{ij} = C_{ijkl}\left(u_{k,l} - \epsilon_{kl}^*\right) \tag{10.8}$$

where $\epsilon_{kl}^* = 0$ for $\mathbf{x} \in D - \Omega$. The substitution of the above stress into the equilibrium equation yields

$$C_{ijkl}u_{k,li} = C_{ijkl}\epsilon_{kl,i}^* \tag{10.9}$$

Green's function technique has been widely used to solve the above type of problem. By using Green's function which satisfies the boundary conditions, the displacement can be written in terms of eigenstrain as follows:

$$u_i(\mathbf{x}) = -\int_D \left[G_{ij}(\mathbf{x}, \mathbf{x}')C_{mjkl}\frac{\partial \epsilon_{kl}^*\left(\mathbf{x}'\right)}{\partial x_m'}\right] d\mathbf{x}' \tag{10.10}$$

Using Gauss' theorem and considering no eigenstrain at the boundary, Equation 10.10 can be rewritten as

$$u_i(\mathbf{x}) = \int_{\Omega} \frac{\partial G_{ij}}{\partial x'_m} C_{mjkl} \epsilon^*_{kl} \, d\mathbf{x}' \tag{10.11}$$

Considering that

$$\frac{\partial \phi}{\partial x'_m} = -\frac{\partial \phi}{\partial x_m}, \frac{\partial \psi}{\partial x'_m} = -\frac{\partial \psi}{\partial x_m}, \frac{\partial \bar{\psi}}{\partial x'_m} = -Q_{mq}\frac{\partial \bar{\psi}}{\partial x_q}, \frac{\partial \bar{\phi}}{\partial x'_m} = -Q_{mq}\frac{\partial \bar{\phi}}{\partial x_q}$$

and defining

$$\Gamma^u_{ijm} = -\frac{\partial G_{ij}}{\partial x'_m}, \tag{10.12}$$

one can obtain

$$\begin{aligned} 4\pi\mu^0 \Gamma^u_{ijm} &= \left[\phi_{,m}\delta_{ij} - \frac{1}{4(1-v^0)}\psi_{,ijm}\right] + Q_{mq}\bar{\phi}_{,q}\delta_{ij} - \frac{(3-4v^0)}{4(1-v^0)} Q_{ik}Q_{jl}Q_{mq}\bar{\psi}_{,klq} \\ &\quad - \frac{x_3}{2(1-v^0)} \begin{bmatrix} Q_{jk}Q_{mq}\bar{\psi}_{,ikq3} + 4(1-v^0)Q_{mq}\delta_{j3}\bar{\phi}_{,iq} \\ +2(1-2v^0)\delta_{i3}Q_{jk}Q_{mq}\bar{\phi}_{,kq} - Q_{jk}Q_{mq}x_3\bar{\phi}_{,ikq} \end{bmatrix} \\ &\quad - 2(1-2v^0)\left(\delta_{i3}\delta_{jk} - \delta_{ik}\delta_{j3}\right)\alpha_{,km'} + (1-2v^0)Q_{jk}\beta_{,ikm'} \end{aligned} \tag{10.13}$$

where $\alpha_{,km'}$ and $\beta_{,ikm'}$ are

$$\alpha_{,km'} = \left[\ln\left(x_3 + x'_3 + \bar{\psi}\right)\right]_{,km'}; \; \beta_{,ikm'} = \left[\left(x_3 + x'_3\right)\ln\left(x_3 + x'_3 + \bar{\psi}\right) - \bar{\psi}\right]_{,ikm'}$$

$\alpha_{,km'}$ and $\beta_{,ikm'}$ can be written as the following line integrals from $(x_1, x_2, +\infty)$ to (x_1, x_2, x_3), which means that the potential at the far-field is zero that leads to an infinitely large constant term but it disappears when taking a derivative.

$$\begin{aligned} \alpha_{,km'} &= \left[\int_{+\infty}^{x_3} \frac{1}{\sqrt{\left(x_1 - x'_1\right)^2 + \left(x_2 - x'_2\right)^2 + \left(t + x'_3\right)^2}} dt\right]_{,km'} \\ &= -\left[\int_{+\infty}^{x_3} Q_{mq}\bar{\phi}_{,q}(\mathbf{x^t}, \bar{\mathbf{x}}')dt\right]_{,k} = -Q_{mq}\left[\int_{+\infty}^{x_3} \bar{\phi}(\mathbf{x^t}, \bar{\mathbf{x}}')dt\right]_{,qk} \\ \beta_{,ikm'} &= \left[\int_{+\infty}^{x_3}\int_{+\infty}^{t} \frac{1}{\sqrt{\left(x_1 - x'_1\right)^2 + \left(x_2 - x'_2\right)^2 + \left(s + x'_3\right)^2}} ds dt\right]_{,ikm'} \\ &= -Q_{mq}\left[\int_{+\infty}^{x_3}\int_{+\infty}^{t} \bar{\phi}(\mathbf{x^s}, \bar{\mathbf{x}}')ds dt\right]_{,qik} \end{aligned}$$

where x^t and x^s mean (x_1, x_2, t) and (x_1, x_2, s), thus $\bar{\phi}(\mathbf{x^t}, \bar{\mathbf{x}}')$ is defined as follows:

$$\bar{\phi}(\mathbf{x^t}, \bar{\mathbf{x}}') = \frac{1}{\sqrt{\left(x_1 - x_1'\right)^2 + \left(x_2 - x_2'\right)^2 + \left(t + x_3'\right)^2}} \tag{10.14}$$

Note the relationship $\frac{\partial \bar{\phi}}{\partial x_m'} = -\frac{\partial \bar{\phi}}{\partial x_q}\frac{\partial x_q}{\partial x_m'} = -Q_{mq}\frac{\partial \bar{\phi}}{\partial x_q}$ is used above. Therefore, we can rewrite Equation 10.13 as

$$\begin{aligned}
4\pi\mu^0\Gamma^u_{ijm} &= \left[\phi_{,m}\delta_{ij} - \frac{1}{4(1-v^0)}\psi_{,ijm}\right] + Q_{mq}\bar{\phi}_{,q}\delta_{ij} - \frac{(3-4v^0)}{4(1-v^0)}Q_{ik}Q_{jl}Q_{mq}\bar{\psi}_{,klq} \\
&\quad - \frac{x_3}{2(1-v^0)}\begin{bmatrix} Q_{jk}Q_{mq}\bar{\psi}_{,ikq3} + 4(1-v^0)Q_{mq}\delta_{j3}\bar{\phi}_{,iq} \\ +2(1-2v^0)\delta_{i3}Q_{jk}Q_{mq}\bar{\phi}_{,kq} - Q_{jk}Q_{mq}x_3\bar{\phi}_{,ikq}\end{bmatrix} \\
&\quad + 2(1-2v^0)\left(\delta_{i3}\delta_{jk} - \delta_{ik}\delta_{j3}\right)Q_{mq}\theta_{,qk} - (1-2v^0)Q_{jk}Q_{mq}\lambda_{,qik}
\end{aligned}$$

where

$$\theta = \left[\int_{+\infty}^{x_3} \bar{\phi}(\mathbf{x^t}, \bar{\mathbf{x}}')dt\right]; \ \lambda = \left[\int_{+\infty}^{x_3}\int_{+\infty}^{t} \bar{\phi}(\mathbf{x^s}, \bar{\mathbf{x}}')dsdt\right]$$

Substituting Equation 10.12 into Equation 10.11, one can obtain

$$u_i(\mathbf{x}) = -\int_{\Omega} \Gamma_{ijm}C_{mjkl}\epsilon^*_{kl}\, d\mathbf{x}' \tag{10.15}$$

where

$$\Gamma_{imn} = \frac{1}{2}\left(\Gamma^u_{imn} + \Gamma^u_{inm}\right) \tag{10.16}$$

in which $C_{nmkl} = C_{mnkl}$ is considered to make $\Gamma_{imn} = \Gamma_{inm}$.

Using the kinematic equation, the strain is written as

$$\epsilon_{ij} = \frac{1}{2}\left(u_{i,j} + u_{j,i}\right) = -\frac{1}{2}\int_{\Omega}\left(\Gamma_{imn,j} + \Gamma_{jmn,i}\right)C_{nmkl}\epsilon^*_{kl}\, d\mathbf{x}' = -\int_{\Omega}\Gamma^e_{ijmn}C_{mnkl}\epsilon^*_{kl}\, d\mathbf{x}' \tag{10.17}$$

where

$$\Gamma^e_{ijmn} = \frac{1}{4}\left(\Gamma^u_{imn,j} + \Gamma^u_{jmn,i} + \Gamma^u_{inm,j} + \Gamma^u_{jnm,i}\right) \tag{10.18}$$

which satisfies both minor and major symmetries of fourth-rank tensors.

Considering that the eigenstrain is a continuous tensor function over the inhomogeneity, one can write it in a polynomial form of the coordinate as follows:

$$\epsilon^*_{ij}(\mathbf{x}') = \begin{cases} E^0_{ij} + E^1_{ijk}\left(x_k' - x_k^c\right) + E^2_{ijkl}\left(x_k' - x_k^c\right)\left(x_l' - x_l^c\right)\ldots, & \mathbf{x}' \in \Omega \\ 0, & \mathbf{x}' \in D - \Omega \end{cases} \tag{10.19}$$

For simplicity, only constant, linear, and quadratic terms are considered in the present paper, that is,

$$\epsilon^{*}_{mn} = E^{0}_{mn} + E^{1}_{mnp}\left(x'_p - x^c_p\right) + E^{2}_{mnst}\left(x'_s - x^c_s\right)\left(x'_t - x^c_t\right) \tag{10.20}$$

Following Mura's book [3], one can integrate Equation 10.17 as

$$\begin{aligned}
\epsilon_{ij} &= -\int_{\Omega} \Gamma^{e}_{ijkl} C_{klmn}\left[E^{0}_{mn} + E^{1}_{mnp}\left(x'_p - x^c_p\right) + E^{2}_{mnst}\left(x'_s - x^c_s\right)\left(x'_t - x^c_t\right)\right] d\mathbf{x}' \\
&= -\frac{1}{4\pi\mu_0}\left(D^{0}_{ijkl} C_{klmn} E^{0}_{mn} + D^{1}_{pijkl} C_{klmn} E^{1}_{mnp} + D^{2}_{stijkl} C_{klmn} E^{2}_{mnst}\right)
\end{aligned} \tag{10.21}$$

where D^0_{ijkl}, D^1_{pijkl}, and D^2_{stijkl} are

$$\begin{cases}
D^{0}_{ijkl} = \dfrac{1}{4}\left(D^{e}_{ijkl} + D^{e}_{jikl} + D^{e}_{ijlk} + D^{e}_{jilk}\right) \\
D^{1}_{pijkl} = \dfrac{1}{4}\left(D^{e}_{pijkl} + D^{e}_{pjikl} + D^{e}_{pijlk} + D^{e}_{pjilk}\right) \\
D^{2}_{stijkl} = \dfrac{1}{4}\left(D^{e}_{stijkl} + D^{e}_{stjikl} + D^{e}_{stijlk} + D^{e}_{stjilk}\right)
\end{cases} \tag{10.22}$$

in which D^e_{ijmn}, D^e_{pijkl}, and D^e_{stijkl} are given as

$$\begin{aligned}
D^{e}_{ijmn} &= \int_{\Omega} \Gamma^{u}_{imn,j}\, d\mathbf{x}' \\
&= \left[\Phi_{,nj}\delta_{im} - \frac{1}{4(1-v^0)}\Psi_{,imnj}\right] + Q_{nq}\bar{\Phi}_{,qj}\delta_{im} - \frac{(3-4v^0)}{4(1-v^0)} Q_{ik}Q_{ml}Q_{nq}\bar{\Psi}_{,klqj} \\
&\quad - \frac{x_3}{2(1-v^0)}\begin{bmatrix} Q_{mk}Q_{nq}\bar{\Psi}_{,ikq3j} + 4(1-v^0)Q_{nq}\delta_{m3}\bar{\Phi}_{,iqj} \\ +2(1-2v^0)\delta_{i3}Q_{mk}Q_{nq}\bar{\Phi}_{,kqj} \\ -Q_{mk}Q_{nq}x_3\bar{\Phi}_{,ikqj} - Q_{mk}Q_{nq}\delta_{3j}\bar{\Phi}_{,ikq} \end{bmatrix} \\
&\quad - \frac{\delta_{3j}}{2(1-v^0)}\begin{bmatrix} Q_{mk}Q_{nq}\bar{\Psi}_{,ikq3} + 4(1-v^0)Q_{nq}\delta_{m3}\bar{\Phi}_{,iq} \\ +2(1-2v^0)\delta_{i3}Q_{mk}Q_{nq}\bar{\Phi}_{,kq} - Q_{mk}Q_{nq}x_3\bar{\Phi}_{,ikq} \end{bmatrix} \\
&\quad + 2(1-2v^0)\left(\delta_{i3}\delta_{mk} - \delta_{ik}\delta_{m3}\right)Q_{nq}\Theta_{,jqk} - (1-2v^0)Q_{mk}Q_{nq}\Lambda_{,qikj}
\end{aligned}$$

$$
\begin{aligned}
D^{e}_{pijmn} &= \int_{\Omega} \Gamma^{u}_{imn,j}\left(x'_p - x^c_p\right) d\mathbf{x}' \\
&= \left[\Phi_{p,nj}\delta_{im} - \frac{1}{4(1-v^0)}\Psi_{p,imnj}\right] + Q_{nq}Q_{pt}\bar{\Phi}_{t,qj}\delta_{im} \\
&\quad - \frac{(3-4v^0)}{4(1-v^0)} Q_{ik}Q_{ml}Q_{nq}Q_{pt}\bar{\Psi}_{t,klqj} \\
&\quad - \frac{x_3}{2(1-v^0)}\begin{bmatrix} Q_{mk}Q_{nq}Q_{pt}\bar{\Psi}_{t,ikq3j} + 4(1-v^0)Q_{nq}\delta_{m3}Q_{pt}\bar{\Phi}_{t,iqj} \\ +2(1-2v^0)\delta_{i3}Q_{mk}Q_{nq}Q_{pt}\bar{\Phi}_{t,kqj} \\ -Q_{mk}Q_{nq}x_3Q_{pt}\bar{\Phi}_{t,ikqj} - Q_{mk}Q_{nq}\delta_{3j}Q_{pt}\bar{\Phi}_{t,ikq} \end{bmatrix} \\
&\quad - \frac{\delta_{3j}}{2(1-v^0)}\begin{bmatrix} Q_{mk}Q_{nq}Q_{pt}\bar{\Psi}_{t,ikq3} + 4(1-v^0)Q_{nq}Q_{pt}\delta_{m3}\bar{\Phi}_{t,iq} \\ +2(1-2v^0)\delta_{i3}Q_{mk}Q_{nq}Q_{pt}\bar{\Phi}_{t,kq} - Q_{mk}Q_{nq}x_3Q_{pt}\bar{\Phi}_{t,ikq} \end{bmatrix} \\
&\quad + 2(1-2v^0)\left(\delta_{i3}\delta_{mk} - \delta_{ik}\delta_{m3}\right)Q_{nq}\Theta_{p,jqk} - (1-2v^0)Q_{mk}Q_{nq}\Lambda_{p,qikj}
\end{aligned}
$$

$$
\begin{aligned}
D^{e}_{stijmn} &= \int_{\Omega} \Gamma^{u}_{imn,j}\left(x'_s - x^c_s\right)\left(x'_t - x^c_t\right) d\mathbf{x}' \\
&= \left[\Phi_{st,nj}\delta_{im} - \frac{1}{4(1-v^0)}\Psi_{st,imnj}\right] + Q_{nq}Q_{ps}Q_{gt}\bar{\Phi}_{pg,qj}\delta_{im} \\
&\quad - \frac{(3-4v^0)}{4(1-v^0)} Q_{ik}Q_{ml}Q_{nq}Q_{ps}Q_{gt}\bar{\Psi}_{pg,klqj} \\
&\quad - \frac{x_3}{2(1-v^0)}\begin{bmatrix} Q_{mk}Q_{nq}Q_{ps}Q_{gt}\bar{\Psi}_{pg,ikq3j} + 4(1-v^0)Q_{nq}\delta_{m3}Q_{ps}Q_{gt}\bar{\Phi}_{pg,iqj} \\ +2(1-2v^0)\delta_{i3}Q_{mk}Q_{nq}Q_{ps}Q_{gt}\bar{\Phi}_{pg,kqj} \\ -Q_{mk}Q_{nq}x_3Q_{ps}Q_{gt}\bar{\Phi}_{pg,ikqj} - Q_{mk}Q_{nq}Q_{ps}Q_{gt}\delta_{3j}\bar{\Phi}_{pg,ikq} \end{bmatrix} \\
&\quad - \frac{\delta_{3j}}{2(1-v^0)}\begin{bmatrix} Q_{mk}Q_{nq}Q_{ps}Q_{gt}\bar{\Psi}_{pg,ikq3} + 4(1-v^0)Q_{nq}\delta_{m3}Q_{ps}Q_{gt}\bar{\Phi}_{pg,iq} \\ +2(1-2v^0)\delta_{i3}Q_{mk}Q_{nq}Q_{ps}Q_{gt}\bar{\Phi}_{pg,kq} - Q_{mk}Q_{nq}x_3Q_{ps}Q_{gt}\bar{\Phi}_{pg,ikq} \end{bmatrix} \\
&\quad + 2(1-2v^0)\left(\delta_{i3}\delta_{mk} - \delta_{ik}\delta_{m3}\right)Q_{nq}\Theta_{st,jqk} - (1-2v^0)Q_{mk}Q_{nq}\Lambda_{st,qikj}
\end{aligned}
$$

where Φ, Ψ, Θ, and Λ denote the integrals of ϕ, ψ, θ, and λ over the ellipsoidal drop centered at (x^c_1, x^c_2, x^c_3); Φ_p, Ψ_p, Θ_p, and Λ_p denote the integrals of $\phi(x'_p - x^c_p)$, $\psi(x'_p - x^c_p)$, $\theta(x'_p - x^c_p)$, and $\lambda(x'_p - x^c_p)$; Φ_{pq}, Ψ_{pq}, Θ_{pq}, and Λ_{pq} denote the integrals of $\phi(x'_p - x^c_p)(x'_q - x^c_q)$, $\psi(x'_p - x^c_p)(x'_q - x^c_q)$, $\theta(x'_p - x^c_p)(x'_q - x^c_q)$, and $\lambda(x'_p - x^c_p)(x'_q - x^c_q)$. These integrals and their derivatives are explicitly provided at the end of this section.

The semi-infinite domain D with stiffness $\mathbf{C}^0$ contains an inclusion Ω with an eigenstrain field $\epsilon^*(\mathbf{x})$. The elastic field has been solved in Equation 10.21. In this section, the subdomain Ω in Figure 10.2 is replaced by an inhomogeneity with a different stiffness $\mathbf{C}^1$. A uniform stress $\boldsymbol{\sigma}^0$ is applied in the far-field. Due to the material mismatch, the local field in the neighborhood of Ω will be disturbed. To solve this problem the concept of Eshelby's equivalent inclusion method (EIM) [5,6] will be used in this section. The particle Ω is firstly filled with the same matrix material of $\mathbf{C}^0$ but an eigenstrain $\epsilon^*(\mathbf{x})$ is applied in Ω to make the stress Ω in the two cases equivalent. Then the BVPs can be equivalently solved.

For one inhomogeneity in an infinite solid subject to a uniform far-field stress, the eigenstrain is uniform [3,5,6]. However, due to the uniform boundary on the semi-infinite solid, the eigenstrain will not be constant anymore. However, the variation of $\boldsymbol{\epsilon}^*(\mathbf{x})$ should be continuous over Ω, which can be written in terms of polynomial of $\mathbf{x} - \mathbf{x}^c$ with the origin at the center of the ellipsoid, such as

$$\epsilon^*_{ij}(\mathbf{x}) = E^0_{ij} + E^1_{ijk}\left(x_k - x^c_k\right) + E^2_{ijkl}\left(x_k - x^c_k\right)\left(x_l - x^c_l\right) + \cdots, \quad \mathbf{x} \in \Omega \tag{10.23}$$

Substituting Equation 10.8 into equilibrium equation yields

$$C_{ijkl}\left(u_{k,li} - \epsilon^*_{kl,i}\right) = C^1_{ijkl}u_{k,li} \tag{10.24}$$

Therefore, if the eigenstrain can be derived from the above equation, one can obtain the elastic field. Using Taylor's expansion, the disturbed strain inside the particle can be written as

$$\begin{aligned}
\epsilon'_{ij}(\mathbf{x}) = &-\frac{1}{4\pi\mu_0}\begin{bmatrix} D^0_{ijkl}\left(\mathbf{x}^c\right)C_{klmn}E^0_{mn} + D^1_{pijkl}\left(\mathbf{x}^c\right)C_{klmn}E^1_{mnp} \\ +D^2_{stijkl}\left(\mathbf{x}^c\right)C_{klmn}E^2_{mnst} \end{bmatrix} \\
&-\frac{1}{4\pi\mu_0}\begin{bmatrix} D^0_{ijkl,r}\left(\mathbf{x}^c\right)C_{klmn}E^0_{mn} + D^1_{pijkl,r}\left(\mathbf{x}^c\right)C_{klmn}E^1_{mnp} \\ +D^2_{stijkl,r}\left(\mathbf{x}^c\right)C_{klmn}E^2_{mnst} \end{bmatrix}\left(x_r - x^c_r\right) \\
&-\frac{1}{8\pi\mu_0}\begin{bmatrix} D^0_{ijkl,yz}\left(\mathbf{x}^c\right)C_{klmn}E^0_{mn} + D^1_{pijkl,yz}\left(\mathbf{x}^c\right)C_{klmn}E^1_{mnp} \\ +D^2_{stijkl,yz}\left(\mathbf{x}^c\right)C_{klmn}E^2_{mnst} \end{bmatrix}\left(x_y - x^c_y\right)\left(x_z - x^c_z\right)
\end{aligned} \tag{10.25}$$

where $D^0_{ijkl,r}$, $D^1_{pijkl,r}$, $D^2_{stijkl,r}$, $D^0_{ijkl,yz}$, $D^1_{pijkl,yz}$, and $D^2_{stijkl,yz}$ are defined as

$$\begin{cases} D^0_{ijkl,r} = \dfrac{1}{4}\left(D^e_{ijkl,r} + D^e_{jikl,r} + D^e_{ijlk,r} + D^e_{jilk,r}\right) \\ D^1_{pijkl,r} = \dfrac{1}{4}\left(D^e_{pijkl,r} + D^e_{pjikl,r} + D^e_{pijlk,r} + D^e_{pjilk,r}\right) \\ D^2_{stijkl,r} = \dfrac{1}{4}\left(D^e_{stijkl,r} + D^e_{stjikl,r} + D^e_{stijlk,r} + D^e_{stjilk,r}\right) \end{cases} \tag{10.26}$$

$$\begin{cases} D^0_{ijkl,yz} = \dfrac{1}{4}\left(D^e_{ijkl,yz} + D^e_{jikl,yz} + D^e_{ijlk,yz} + D^e_{jilk,yz}\right) \\ D^1_{pijkl,yz} = \dfrac{1}{4}\left(D^e_{pijkl,yz} + D^e_{pjikl,yz} + D^e_{pijlk,yz} + D^e_{pjilk,yz}\right) \\ D^2_{stijkl,yz} = \dfrac{1}{4}\left(D^e_{stijkl,yz} + D^e_{stjikl,yz} + D^e_{stijlk,yz} + D^e_{stjilk,yz}\right) \end{cases} \tag{10.27}$$

in which $D^e_{ijmn,r}$, $D^e_{pijmn,r}$ $D^e_{stijmn,r}$ $D^e_{ijmn,yz}$, $D^e_{pijmn,yz}$, and $D^e_{stijmn,yz}$ are given at the end of this section. When a uniform force is applied on the boundary of the semi-infinite domain, a uniform average stress σ^0 will be caused. For the equivalent inclusion problem, a uniform strain $\epsilon^0_{ij} = C^{-1}_{ijkl}\sigma^0_{kl}$ will be induced, the equivalent inclusion condition is written as

$$C^*_{ijkl}\left(\epsilon^0_{kl} + \epsilon'_{kl}\right) = C_{ijkl}\left(\epsilon^0_{kl} + \epsilon'_{kl} - \epsilon^*_{kl}\right) \tag{10.28}$$

which can be rewritten as

$$\Delta\mathbf{C} : \left(\epsilon^0 + \epsilon'\right) = -\mathbf{C} : \epsilon^* \tag{10.29}$$

where $\Delta C = C^* - C$. The substitution of Equation 10.25 into Equation 10.29 yields,

$$\left(\frac{C_{ijmn} - C^*_{ijmn}}{8\pi\mu_0}\right)\left\{\begin{array}{l} 8\pi\mu_0\epsilon^0_{mn} - 2\begin{pmatrix} D^0_{ijkl}\left(\mathbf{x}^c\right)C_{klmn}E^0_{mn} \\ +D^1_{pijkl}\left(\mathbf{x}^c\right)C_{klmn}E^1_{mnp} \\ s + D^2_{stijkl}(\mathbf{x}^c)C_{klmn}E^2_{mnst} \end{pmatrix} \\ -2\begin{pmatrix} D^0_{ijkl,r}\left(\mathbf{x}^c\right)C_{klmn}E^0_{mn} \\ +D^1_{pijkl,r}\left(\mathbf{x}^c\right)C_{klmn}E^1_{mnp} \\ +D^2_{stijkl,r}(\mathbf{x}^c)C_{klmn}E^2_{mnst} \end{pmatrix}(x_r - x^c_r) \\ -\begin{pmatrix} D^0_{ijkl,yz}\left(\mathbf{x}^c\right)C_{klmn}E^0_{mn} \\ +D^1_{pijkl,yz}\left(\mathbf{x}^c\right)C_{klmn}E^1_{mnp} \\ +D^2_{stijkl,yz}(\mathbf{x}^c)C_{klmn}E^2_{mnst} \end{pmatrix}(x_y - x^c_y)(x_z - x^c_z) \end{array}\right\}$$

$$= C_{ijmn}\left[\begin{array}{c} E^0_{mn} + E^1_{mnp}(x_p - x^c_p) \\ +E^2_{mnst}(x_s - x^c_s)(x_t - x^c_t) \end{array}\right] \tag{10.30}$$

The coefficients of the constant, linear, and quadratic terms at left side should be equal to those on the right side, that is,

$$\left\{\begin{array}{l} \left[\begin{array}{c} \left[\left(C_{ijkl} - C^*_{ijkl}\right)D_{klst}\left(\mathbf{x}^c\right)C_{stmn} + 4\pi\mu_0 C_{ijmn}\right]E^0_{mn} \\ +\left(C_{ijkl} - C^*_{ijkl}\right)D_{pklst}\left(\mathbf{x}^c\right)C_{stmn}E^1_{mnp} \\ +\left(C_{ijkl} - C^*_{ijkl}\right)D_{pqklst}\left(\mathbf{x}^c\right)C_{stmn}E^2_{mnpq} \end{array}\right] = 4\pi\mu_0(C_{ijkl} - C^*_{ijkl})\epsilon^0_{kl} \\ \left[\begin{array}{c} \left[\left(C_{ijkl} - C^*_{ijkl}\right)D_{pklst,r}\left(\mathbf{x}^c\right)C_{stmn} + 4\pi\mu_0 C_{ijmn}\delta_{pr}\right]E^1_{mnp} \\ +\left(C_{ijkl} - C^*_{ijkl}\right)D_{klst,r}\left(\mathbf{x}^c\right)C_{stmn}E^0_{mn} \\ +\left(C_{ijkl} - C^*_{ijkl}\right)D_{pqklst,r}\left(\mathbf{x}^c\right)C_{stmn}E^2_{mnpq} \end{array}\right] = 0 \\ \left[\begin{array}{c} \left(C_{ijkl} - C^*_{ijkl}\right)D_{klst,yz}\left(\mathbf{x}^c\right)C_{stmn}E^0_{mn} \\ +\left(C_{ijkl} - C^*_{ijkl}\right)D_{pklst,yz}\left(\mathbf{x}^c\right)C_{stmn}E^1_{mnp} \\ +\left[\left(C_{ijkl} - C^*_{ijkl}\right)D_{pqklst,yz}\left(\mathbf{x}^c\right)C_{stmn} + 8\pi\mu_0 C_{ijmn}\delta_{py}\delta_{qz}\right]E^2_{mnpq} \end{array}\right] = 0 \end{array}\right. \tag{10.31}$$

Solving this linear equation system, one can obtain the eigenstrain terms of $\mathbf{E}^0$, $\mathbf{E}^1$, and $\mathbf{E}^2$, with which the displacement is written as

$$u_i(\mathbf{x}) = -\int_{\Omega} \Gamma_{imn} C_{mnkl} \left(E^0_{kl} + E^1_{klp}(x'_p - x^c_p) + E^2_{klpq}(x'_p - x^c_p)(x'_q - x^c_q) \right) d\mathbf{x}'$$

$$= -\frac{1}{4\pi\mu_0} \left(M^0_{imn} C_{mnkl} E^0_{kl} + N^1_{pimn} C_{mnkl} E^1_{klp} + T^2_{pqimn} C_{mnkl} E^2_{klpq} \right) \quad (10.32)$$

where M^0_{imn}, N^1_{pimn}, and T^2_{pqimn} are defined as

$$M^0_{imn} = \frac{1}{2}\left(M^u_{imn} + M^u_{inm}\right); \; N^1_{pimn} = \frac{1}{2}\left(N^u_{pimn} + N^u_{pinm}\right); \; T^2_{pqimn} = \frac{1}{2}\left(T^u_{pqimn} + T^u_{pqinm}\right)$$

where the explicit forms of M^u_{imn}, N^u_{pimn}, and T^u_{pqimn} are given at the end of this section.

To investigate the accuracy of the present formulation, the finite element method (FEM) for the axisymmetric cases of one particle embedded in a semi-infinite domain will be studied. The boundary effects on the local field and particle interactions are also discussed. An aluminum particle with radius $a = 10\,\mu$m is embedded in a semi-infinite HDPE at $(0, 0, h)$, where $h = 11$, 20, and 30 μm to the boundary. Young's moduli and Poisson's ratios for aluminum and HDPE are $E_1 = 69$ GPa, $E_0 = 0.8$ GPa, $\nu_1 = 0.334$, and $\nu_0 = 0.38$, respectively [74,77]. The strain components corresponding to the uniform load are $\epsilon^0_{33} = 0.001$, $\epsilon^0_{11} = \epsilon^0_{22} = -\nu_0\epsilon^0_{33}$ is applied to the system to represent a far-field tension $\sigma^0_{33} = 0.8$ MPa. Under this far-field tension, the elastic field is investigated. Finite element models (FEMs) are used to verify our analytical solution. An axisymmetric domain with 0.5 mm × 1 mm is used to represent the semi-infinite domain with a free boundary at $x_3 = 0$ in the FEM simulation.

Figure 10.3a illustrates the variation of stress σ_{33} along the axis x_3 ($x_1 = x_2 = 0$), when $h = 11\,\mu$m. One can observe a considerable difference of the results predicted by EIM with a constant eigenstrain from the FEM result. When linear terms are introduced, much better results are obtained.When quadratic terms are considered, the results almost overlaps with the FEM results, which indicates that the quadratic form of eigenstrains are sufficient for the present elastic analysis when the spacing to the boundary is higher than 5% of particle's size. When the particle is embedded at $h = 20\,\mu$m (Figure 10.3b), the results obtained by uniformly and linearly distributed eigenstrain are much closer to the FEM results than those in the previous case because the boundary effect decays with h. With the distance to the boundary h increasing, the difference of the results from uniform, linear, and quadratic distributed eigenstrain decreases. When h is large enough, the model reduces to one particle embedded into an infinite domain [5,6], in which the eigenstrain is uniform inside the inhomogeneity and higher-order terms disappear.

In Figure 10.3a, σ_{33} is equal to the far-field tension at the boundary $x_3 = 0$, and decreases with x_3 to the bottom of the particle. Inside the particle, it increases with x_3 to the top of the particle. Then, a sudden change of the slope is observed and σ_{33} increases quickly to its maximum value, which is about $2\sigma^0_{33}$. Then, it decreases with x_3 and finally converges to the

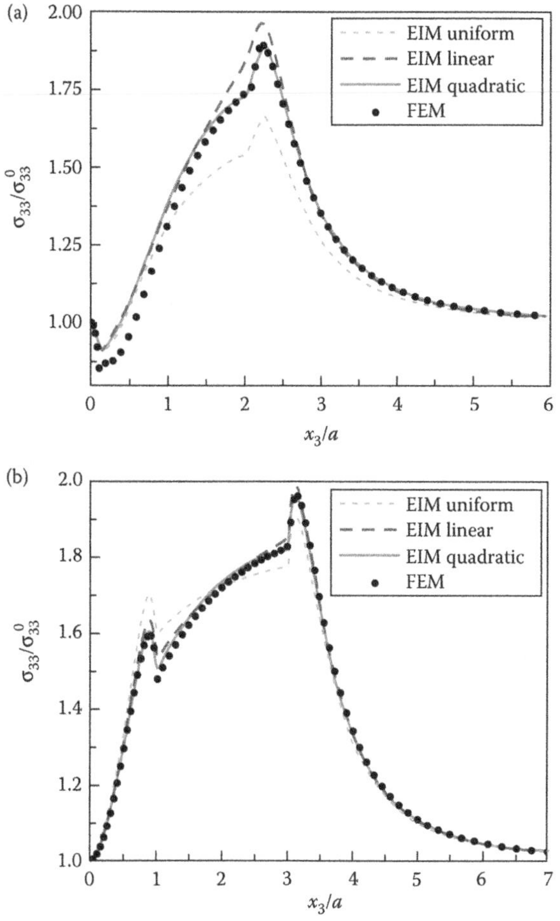

FIGURE 10.3 Variation of stress σ_{33} with x_3: (a) $h = 1.1a$ and (b) $h = 2a$.

far-field tension σ_{33}^0. Figure 10.3b illustrates the variation of σ_{33} with x_3, when the aluminum particle is embedded at $h = 20\ \mu\text{m}$, whose trend is quite different from the one for $h = 11\ \mu\text{m}$. Stress σ_{33} is equal to the far-field tension σ_{33}^0 at the boundary, and then increases with x_3 and reaches the peak value near the bottom of the particle. Then, it decreases with x_3 to the bottom of the particle. Inside the particle, σ_{33} increases with x_3. One can observe here that the change of σ_{33} inside the particle is much smaller than that for $h = 11\ \mu\text{m}$. Then, it goes outside of the particle and increases to the maximum value quickly near the top of particle. Finally, it decreases with x_3 and converges to the far-field load. In Figure 10.3b, the stress distribution on the particle is not symmetric either, and the peak values near the bottom of particle is smaller than that near the top. Actually, when the distance between the particle and boundary is large enough, the boundary effect is negligible so that the result will reduce to Eshelby's original problem for one particle embedded in an infinite domain, with which the curve will become absolutely symmetric and the stress inside the particle is uniform.

The Integrals and Derivatives of Vectors and Tensors in this Section

$$D^e_{ijmn,\gamma}$$

$$= \left[\Phi_{,nj\gamma}\delta_{im} - \frac{1}{4(1-v^0)}\Psi_{,imnj\gamma}\right] + Q_{nq}\bar{\Phi}_{,qj\gamma}\delta_{im} - \frac{(3-4v^0)}{4(1-v^0)}Q_{ik}Q_{ml}Q_{nq}\bar{\Psi}_{,klqj\gamma}$$

$$- \frac{x_3}{2(1-v^0)}\begin{bmatrix} Q_{mk}Q_{nq}\bar{\Psi}_{,ikq3j\gamma} + 4(1-v^0)Q_{nq}\delta_{m3}\bar{\Phi}_{,iqj\gamma} + 2(1-2v^0)\delta_{i3}Q_{mk}Q_{nq}\bar{\Phi}_{,kqj\gamma} \\ -Q_{mk}Q_{nq}x_3\bar{\Phi}_{,ikqj\gamma} - Q_{mk}Q_{nq}\delta_{3\gamma}\bar{\Phi}_{,ikqj} - Q_{mk}Q_{nq}\delta_{3j}\bar{\Phi}_{,ikq\gamma} \end{bmatrix}$$

$$- \frac{\delta_{3\gamma}}{2(1-v^0)}\begin{bmatrix} Q_{mk}Q_{nq}\bar{\Psi}_{,ikq3j} + 4(1-v^0)Q_{nq}\delta_{m3}\bar{\Phi}_{,iqj} + 2(1-2v^0)\delta_{i3}Q_{mk}Q_{nq}\bar{\Phi}_{,kqj} \\ -Q_{mk}Q_{nq}x_3\bar{\Phi}_{,ikqj} - Q_{mk}Q_{nq}\delta_{3j}\bar{\Phi}_{,ikq} \end{bmatrix}$$

$$- \frac{\delta_{3j}}{2(1-v^0)}\begin{bmatrix} Q_{mk}Q_{nq}\bar{\Psi}_{,ikq3\gamma} + 4(1-v^0)Q_{nq}\delta_{m3}\bar{\Phi}_{,iq\gamma} + 2(1-2v^0)\delta_{i3}Q_{mk}Q_{nq}\bar{\Phi}_{,kq\gamma} \\ -Q_{mk}Q_{nq}x_3\bar{\Phi}_{,ikq\gamma} - Q_{mk}Q_{nq}\delta_{3\gamma}\bar{\Phi}_{,ikq} \end{bmatrix}$$

$$+ 2(1-2v^0)\left(\delta_{i3}\delta_{mk} - \delta_{ik}\delta_{m3}\right)Q_{nq}\Theta_{,jqk\gamma} - (1-2v^0)Q_{mk}Q_{nq}\Lambda_{,qikj\gamma}$$

$$D^e_{pijmn,\gamma}$$

$$= \left[\Phi_{p,nj\gamma}\delta_{im} - \frac{1}{4(1-v^0)}\Psi_{p,imnj\gamma}\right]$$

$$+ Q_{nq}Q_{pt}\bar{\Phi}_{t,qj\gamma}\delta_{im} - \frac{(3-4v^0)}{4(1-v^0)}Q_{ik}Q_{ml}Q_{nq}Q_{pt}\bar{\Psi}_{t,klqj\gamma}$$

$$- \frac{x_3}{2(1-v^0)}\begin{bmatrix} Q_{mk}Q_{nq}Q_{pt}\bar{\Psi}_{t,ikq3j\gamma} + 4(1-v^0)Q_{nq}\delta_{m3}Q_{pt}\bar{\Phi}_{t,iqj\gamma} \\ +2(1-2v^0)\delta_{i3}Q_{mk}Q_{nq}Q_{pt}\bar{\Phi}_{t,kqj\gamma} - Q_{mk}Q_{nq}x_3Q_{pt}\bar{\Phi}_{t,ikqj\gamma} \\ -Q_{mk}Q_{nq}\delta_{3\gamma}Q_{pt}\bar{\Phi}_{t,ikqj} - Q_{mk}Q_{nq}\delta_{3j}Q_{pt}\bar{\Phi}_{t,ikq\gamma} \end{bmatrix}$$

$$- \frac{\delta_{3\gamma}}{2(1-v^0)}\begin{bmatrix} Q_{mk}Q_{nq}Q_{pt}\bar{\Psi}_{t,ikq3j} + 4(1-v^0)Q_{nq}\delta_{m3}Q_{pt}\bar{\Phi}_{t,iqj} \\ +2(1-2v^0)\delta_{i3}Q_{mk}Q_{nq}Q_{pt}\bar{\Phi}_{t,kqj} \\ -Q_{mk}Q_{nq}x_3Q_{pt}\bar{\Phi}_{t,ikqj} - Q_{mk}Q_{nq}\delta_{3j}Q_{pt}\bar{\Phi}_{t,ikq} \end{bmatrix}$$

$$- \frac{\delta_{3j}}{2(1-v^0)}\begin{bmatrix} Q_{mk}Q_{nq}Q_{pt}\bar{\Psi}_{t,ikq3\gamma} + 4(1-v^0)Q_{nq}Q_{pt}\delta_{m3}\bar{\Phi}_{t,iq\gamma} \\ +2(1-2v^0)\delta_{i3}Q_{mk}Q_{nq}Q_{pt}\bar{\Phi}_{t,kq\gamma} \\ -Q_{mk}Q_{nq}x_3Q_{pt}\bar{\Phi}_{t,ikq\gamma} - Q_{mk}Q_{nq}\delta_{3\gamma}Q_{pt}\bar{\Phi}_{t,ikq} \end{bmatrix}$$

$$+ 2(1-2v^0)\left(\delta_{i3}\delta_{mk} - \delta_{ik}\delta_{m3}\right)Q_{nq}\Theta_{p,jqk\gamma} - (1-2v^0)Q_{mk}Q_{nq}\Lambda_{p,qikj\gamma}$$

$$D^e_{stijmn,\gamma}$$

$$= \left[\Phi_{st,nj\gamma}\delta_{im} - \frac{1}{4(1-v^0)}\Psi_{st,imnj\gamma}\right]$$

$$+ Q_{nq}Q_{ps}Q_{gt}\bar{\Phi}_{pg,qj\gamma}\delta_{im} - \frac{(3-4v^0)}{4(1-v^0)}Q_{ik}Q_{ml}Q_{nq}Q_{ps}Q_{gt}\bar{\Psi}_{pg,klqj\gamma}$$

$$- \frac{x_3}{2(1-v^0)}\begin{bmatrix} Q_{mk}Q_{nq}Q_{ps}Q_{gt}\bar{\Psi}_{pg,ikq3j\gamma} + 4(1-v^0)Q_{nq}\delta_{m3}Q_{ps}Q_{gt}\bar{\Phi}_{pg,iqj\gamma} \\ +2(1-2v^0)\delta_{i3}Q_{mk}Q_{nq}Q_{ps}Q_{gt}\bar{\Phi}_{pg,kqj\gamma} - Q_{mk}Q_{nq}x_3Q_{ps}Q_{gt}\bar{\Phi}_{pg,ikqj\gamma} \\ -Q_{mk}Q_{nq}\delta_{3\gamma}Q_{ps}Q_{gt}\bar{\Phi}_{pg,ikqj} - Q_{mk}Q_{nq}Q_{ps}Q_{gt}\delta_{3j}\bar{\Phi}_{pg,ikq\gamma} \end{bmatrix}$$

$$-\frac{\delta_{3\gamma}}{2(1-v^0)}\left[\begin{array}{c} Q_{mk}Q_{nq}Q_{ps}Q_{gt}\bar{\Psi}_{pg,ikq3j}+4(1-v^0)Q_{nq}\delta_{m3}Q_{ps}Q_{gt}\bar{\Phi}_{pg,iqj} \\ +2(1-2v^0)\delta_{i3}Q_{mk}Q_{nq}Q_{ps}Q_{gt}\bar{\Phi}_{pg,kqj} \\ -Q_{mk}Q_{nq}x_3Q_{ps}Q_{gt}\bar{\Phi}_{pg,ikqj}-Q_{mk}Q_{nq}Q_{ps}Q_{gt}\delta_{3j}\bar{\Phi}_{pg,ikq} \end{array}\right]$$

$$-\frac{\delta_{3j}}{2(1-v^0)}\left[\begin{array}{c} Q_{mk}Q_{nq}Q_{ps}Q_{gt}\bar{\Psi}_{pg,ikq3\gamma}+4(1-v^0)Q_{nq}\delta_{m3}Q_{ps}Q_{gt}\bar{\Phi}_{pg,iq\gamma} \\ +2(1-2v^0)\delta_{i3}Q_{mk}Q_{nq}Q_{ps}Q_{gt}\bar{\Phi}_{pg,kq\gamma} \\ -Q_{mk}Q_{nq}x_3Q_{ps}Q_{gt}\bar{\Phi}_{pg,ikq\gamma}-Q_{mk}Q_{nq}\delta_{3\gamma}Q_{ps}Q_{gt}\bar{\Phi}_{pg,ikq} \end{array}\right]$$

$$+2(1-2v^0)\left(\delta_{i3}\delta_{mk}-\delta_{ik}\delta_{m3}\right)Q_{nq}\Theta_{st,jqk\gamma}-(1-2v^0)Q_{mk}Q_{nq}\Lambda_{st,qikj\gamma}$$

$D^e_{ijmn,\alpha\beta}$

$$=\left[\Phi_{,nj\alpha\beta}\delta_{im}-\frac{1}{4(1-v^0)}\Psi_{,imnj\alpha\beta}\right]$$

$$+Q_{nq}\bar{\Phi}_{,qj\alpha\beta}\delta_{im}-\frac{(3-4v^0)}{4(1-v^0)}Q_{ik}Q_{ml}Q_{nq}\bar{\Psi}_{,klqj\alpha\beta}$$

$$-\frac{x_3}{2(1-v^0)}\left[\begin{array}{c} Q_{mk}Q_{nq}\bar{\Psi}_{,ikq3j\alpha\beta}+4(1-v^0)Q_{nq}\delta_{m3}\bar{\Phi}_{,iqj\alpha\beta} \\ +2(1-2v^0)\delta_{i3}Q_{mk}Q_{nq}\bar{\Phi}_{,kqj\alpha\beta}-Q_{mk}Q_{nq}x_3\bar{\Phi}_{,ikqj\alpha\beta} \\ -Q_{mk}Q_{nq}\delta_{3\beta}\bar{\Phi}_{,ikqj\alpha}-Q_{mk}Q_{nq}\delta_{3\alpha}\bar{\Phi}_{,ikqj}-Q_{mk}Q_{nq}\delta_{3j}\bar{\Phi}_{,ikq\alpha} \end{array}\right]$$

$$-\frac{\delta_{3\beta}}{2(1-v^0)}\left[\begin{array}{c} Q_{mk}Q_{nq}\bar{\Psi}_{,ikq3j\alpha}+4(1-v^0)Q_{nq}\delta_{m3}\bar{\Phi}_{,iqj\alpha} \\ +2(1-2v^0)\delta_{i3}Q_{mk}Q_{nq}\bar{\Phi}_{,kqj\alpha}-Q_{mk}Q_{nq}x_3\bar{\Phi}_{,ikqj\alpha} \\ -Q_{mk}Q_{nq}\delta_{3\alpha}\bar{\Phi}_{,ikqj}-Q_{mk}Q_{nq}\delta_{3j}\bar{\Phi}_{,ikq\alpha} \end{array}\right]$$

$$-\frac{\delta_{3\alpha}}{2(1-v^0)}\left[\begin{array}{c} Q_{mk}Q_{nq}\bar{\Psi}_{,ikq3j\beta}+4(1-v^0)Q_{nq}\delta_{m3}\bar{\Phi}_{,iqj\beta} \\ +2(1-2v^0)\delta_{i3}Q_{mk}Q_{nq}\bar{\Phi}_{,kqj\beta}-Q_{mk}Q_{nq}x_3\bar{\Phi}_{,ikqj\beta} \\ -Q_{mk}Q_{nq}\delta_{3\beta}\bar{\Phi}_{,ikqj}-Q_{mk}Q_{nq}\delta_{3j}\bar{\Phi}_{,ikq\beta} \end{array}\right]$$

$$-\frac{\delta_{3j}}{2(1-v^0)}\left[\begin{array}{c} Q_{mk}Q_{nq}\bar{\Psi}_{,ikq3\alpha\beta}+4(1-v^0)Q_{nq}\delta_{m3}\bar{\Phi}_{,iq\alpha\beta} \\ +2(1-2v^0)\delta_{i3}Q_{mk}Q_{nq}\bar{\Phi}_{,kq\alpha\beta}-Q_{mk}Q_{nq}x_3\bar{\Phi}_{,ikq\alpha\beta} \\ -Q_{mk}Q_{nq}\delta_{3\beta}\bar{\Phi}_{,ikq\alpha}-Q_{mk}Q_{nq}\delta_{3\alpha}\bar{\Phi}_{,ikq\beta} \end{array}\right]$$

$$+2(1-2v^0)\left(\delta_{i3}\delta_{mk}-\delta_{ik}\delta_{m3}\right)Q_{nq}\Theta_{,jqk\alpha\beta}-(1-2v^0)Q_{mk}Q_{nq}\Lambda_{,qikj\alpha\beta}$$

$D^e_{pijmn,\alpha\beta}$

$$=\left[\Phi_{p,nj\alpha\beta}\delta_{im}-\frac{1}{4(1-v^0)}\Psi_{p,imnj\alpha\beta}\right]$$

$$+Q_{nq}Q_{pt}\bar{\Phi}_{t,qj\alpha\beta}\delta_{im}-\frac{(3-4v^0)}{4(1-v^0)}Q_{ik}Q_{ml}Q_{nq}Q_{pt}\bar{\Psi}_{t,klqj\alpha\beta}$$

$$-\frac{x_3}{2(1-v^0)}\left[\begin{array}{c} Q_{mk}Q_{nq}Q_{pt}\bar{\Psi}_{t,ikq3j\alpha\beta}+4(1-v^0)Q_{nq}\delta_{m3}Q_{pt}\bar{\Phi}_{t,iqj\alpha\beta} \\ +2(1-2v^0)\delta_{i3}Q_{mk}Q_{nq}Q_{pt}\bar{\Phi}_{t,kqj\alpha\beta}-Q_{mk}Q_{nq}x_3Q_{pt}\bar{\Phi}_{t,ikqj\alpha\beta} \\ -Q_{mk}Q_{nq}\delta_{3\beta}Q_{pt}\bar{\Phi}_{t,ikqj\alpha}-Q_{mk}Q_{nq}\delta_{3\alpha}Q_{pt}\bar{\Phi}_{t,ikqj\beta} \\ -Q_{mk}Q_{nq}\delta_{3j}Q_{pt}\bar{\Phi}_{t,ikq\alpha\beta} \end{array}\right]$$

$$
- \frac{\delta_{3\beta}}{2(1-v^0)} \left[\begin{array}{c} Q_{mk}Q_{nq}Q_{pt}\bar{\Psi}_{t,ikq3j\alpha} + 4(1-v^0)Q_{nq}\delta_{m3}Q_{pt}\bar{\Phi}_{t,iqj\alpha} \\ +2(1-2v^0)\delta_{i3}Q_{mk}Q_{nq}Q_{pt}\bar{\Phi}_{t,kqj\alpha} - Q_{mk}Q_{nq}x_3Q_{pt}\bar{\Phi}_{t,ikqj\alpha} \\ -Q_{mk}Q_{nq}\delta_{3\alpha}Q_{pt}\bar{\Phi}_{t,ikqj} - Q_{mk}Q_{nq}\delta_{3j}Q_{pt}\bar{\Phi}_{t,ikq\alpha} \end{array} \right]
$$

$$
- \frac{\delta_{3\alpha}}{2(1-v^0)} \left[\begin{array}{c} Q_{mk}Q_{nq}Q_{pt}\bar{\Psi}_{t,ikq3j\beta} + 4(1-v^0)Q_{nq}\delta_{m3}Q_{pt}\bar{\Phi}_{t,iqj\beta} \\ +2(1-2v^0)\delta_{i3}Q_{mk}Q_{nq}Q_{pt}\bar{\Phi}_{t,kqj\beta} - Q_{mk}Q_{nq}x_3Q_{pt}\bar{\Phi}_{t,ikqj\beta} \\ -Q_{mk}Q_{nq}\delta_{3\beta}Q_{pt}\bar{\Phi}_{t,ikqj} - Q_{mk}Q_{nq}\delta_{3j}Q_{pt}\bar{\Phi}_{t,ikq\beta} \end{array} \right]
$$

$$
- \frac{\delta_{3j}}{2(1-v^0)} \left[\begin{array}{c} Q_{mk}Q_{nq}Q_{pt}\bar{\Psi}_{t,ikq3\alpha\beta} + 4(1-v^0)Q_{nq}Q_{pt}\delta_{m3}\bar{\Phi}_{t,iq\alpha\beta} \\ +2(1-2v^0)\delta_{i3}Q_{mk}Q_{nq}Q_{pt}\bar{\Phi}_{t,kq\alpha\beta} - Q_{mk}Q_{nq}x_3Q_{pt}\bar{\Phi}_{t,ikq\alpha\beta} \\ -Q_{mk}Q_{nq}\delta_{3\beta}Q_{pt}\bar{\Phi}_{t,ikq\alpha} - Q_{mk}Q_{nq}\delta_{3\alpha}Q_{pt}\bar{\Phi}_{t,ikq\beta} \end{array} \right]
$$

$$
+ 2(1-2v^0)\left(\delta_{i3}\delta_{mk} - \delta_{ik}\delta_{m3}\right)Q_{nq}\Theta_{p,jqk\alpha\beta} - (1-2v^0)Q_{mk}Q_{nq}\Lambda_{p,qikj\alpha\beta}
$$

and

$$
\begin{aligned}
D^e_{stijmn,\alpha\beta}
&= \left[\Phi_{st,nj\alpha\beta}\delta_{im} - \frac{1}{4(1-v^0)}\Psi_{st,imnj\alpha\beta}\right] \\
&+ Q_{nq}Q_{ps}Q_{gt}\bar{\Phi}_{pg,qj\alpha\beta}\delta_{im} - \frac{(3-4v^0)}{4(1-v^0)}Q_{ik}Q_{ml}Q_{nq}Q_{ps}Q_{gt}\bar{\Psi}_{pg,klqj\alpha\beta}
\end{aligned}
$$

$$
- \frac{x_3}{2(1-v^0)} \left[\begin{array}{c} Q_{mk}Q_{nq}Q_{ps}Q_{gt}\bar{\Psi}_{pg,ikq3j\alpha\beta} + 4(1-v^0)Q_{nq}\delta_{m3}Q_{ps}Q_{gt}\bar{\Phi}_{pg,iqj\alpha\beta} \\ +2(1-2v^0)\delta_{i3}Q_{mk}Q_{nq}Q_{ps}Q_{gt}\bar{\Phi}_{pg,kqj\alpha\beta} \\ -Q_{mk}Q_{nq}x_3Q_{ps}Q_{gt}\bar{\Phi}_{pg,ikqj\alpha\beta} - Q_{mk}Q_{nq}\delta_{3\beta}Q_{ps}Q_{gt}\bar{\Phi}_{pg,ikqj\alpha} \\ -Q_{mk}Q_{nq}\delta_{3\alpha}Q_{ps}Q_{gt}\bar{\Phi}_{pg,ikqj\beta} - Q_{mk}Q_{nq}Q_{ps}Q_{gt}\delta_{3j}\bar{\Phi}_{pg,ikq\alpha\beta} \end{array} \right]
$$

$$
- \frac{\delta_{3\beta}}{2(1-v^0)} \left[\begin{array}{c} Q_{mk}Q_{nq}Q_{ps}Q_{gt}\bar{\Psi}_{pg,ikq3j\alpha} + 4(1-v^0)Q_{nq}\delta_{m3}Q_{ps}Q_{gt}\bar{\Phi}_{pg,iqj\alpha} \\ +2(1-2v^0)\delta_{i3}Q_{mk}Q_{nq}Q_{ps}Q_{gt}\bar{\Phi}_{pg,kqj\alpha} - Q_{mk}Q_{nq}x_3Q_{ps}Q_{gt}\bar{\Phi}_{pg,ikqj\alpha} \\ -Q_{mk}Q_{nq}\delta_{3\alpha}Q_{ps}Q_{gt}\bar{\Phi}_{pg,ikqj} - Q_{mk}Q_{nq}Q_{ps}Q_{gt}\delta_{3j}\bar{\Phi}_{pg,ikq\alpha} \end{array} \right]
$$

$$
- \frac{\delta_{3\alpha}}{2(1-v^0)} \left[\begin{array}{c} Q_{mk}Q_{nq}Q_{ps}Q_{gt}\bar{\Psi}_{pg,ikq3j\beta} + 4(1-v^0)Q_{nq}\delta_{m3}Q_{ps}Q_{gt}\bar{\Phi}_{pg,iqj\beta} \\ +2(1-2v^0)\delta_{i3}Q_{mk}Q_{nq}Q_{ps}Q_{gt}\bar{\Phi}_{pg,kqj\beta} - Q_{mk}Q_{nq}x_3Q_{ps}Q_{gt}\bar{\Phi}_{pg,ikqj\beta} \\ -Q_{mk}Q_{nq}\delta_{3\beta}Q_{ps}Q_{gt}\bar{\Phi}_{pg,ikqj} - Q_{mk}Q_{nq}Q_{ps}Q_{gt}\delta_{3j}\bar{\Phi}_{pg,ikq\beta} \end{array} \right]
$$

$$
- \frac{\delta_{3j}}{2(1-v^0)} \left[\begin{array}{c} Q_{mk}Q_{nq}Q_{ps}Q_{gt}\bar{\Psi}_{pg,ikq3\alpha\beta} + 4(1-v^0)Q_{nq}\delta_{m3}Q_{ps}Q_{gt}\bar{\Phi}_{pg,iq\alpha\beta} \\ +2(1-2v^0)\delta_{i3}Q_{mk}Q_{nq}Q_{ps}Q_{gt}\bar{\Phi}_{pg,kq\alpha\beta} - Q_{mk}Q_{nq}x_3Q_{ps}Q_{gt}\bar{\Phi}_{pg,ikq\alpha\beta} \\ -Q_{mk}Q_{nq}\delta_{3\beta}Q_{ps}Q_{gt}\bar{\Phi}_{pg,ikq\alpha} - Q_{mk}Q_{nq}\delta_{3\alpha}Q_{ps}Q_{gt}\bar{\Phi}_{pg,ikq\beta} \end{array} \right]
$$

$$
+ 2(1-2v^0)\left(\delta_{i3}\delta_{mk} - \delta_{ik}\delta_{m3}\right)Q_{nq}\Theta_{st,jqk\alpha\beta} - (1-2v^0)Q_{mk}Q_{nq}\Lambda_{st,qikj\alpha\beta}
$$

Moreover, the displacement is written as

$$u_i(\mathbf{x}) = -\int_{\Omega} \Gamma_{imn} C_{mnkl} \left[E^0_{kl} + E^1_{klp}\left(x'_p - x^c_p\right) + E^2_{klpq}\left(x'_p - x^c_p\right)\left(x'_q - x^c_q\right)\right] d\mathbf{x}'$$

$$= -\frac{1}{4\pi\mu_0}\left(M_{imn} C_{mnkl} E^0_{kl} + N_{pimn} C_{mnkl} E^1_{klp} + T_{pqimn} C_{mnkl} E^2_{klpq}\right)$$

The tensors $\mathbf{M}^u$, $\mathbf{N}^u$, and $\mathbf{T}^u$ are provided as

$$M^u_{imn} = \left[\Phi_{,n}\delta_{im} - \frac{1}{4(1-v^0)}\Psi_{,imn}\right] + Q_{nq}\bar{\Phi}_{,q}\delta_{im} - \frac{(3-4v^0)}{4(1-v^0)} Q_{ik} Q_{ml} Q_{nq} \bar{\Psi}_{,klq}$$
$$- \frac{x_3}{2(1-v^0)}\left[\begin{array}{c} Q_{mk}Q_{nq}\bar{\Psi}_{,ikq3} + 4(1-v^0)Q_{nq}\delta_{m3}\bar{\Phi}_{,iq} \\ +2(1-2v^0)\delta_{i3}Q_{mk}Q_{nq}\bar{\Phi}_{,kq} - Q_{mk}Q_{nq}x_3\bar{\Phi}_{,ikq}\end{array}\right]$$
$$+ 2(1-2v^0)\left(\delta_{i3}\delta_{mk} - \delta_{ik}\delta_{m3}\right) Q_{nq}\Theta_{,qk} - (1-2v^0) Q_{mk} Q_{nq} \Lambda_{,qik}$$

$$N^u_{pimn} = \left[\Phi_{p,n}\delta_{im} - \frac{1}{4(1-v^0)}\Psi_{p,imn}\right] + Q_{tp}Q_{nq}\bar{\Phi}_{t,q}\delta_{im} - \frac{(3-4v^0)}{4(1-v^0)} Q_{ik} Q_{ml} Q_{nq} Q_{pt} \bar{\Psi}_{t,klq}$$
$$- \frac{x_3}{2(1-v^0)}\left[\begin{array}{c} Q_{mk}Q_{nq}Q_{pt}\bar{\Psi}_{t,ikq3} + 4(1-v^0)Q_{nq}Q_{pt}\delta_{m3}\bar{\Phi}_{t,iq} \\ +2(1-2v^0)\delta_{i3}Q_{mk}Q_{nq}Q_{pt}\bar{\Phi}_{t,kq} - Q_{mk}Q_{nq}Q_{pt}x_3\bar{\Phi}_{t,ikq}\end{array}\right]$$
$$+ 2(1-2v^0)\left(\delta_{i3}\delta_{mk} - \delta_{ik}\delta_{m3}\right) Q_{nq}\Theta_{p,qk} - (1-2v^0) Q_{mk} Q_{nq} \Lambda_{p,qik}$$

and

$$T^u_{pqimn} = \left[\Phi_{pq,n}\delta_{im} - \frac{1}{4(1-v^0)}\Psi_{pq,imn}\right] + Q_{ng}Q_{ps}Q_{qt}\bar{\Phi}_{st,g}\delta_{im}$$
$$- \frac{(3-4v^0)}{4(1-v^0)} Q_{ik} Q_{ml} Q_{ng} Q_{ps} Q_{qt} \bar{\Psi}_{st,klg}$$
$$- \frac{x_3}{2(1-v^0)}\left[\begin{array}{c} Q_{mk}Q_{ng}Q_{ps}Q_{qt}\bar{\Psi}_{st,ikg3} + 4(1-v^0)Q_{ng}Q_{ps}Q_{qt}\delta_{m3}\bar{\Phi}_{st,ig} \\ +2(1-2v^0)\delta_{i3}Q_{mk}Q_{ng}Q_{ps}Q_{qt}\bar{\Phi}_{st,kg} - Q_{mk}Q_{ng}Q_{ps}Q_{qt}x_3\bar{\Phi}_{st,ikg}\end{array}\right]$$
$$+ 2(1-2v^0)\left(\delta_{i3}\delta_{mk} - \delta_{ik}\delta_{m3}\right) Q_{ng}\Theta_{pq,gk} - (1-2v^0) Q_{mk} Q_{ng} \Lambda_{pq,gik}$$

The volume integrals of θ and λ are given as

$$\Theta(\mathbf{x}, \mathbf{x}^c) = \int_{\Omega} \left[\ln\left(x_3 + x'_3 + \bar{\psi}\right)\right] d\mathbf{x}' = \frac{4\pi a^3}{3} \ln \widetilde{\psi}(\mathbf{x}, \mathbf{x}^c)$$

and

$$\Lambda(\mathbf{x},\mathbf{x}^c) = \int_\Omega \left[\left(x_3 + x_3'\right)\ln\left(x_3 + x_3' + \bar{\psi}\right) - \bar{\psi}\right] d\mathbf{x}'$$

$$= \frac{4\pi a^3}{3}\left[\left(x_3 + x_3^c\right)\ln\widetilde{\psi}(\mathbf{x},\mathbf{x}^c) - \bar{\psi}(\mathbf{x},\mathbf{x}^c)\right]$$

where $\widetilde{\psi}(\mathbf{x},\mathbf{x}^c) = \bar{\psi}(\mathbf{x},\mathbf{x}^c) + x_3 + x_3^c$.

The volume integrals of $\theta\left(x_p' - x_p^c\right)$ and $\lambda\left(x_p' - x_p^c\right)$ are given as

$$\Theta_p(\mathbf{x},\mathbf{x}^c) = \int_\Omega \left[\ln\left(x_3 + x_3' + \bar{\psi}\right)\right]\left(x_p' - x_p^c\right) d\mathbf{x}'$$

$$= \begin{bmatrix} -\dfrac{4\pi}{15}a^5\left(\ln\widetilde{\psi}(\mathbf{x},\mathbf{x}^c)\right)_{,p} \; (p = 1, 2) \\ \dfrac{4\pi}{15}a^5\bar{\phi}(\mathbf{x},\mathbf{x}^c)\; (p = 3) \end{bmatrix}$$

$$\Lambda_p(\mathbf{x},\mathbf{x}^c) = \int_\Omega \left[\left(x_3 + x_3'\right)\ln\left(x_3 + x_3' + \bar{\psi}\right) - \bar{\psi}\right]\left(x_p' - x_p^c\right) d\mathbf{x}'$$

$$= \begin{bmatrix} \dfrac{4\pi a^5}{15}\left(x_p - x_p^c\right)\widetilde{\phi}(\mathbf{x},\mathbf{x}^c)\; (p = 1, 2) \\ \dfrac{4\pi}{15}a^5\ln\widetilde{\psi}(x,x^c)\; (p = 3) \end{bmatrix}$$

where $\widetilde{\phi}(\mathbf{x},\mathbf{x}^c) = \frac{1}{\bar{\psi}(\mathbf{x},\mathbf{x}^c)+x_3+x_3^c}$. Note that

$$\widetilde{\psi}_{,i} = \bar{\psi}_{,i} + \delta_{3i} \quad \widetilde{\psi}_{,ij\ldots} = \bar{\psi}_{,ij\ldots};$$

The volume integrals of $\theta\left(x_m' - x_m^c\right)\left(x_n' - x_n^c\right)$, $\lambda\left(x_m' - x_m^c\right)\left(x_n' - x_n^c\right)$ are given as

$$\Theta_{mn}(\mathbf{x},\mathbf{x}^c)$$

$$= \int_\Omega \left[\ln\left(x_3 + x_3' + \bar{\psi}\right)\right]\left(x_m' - x_m^c\right)\left(x_n' - x_n^c\right) d\mathbf{x}'$$

$$= \begin{cases} \dfrac{4\pi a^5}{105}\left\{ \begin{array}{c} -a^2\left[\left(\ln\bar{\psi}\right)_{,m}\left(\ln\widetilde{\psi}\right)_{,n} + \left(\ln\widetilde{\psi}\right)_{,m}\left(\ln\widetilde{\psi}\right)_{,n}\right] \\ +\delta_{mn}\left(7\ln\widetilde{\psi} + a^2\widetilde{\phi}\,\bar{\phi}\right) \end{array} \right\} & (m \neq 3; n \neq 3). \\ -\dfrac{4\pi a^7}{105}\bar{\phi}_{,m} & (m \neq 3; n = 3). \\ \dfrac{4\pi a^5}{105}\left(a^2\bar{\phi}_{,3} + 7\ln\widetilde{\psi}\right) & (m = 3; n = 3). \end{cases}$$

and

$$\Lambda_{mn}(\mathbf{x},\mathbf{x}^c)$$
$$= \int_\Omega \left[(x_3 + x_3') \ln (x_3 + x_3' + \bar{\psi}) - \bar{\psi} \right] (x_m' - x_m^c) (x_n' - x_n^c) \, d\mathbf{x}'$$
$$= \begin{cases} \dfrac{4\pi a^5}{105} \left\{ \begin{array}{c} -a^2 (x_m - x_m^c) \widetilde{\phi}_{,n} \\ +\delta_{mn} \left[7 \left((x_3 + x_3^c) \ln \widetilde{\psi}(\mathbf{x},\mathbf{x}^c) - \bar{\psi}(\mathbf{x},\mathbf{x}^c) \right) - a^2 \widetilde{\phi} \right] \end{array} \right\} & (m \neq 3; n \neq 3). \\ -\dfrac{4\pi a^7}{105} \left(\ln \widetilde{\psi} \right)_{,m} & (m \neq 3; n = 3). \\ \dfrac{4\pi a^5}{105} \left\{ a^2 \bar{\phi} + 7 \left[(x_3 + x_3^c) \ln \widetilde{\psi}(\mathbf{x},\mathbf{x}^c) - \bar{\psi}(\mathbf{x},\mathbf{x}^c) \right] \right\} & (m = 3; n = 3). \end{cases}$$

10.3 ELASTIC SOLUTION FOR MULTIPLE PARTICLES IN A SEMI-INFINITE DOMAIN

The above formulation for a single particle can be extended to multiple particles in a very straightforward way. When many particles are inserted in the semi-infinite domain, their interactions will significantly affect the elastic fields. Figure 10.1 illustrates the whole system with many particles. To formulate this problem, a global coordinate system has been set up in Figure 10.1, and each particle, say Ω^I, is centered at $\mathbf{x}^I = (x_1^I, x_2^I, x_3^I)$. All particles being treated as including the same material properties with the matrix, the eigenstrain can be introduced with a polynomial form as:

$$\epsilon_{ij}^{*,I}(\mathbf{x}) = E_{ij}^{0,I} + E_{ijk}^{1,I} \left(x_k - x_k^I\right) + E_{ijkl}^{2,I} \left(x_k - x_k^I\right) \left(x_l - x_l^I\right) + \cdots, \quad \mathbf{x} \in \Omega^I \tag{10.33}$$

where no summation is applied to the superscript, while repeated subscripts still mean summation. Then, the equivalent condition Equation 10.24 for one of the particles, say Ω^J, reads

$$C_{ijkl} \left(u_{k,li} - \epsilon_{kl,i}^{*,J} \right) = C_{ijkl}^1 u_{k,li}, \; \mathbf{x} \in \Omega^J \tag{10.34}$$

The displacement of any point in the domain is caused by the eigenstrain of all particles, which can be obtained by the superposition as follows:

$$u_i(\mathbf{x}) = -\frac{1}{4\pi\mu_0} \sum_{I=1}^{N} \left(M_{imn}^{0,I} C_{mnkl} E_{kl}^{0,I} + N_{pimn}^{1,I} C_{mnkl} E_{klp}^{1,I} + T_{pqimn}^{2,I} C_{mnkl} E_{klpq}^{2,I} \right) \tag{10.35}$$

The disturbed strain on particle Ω^J can be written in terms of Taylor's expansion as

$$\epsilon'_{ij} = -\frac{1}{4\pi\mu_0}\sum_{I=1}^{N}\left[\begin{array}{c} D^{0,I}_{ijkl}\left(\mathbf{x}^J\right)C_{klmn}E^0_{mn} + D^{1,I}_{pijkl}\left(\mathbf{x}^J\right)C_{klmn}E^1_{mnp} \\ D^{2,I}_{stijkl}\left(\mathbf{x}^J\right)C_{klmn}E^2_{mnst}\end{array}\right]$$
$$+\left[\begin{array}{c} D^{0,I}_{ijkl,r}\left(\mathbf{x}^J\right)C_{klmn}E^0_{mn} + D^{1,I}_{pijkl,r}\left(\mathbf{x}^J\right)C_{klmn}E^1_{mnp} \\ +D^{2,I}_{stijkl,r}\left(\mathbf{x}^J\right)C_{klmn}E^2_{mnst}\end{array}\right]\left(x_r - x^I_r\right)$$
$$+\frac{1}{2!}\left[\begin{array}{c} D^{0,I}_{ijkl,yz}\left(\mathbf{x}^J\right)C_{klmn}E^0_{mn} + D^{1,I}_{pijkl,yz}\left(\mathbf{x}^J\right)C_{klmn}E^1_{mnp} \\ +D^{2,I}_{stijkl,yz}\left(\mathbf{x}^J\right)C_{klmn}E^2_{mnst}\end{array}\right]\left(x_y - x^I_y\right)\left(x_z - x^I_z\right) \tag{10.36}$$

Comparing the coefficient for different orders of $(x_r - x^J_r)$ terms of the equivalent inclusion condition Equation 10.37, one can also obtain

$$\left\{\begin{array}{l}
\left[\begin{array}{c} \sum_{I=1}^{N}\left(C_{ijkl} - C^*_{ijkl}\right)D^0_{klst}\left(\mathbf{x}^c\right)C_{stmn}E^{0,I}_{mn} + 4\pi\mu_0 C_{ijmn}E^{0,J}_{mn} \\ +\sum_{I=1}^{N}\left(C_{ijkl} - C^*_{ijkl}\right)D^{1,I}_{pklst}\left(\mathbf{x}^c\right)C_{stmn}E^{1,I}_{mnp} \\ +\sum_{I=1}^{N}\left(C_{ijkl} - C^*_{ijkl}\right)D^{2,I}_{pqklst}\left(\mathbf{x}^c\right)C_{stmn}E^{2,I}_{mnpq}\end{array}\right] = 4\pi\mu_0(C_{ijkl} - C^*_{ijkl})\epsilon^0_{kl} \\
\left[\begin{array}{c} \sum_{I=1}^{N}\left(C_{ijkl} - C^*_{ijkl}\right)D^{1,I}_{pklst,r}\left(\mathbf{x}^c\right)C_{stmn}E^{1,I}_{mnp} + 4\pi\mu_0 C_{ijmn}\delta_{pr}E^{1,J}_{mnp} \\ +\sum_{I=1}^{N}\left(C_{ijkl} - C^*_{ijkl}\right)D^{0,I}_{klst,r}\left(\mathbf{x}^c\right)C_{stmn}E^{0,I}_{mn} \\ +\sum_{I=1}^{N}\left(C_{ijkl} - C^*_{ijkl}\right)D^{2,I}_{pqklst,r}\left(\mathbf{x}^c\right)C_{stmn}E^{2,I}_{mnpq}\end{array}\right] = 0 \\
\left[\begin{array}{c} \sum_{I=1}^{N}\left(C_{ijkl} - C^*_{ijkl}\right)D^{0,I}_{klst,yz}\left(\mathbf{x}^c\right)C_{stmn}E^{0,I}_{mn} \\ +\sum_{I=1}^{N}\left(C_{ijkl} - C^*_{ijkl}\right)D^{1,I}_{pklst,yz}\left(\mathbf{x}^c\right)C_{stmn}E^{1,I}_{mnp} \\ +\sum_{I=1}^{N}\left(C_{ijkl} - C^*_{ijkl}\right)D^{2,I}_{pqklst,yz}\left(\mathbf{x}^c\right)C_{stmn}E^{2,I}_{mnpq} + 8\pi\mu_0 C_{ijmn}\delta_{py}\delta_{qz}E^{2,J}_{mnpq}\end{array}\right] = 0
\end{array}\right. \tag{10.37}$$

Due to symmetry, there are total 6, 24, and 60 independent components of the eigenstrain in each particle when constant, linear, and quadratic terms are considered, respectively. Using the quadratic terms as an example, Equation 10.37 can be rewritten in the matrix notation,

$$[A]^J_{60\times 60n}[E]_{60n} = [F]^J_{60} \tag{10.38}$$

where the eigenstrain terms of all the particles are represented by a vector $[E]$ whose length is $60n$. $[A]^J$ is the local coefficient matrix with dimensions $60 \times 60n$ whose elements are the coefficients in Equation 10.37. $[F]^J$ is the local force vector with length 60, whose elements are the summation of the body force multiplying the corresponding coefficients.

For each particle, one can write one set of equations as the above. When a higher rank of the polynomial form of eigenstrain is used, more equation can be obtained as the above.

Rewriting all the equations in the matrix notation, assembling all the $[A]^J$ into the global coefficient matrix, and all the $[F]^J$ into global force vector, one obtains

$$[A]_{60n\times 60n}[E]_{60n} = [F]_{60n} \tag{10.39}$$

Solving this global linear equations of system, one can obtain the eigenstrain terms of all particles. Substituting them into Equation 10.35, one can also obtain the displacement field and thus the strain and stress fields.

Because we consider the boundary value problem (BVP) through a continuum mechanics based approach, the particle interactions have been implicitly included in the analysis by solving the above BVP. This kind of pairwise interaction exists between any pair of particles and each particle is subject to the effects of all other particles, which generally decay with the particle–particle distance rapidly. Therefore, one can set a cut-off distance when $[A]^J_{60\times 60n}$ in Equation 10.38 is calculated, which will make the global coefficient matrix $[A]_{60n\times 60n}$ a sparse matrix for a large scale particle system.

To demonstrate the accuracy of this method, we follow the case in Section 10.2 and add one more particle on the top of the existing particle with a certain spacing. When two aluminum particles are embedded into the same semi-infinite HDPE domain, the problem becomes more complicated due to the coupling of the particle–boundary interaction and the particle–particle interaction. For the top-down case, the two particles are located at $(0, 0, h)$ and $(0, 0, h + s)$, where s is the center–center distance of two particles. Obviously, s should be higher than $2a$ to avoid the particles' overlap. The same uniaxial load of $\sigma^0_{33} = 0.8$ MPa is applied to the system corresponding to $\epsilon^0_{33} = 0.001$, $\epsilon^0_{11} = \epsilon^0_{22} = -\nu_0\epsilon^0_{33}$.

First, an extreme case is considered first with $h = 1.1a$, $s = 2.5a$. The results from EIM using uniform, linear, and quadratic forms of eigenstrains are compared with the FEM results in Figure 10.4a. One can find that σ_{33} increases with x_3 from boundary to specific peak value and then decrease a little bit when it reaches the bottom of the first particle. Then it increases again until it reaches the mid-point of the two particles, that is, $(0, 0, h + s/2)$. Then it decreases with x_3 until the top of the second particle. After a slight increment, it decreases again and finally converges to the far-field tension load. The four curves are fairly different. EIM with uniform and linear distributed eigenstrain cannot address the stress distribution accurately. The maximum stress from uniform distributed eigenstrain has relative error of 30%. However, the EIM with the quadratic form of eigenstrain agrees with the FEM results very well, in which the difference of the maximum stress is less than 3%.

When the distance to the boundary is larger, the eigenstrain is simpler due to the decay of boundary effect. Figure 10.4b illustrates the variation of σ_{33} with x_3 when $h = 2a$. One can see that the difference of the four curves is smaller although the EIM results for uniform eigenstrain are still fairly distinct from the other three due to the particle interaction. When quadratic terms are introduced in the eigenstrain, much better agreement with the FEM results is obtained, which highlights the advantage of using the quadratic form of eigenstrain.

Notice that when two particles are closer to the boundary in Figure 10.4a, the maximum stress locates in the middle of the two particles; whereas when they move far from

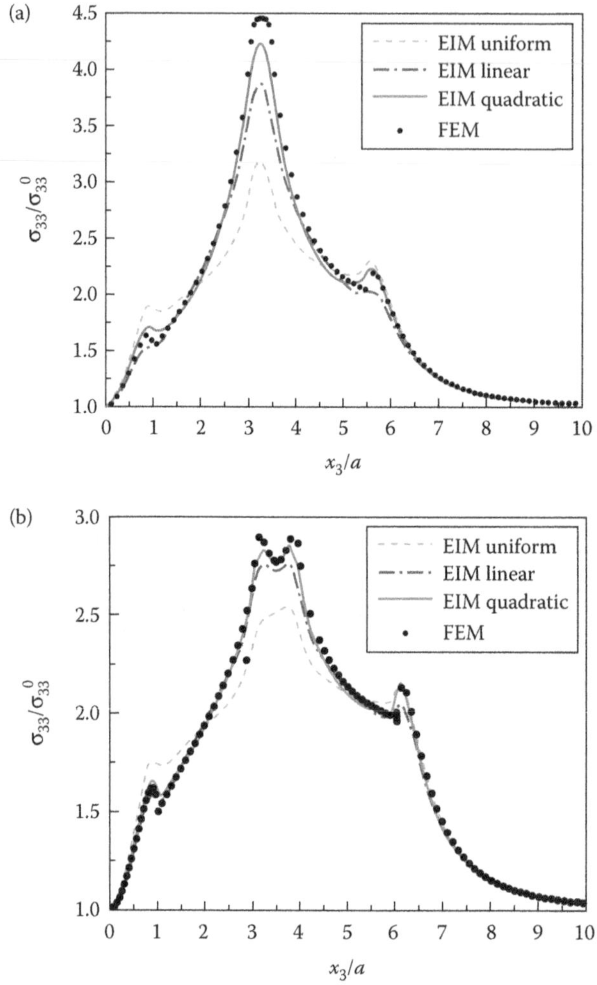

FIGURE 10.4 Variation of stress σ_{33} with x_3 when two top-down particles are embedded into a semi-infinite domain: (a) $h = 1.1a$ and (b) $h = 2a$.

the boundary in Figure 10.4b, the maximum stress locates close to the boundary of the two particles.

10.4 BOUNDARY EFFECTS ON EFFECTIVE ELASTICITY OF A PERIODIC COMPOSITE

The formulation has been extended to a semi-infinite domain containing many particles, which can represent a general composite material. The boundary effect may change the local field in the composite considerably and lead to different effective material behavior, which has not been studied in the literature yet. To simplify the problem, we consider a simple cubic lattice composite illustrated in Figure 10.5a. Because the microstructure is periodic, using one unit cell containing one particle, one can calculate the average stress and strain under a load on the boundary. The relation between the average stress and strain can represent the effective elasticity of the composite. In general cases, periodic boundary condition is used to

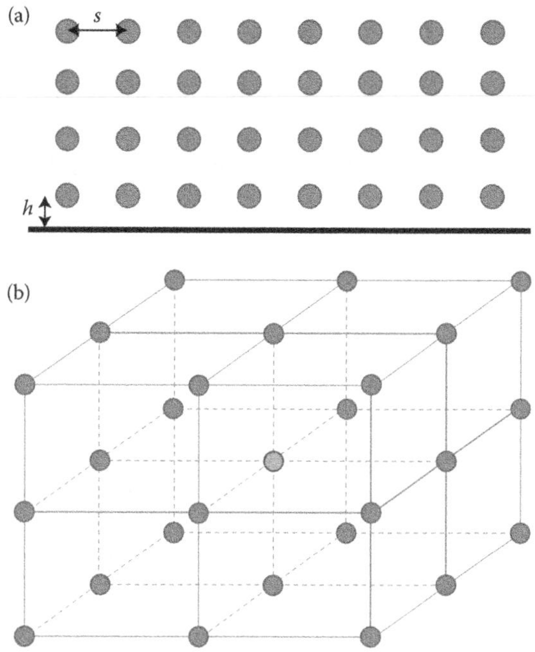

FIGURE 10.5 (a) Configuration of a simple cubic lattice in semi-infinite domain. (b) The configuration of neighboring particles which effect are considered.

solve for effective elasticity of a periodic composite [4]. However, due to the boundary effect, the periodic boundary condition in the depth direction is not valid any more. Therefore, the effective elasticity of the unit cells at different depths will be different. In the following, we number the unit cell starting with the traction free boundary, that is, the bottom of first unit cell at the boundary.

Given a simple lattice composite in Figure 10.5a, an aluminum particle with radius $a = 10\,\mu\text{m}$ is embedded in a semi-infinite HDPE. Young's moduli and Poisson's ratios for aluminum and HDPE are $E_1 = 69\,\text{GPa}$, $E_0 = 0.8\,\text{GPa}$, $\nu_1 = 0.334$, and $\nu_0 = 0.38$, respectively. The distance between the bottom and the center of first layer particle is $h = 2a$. The center–center distance of neighboring particle is $s = 4a$. Due to the geometry of this microstructure, the elastic field in the plane in parallel to the boundary surface will exhibit a periodic pattern for unit cells, so that the particles in the plane will share the same eigenstrain distributions. Therefore, the eigenstrain distribution on a particle will only change along the depth direction. From the above discussion, when the center–center distance is large, the particle interaction decays significantly. For simplicity, only 26 particles are taken into account to solve the eigenstrain, that is, the cut-off of the effective zone is $2\sqrt{3}a$.

For each layer, equivalent stress formula is built as Equation 10.37. Using the particle-particle interaction cut-off, the number of neighboring particles considered in each formula is only 26 as shown in Figure 10.5b. It is expected that the boundary effect will decay along the depth direction. Therefore, eigenstrain will converge to a stable value, when the distance of the unit cell to the boundary is far enough. Then, we can assume the eigenstrains of n layer and $n+1$ layer are the same.

Assembling the equations of all the layers into a linear equation system, the eigenstrain can be obtained. With the eigenstrain, the local displacement and strain can be obtained from Equation 10.21 and Equation 10.10. Furthermore, using the homogenization technique, the average stress and strain can be calculated and therefore the effective elastic moduli can be derived. Uniform uniaxial tension and simple shear along the surface will be considered to compute the effective Young's modulus, Poisson's ratio, and the shear modulus, respectively.

10.4.1 Uniaxial Tensile Loading on the Boundary

Consider a unit cell at a certain distance to the surface containing an inclusion Ω with an eigenstrain ϵ^*_{ij} embedded in a large domain D with free stress boundary condition. The stress and displacement field in the domain is continuous. The average stress can be written as

$$\langle\sigma_{ij}\rangle_D = \frac{1}{V_D}\int_D \sigma_{ij}(\mathbf{x})\,d\mathbf{x} = \frac{1}{V_D}\int_{\partial D} \sigma_{ik}(\mathbf{x})n_k x_j\,d\mathbf{x} = \frac{\sigma^0_{ik}}{V_D}\int_{\partial D} n_k x_j\,d\mathbf{x} = \sigma^0_{ij} \tag{10.40}$$

where the symmetric and periodic boundary conditions along the opposite lateral surfaces and equilibrium in the vertical direction are used.

Similarly, the average strain can be written as

$$\begin{aligned}\langle\epsilon_{ij}\rangle_D &= \frac{1}{V_D}\int_D \epsilon_{ij}(\mathbf{x})\,d\mathbf{x}\\ &= \frac{1}{V_D}\int_D \left(\epsilon_{ij}-\epsilon^*_{ij}\right)(\mathbf{x})\,d\mathbf{x} + \frac{1}{V_D}\int_D \epsilon^*_{ij}(\mathbf{x})\,d\mathbf{x}\\ &= \frac{C^{-1}_{ijkl}}{V_D}\int_D \sigma_{kl}(\mathbf{x})\,d\mathbf{x} + \frac{1}{V_D}\int_D \epsilon^*_{ij}\,d\mathbf{x}\\ &= C^{-1}_{ijkl}\sigma^0_{kl} + \eta\langle\epsilon^*_{ij}\rangle\end{aligned} \tag{10.41}$$

where η is the volume fraction of the inhomogeneity. Recall that there is only one particle in one unit cell. Therefore, by substituting Equation 10.33 into Equation 10.41, one can obtain

$$\begin{aligned}\langle\epsilon_{ij}\rangle^I_D &= C^{-1}_{ijkl}\sigma^0_{kl} + \frac{1}{V_D}\int_\Omega \left[E^{0,I}_{ij} + E^{1,I}_{ijk}\left(x_k - x^I_k\right) + E^{2,I}_{ijkl}\left(x_k - x^I_k\right)\left(x_l - x^I_l\right)\right]d\mathbf{x}\\ &= C^{-1}_{ijkl}\sigma^0_{kl} + \eta\left(E^{0,I}_{ij} + \frac{a^2_I}{5}E^{2,I}_{ijkk}\right)\end{aligned} \tag{10.42}$$

where a_I is the radius of particle I.

When the composite material is under uniaxial tension, Equations 10.40 and 10.41 can be rewritten as follows:

$$\langle \sigma_{33} \rangle_D = \sigma_{33}^0 \tag{10.43}$$

$$\left\langle \epsilon_{ij} \right\rangle_D^I = C_{ij33}^{-1} \sigma_{33}^0 + \eta \left(E_{ij}^{0,I} + \frac{a_I^2}{5} E_{ijkk}^{2,I} \right) \tag{10.44}$$

Therefore, the effective Young's modulus and Poisson's ratio of unit cell I for the present loading condition can be written as

$$\langle E \rangle = \sigma_{33}^0 / \left\langle \epsilon_{33} \right\rangle_D^I \tag{10.45}$$

and

$$\langle v \rangle = - \left\langle \epsilon_{11} \right\rangle_D / \left\langle \epsilon_{33} \right\rangle_D \tag{10.46}$$

Once the testing load is given, the elastic moduli only depends on the eigenstrain.

Under the far-field tension $\sigma_{33}^0 = 0.8$ MPa, the eigenstrain component ϵ_{33}^* distribution at x_1–x_3 plane are illustrated in Figure 10.6. One can observe that due to the boundary effect, the ϵ_{33}^* inside the unit cell 1 is much more different from the others. The distribution along x_1 direction is symmetric, but the symmetry about x_3 axis does not exist due to the boundary effect. Noted that in all the unit cells, no eigenstrain exists outside the particle. In Unit Cell 1,

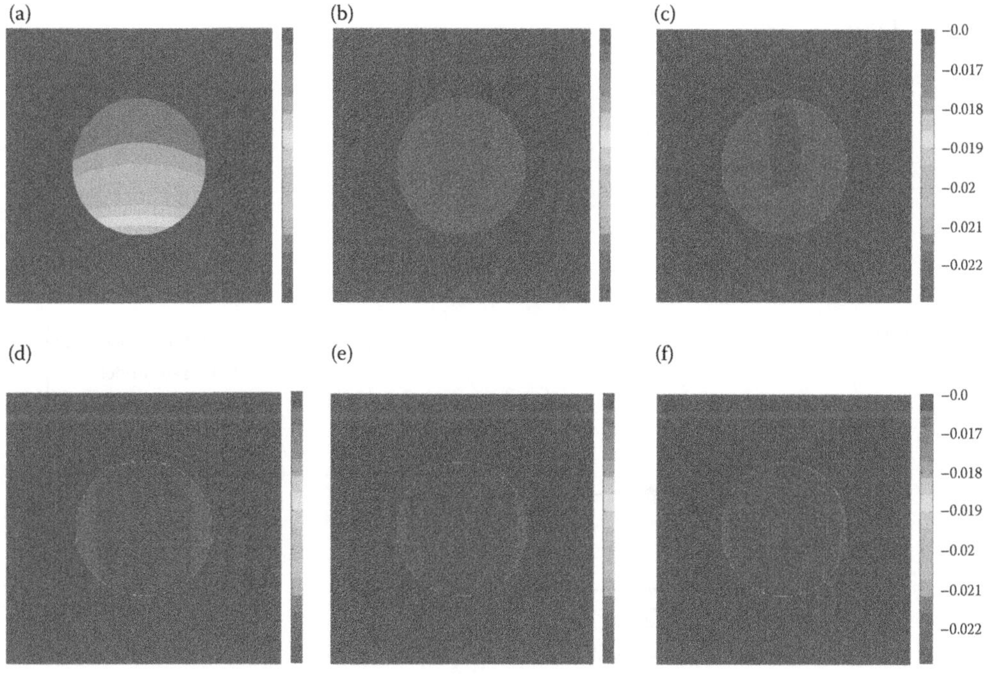

FIGURE 10.6 Distribution of ϵ_{33}^* in unit cells: (a) Unit Cell 1; (b) Unit Cell 2; (c) Unit Cell 3; (d) Unit Cell 4; (e) Unit Cell 5; (f) Unit Cell 6.

ϵ^*_{33} is highest at the bottom, and then decreases along x_3 direction. The difference of ϵ^*_{33} at bottom and top are fairly large. For Unit Cell 2, ϵ^*_{33} is also decreasing along x_3 direction, but the difference is much smaller compared to that of Unit Cell 1. From Unit Cell 3 to Unit Cell 6, the boundary effect becomes smaller and smaller, and therefore the stress along x_3 axis becomes nearly symmetric. Noted that with boundary effect decaying, not only the symmetry along x_3 is gradually recovered, the distribution is uniform. This is because the linear and quadratic terms of eigenstrain become smaller, that is, the uniform component plays a more dominant role.

The effective Young's modulus and Poisson's ratio are predicted by Equations 10.45 and 10.46. Figure 10.7 illustrates the effective material properties of the composite materials. The effective Young's modulus of first unit cell is much lower than those of other unit cells, which indicates that the boundary effect causes the lower stress in the particle of the first unit cell, which leads to a lower effective Young's modulus. One can conclude that the free surface

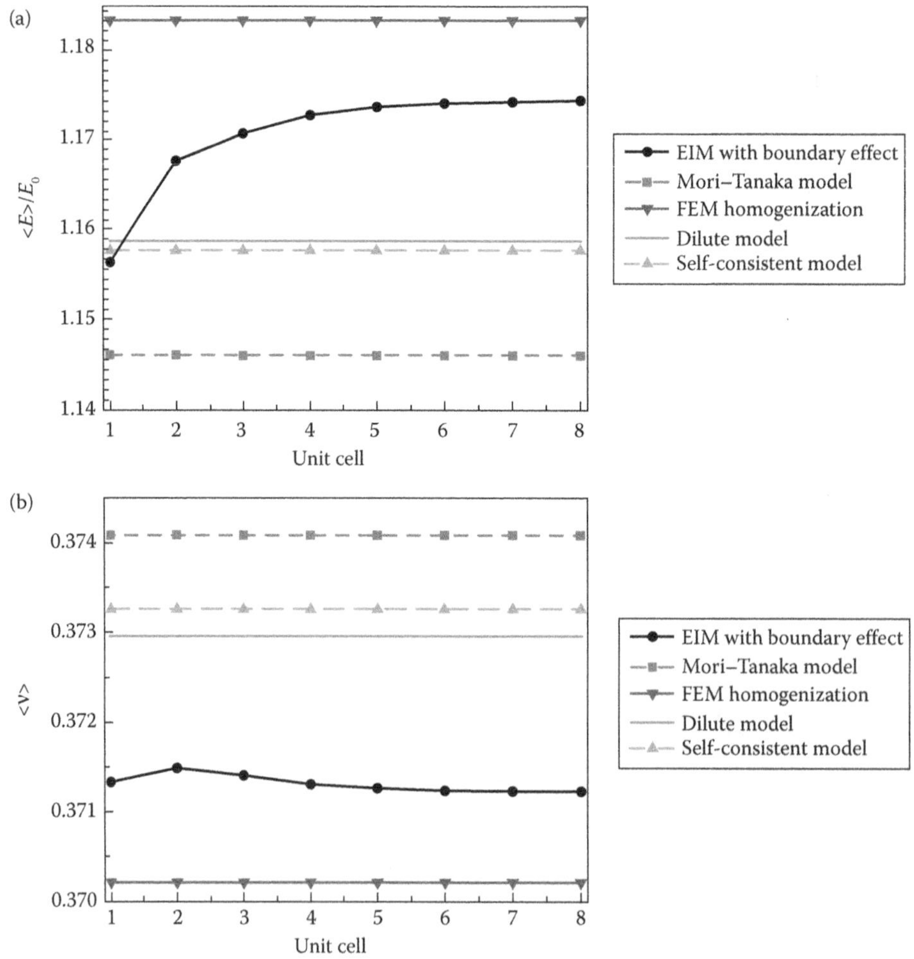

FIGURE 10.7 Variation of effective material properties for unit cells: (a) Young's modulus; (b) Possion's ratio.

will soften the composites with particulate reinforcements. Then the effective Young's modulus increases along the depth as it approaches a stable value after a distance of 5–6 unit cells, which means that the effect of boundary disappears and the effective Young's modulus converges to the effective Young's modulus of a composite material with the same volume fraction but no boundary effect.

One can observe that the convergence value is only lower than the value predicted by the FEM homogenization [52]. It is higher than the dilute model, the self-consistent model, and the Mori–Tanaka model [4]. The predicted effective Young's modulus by the dilute model and the self-consistent model are fairly close.

Figure 10.7b illustrates the effective Poisson's ratio of the composite material. One can observe that the Poisson's ratio of the first unit cell is smaller than the others. However, the second unit cell shows a higher Poisson's ratio. Finally, the effective Poisson's ratio converges to a stable value, which is between the values of two pure materials. The converged value is higher than the one predicted by FEM homogenization [52], but lower than the Mori–Tanaka model [4], the dilute model and the self-consistent model. One may observe a trend among the five predictions that the model with a higher Young's modulus prediction provides a lower Poisson's ratio prediction.

10.4.2 Uniform Simple Shear Loading on the Boundary

For a composite material with simple lattice subject to an uniform shear loading σ_{31}^0 at the boundary. The average stress and average strain can be obtained by rewriting Equations 10.40 and 10.41, that is,

$$\langle \sigma_{31} \rangle_D = \sigma_{31}^0 \tag{10.47}$$

$$\langle \epsilon_{ij} \rangle_D^I = C_{ij31}^{-1} \sigma_{31}^0 + \eta \left(E_{ij}^{0,I} + \frac{a^2}{5} E_{ijkk}^{2,I} \right) \tag{10.48}$$

The effective shear modulus $\langle \mu \rangle$ can be calculated as

$$\langle \mu \rangle = \sigma_{31} / 2 \langle \epsilon_{31} \rangle \tag{10.49}$$

Under a uniform shear loading $\sigma_{31}^0 = 0.58$ MPa, the eigenstrain component ϵ_{31}^* distributions at x_1–x_3 plane are illustrated in Figure 10.8. The eigenstrain ϵ_{31}^* in Unit Cell 1 is quite different from that of other unit cells. ϵ_{31}^* is symmetric about the x_1 axis. But the symmetry about the x_3 axis does not exist due to the boundary effect. Eigenstrain ϵ_{31}^* reaches the highest at the bottom of the particle, and then decreases along x_3 direction. From Unit Cell 2 to Unit Cell 6, the symmetry about x_3 is gradually recovering. This indicates that the boundary effect is decaying with the increasing distance to the boundary, as for Unit Cell 6, ϵ_{31}^* is almost symmetric along the x_3 axis. Eigenstrain distributions in Unit Cell 5 and Unit Cell 6 are almost the same, which means that the eigenstrain is convergent.

Figure 10.9 illustrates the effective shear modulus of the unit cells. Similar to the effective Young's modulus, the effective shear modulus of the first unit cell is much lower than others,

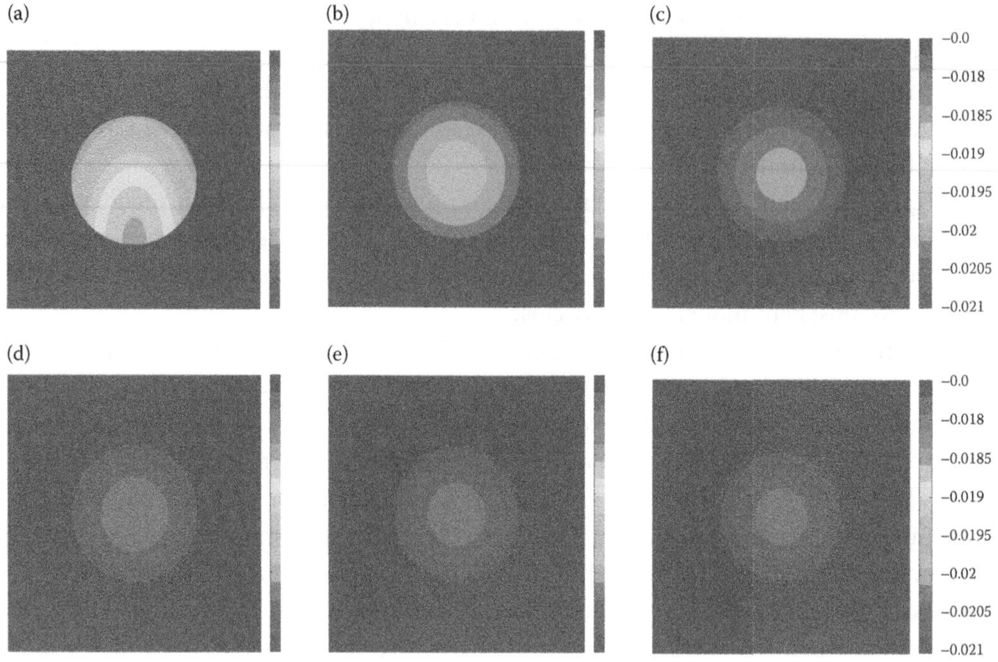

FIGURE 10.8 Distribution of ϵ_{31}^* in unit cells: (a) Unit Cell 1; (b) Unit Cell 2; (c) Unit Cell 3; (d) Unit Cell 4; (e) Unit Cell 5; (f) Unit Cell 6.

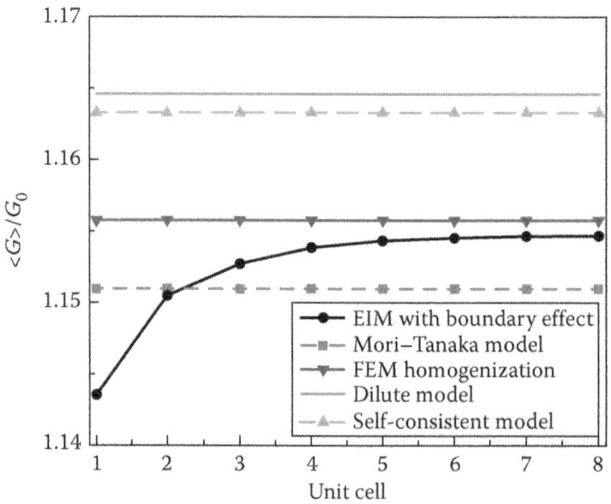

FIGURE 10.9 Variation of shear modulus for unit cell.

which again highlights that the boundary effect can soften the particle reinforced composites. Then, the effective shear modulus increases and finally converges to a stable value. The convergence value is between the prediction of the FEM homogenization [52] and the Mori–Tanaka model [4]. The predictions of the dilute model and the self-consistent model are also illustrated.

Parametric Study

Given the size of a cubic unit cell, say 40 μ each side, if the particle size becomes larger, the volume fraction of the particle will be much higher. It is expected that the effective Young's modulus and shear modulus will also increase with the particle size, since $E_1 > E_0$ and $G_1 > G_0$. Figure 10.10a illustrates the variation of effective Young's modulus with unit cells for the composites with different aluminum particle sizes, say $a = 10$ μm, $a = 12.5$ μm, and $a = 15$ μm, respectively. As expected, the composite has larger particle size shows higher effective Young's modulus. Moreover, the effective Young's moduli of all the three cases increase along x_3 direction. Furthermore, the boundary effect of the case $a = 15$ μm is much more significant than that of the other two, which causes higher variation of the effective Young's modulus versus unit cell number. Actually, when the center of a particle remains the same at the center of unit cells but its size increases, the spacing between the particle to the free boundary will decrease, so that the boundary effect is amplified.

Figure 10.10b illustrates the variation of shear modulus with unit cell number along depth direction for the composites with different particle sizes. Similar to Young's modulus, the

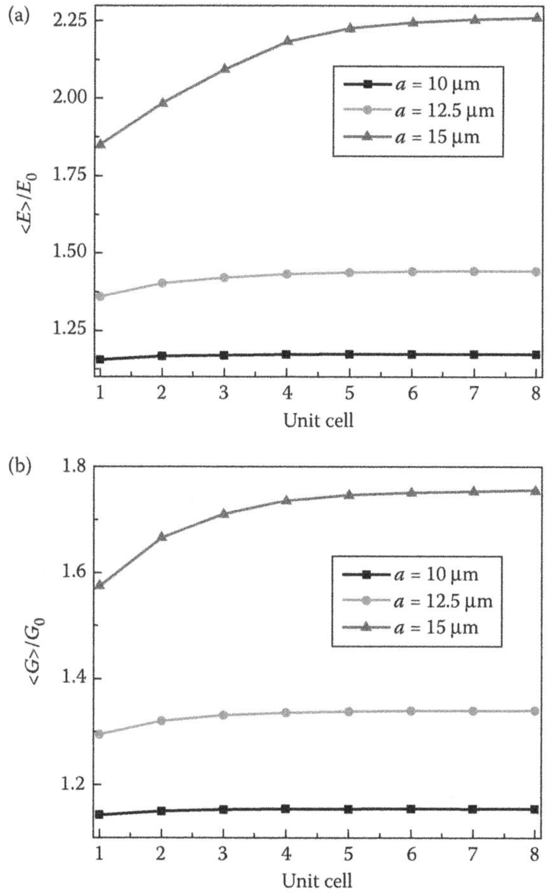

FIGURE 10.10 Variation of effective material properties for unit cells for different particle size: (a) Young's modulus; (b) Shear modulus.

composite with a larger particle size not only shows higher effective shear modulus, but also exhibits more significant variation in the depth direction due to the boundary effect. Noted that the effect of particle–particle interaction will also increase when particle size becomes larger. If the particle size is very large, to reach the equivalent accuracy, a longer cut-off is needed and more neighboring particles should be considered.

Note that the effective material properties should vary with the volume fraction of inhomogeneity. Figure 10.11 illustrates the variation of the effective moduli with the volume fraction of the aluminum for the unit cell at the boundary and far from the boundary. The effective Young's modulus increase with the volume fraction for both the unit cell at the boundary or far from boundary, as illustrated in Figure 10.11a. However, the effective Young's modulus of the unit cell far from boundary increases much faster than that of the unit cell at the boundary. Therefore, the difference of the effective material property of the

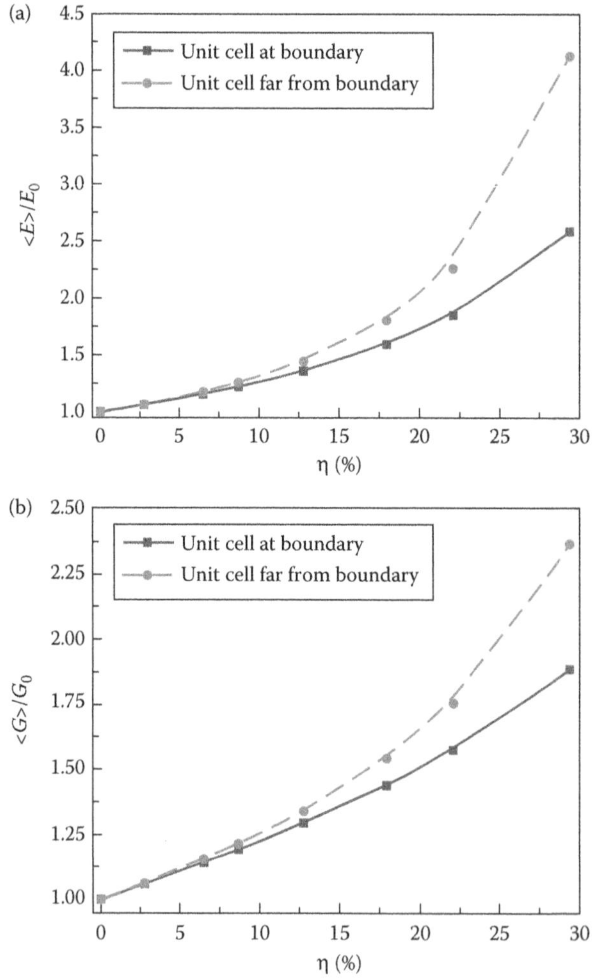

FIGURE 10.11 Variation of effective material properties with volume fraction: (a) Young's modulus; (b) Shear modulus.

two unit cells is also increasing with the volume fraction. This implies that the particulate reinforcement is limited by the boundary.

The similar trend has also been observed in Figure 10.11b for the effective shear modulus changing with volume fractions. Although the same geometry and volume fraction of particle are in the unit cells, the effective elastic moduli of the unit cells is lower when it is close to the boundary due to the boundary effect.

The effective Young's modulus, Poisson's ratio, and shear modulus of the composite material are analyzed above with the aid of the equivalent inclusion method. Note that although only the example of the composites with a simple cubic lattice is analyzed, this method can be applied to composites with other periodic microstructures, for example, face centered cubic (FCC) or body centered cubic (BCC). Moreover, the model can be applied to the composite with different particle shapes, say ellipsoids, and different constitutive properties, say viscoelastic or elastoplastic materials.

10.5 INCLUSION-BASED BOUNDARY ELEMENT METHOD FOR VIRTUAL EXPERIMENTS OF A COMPOSITE SAMPLE

Given a composite specimen, one can apply a test load on the boundary, which is generally a uniform stress or displacement, to characterize the effective mechanical properties. This procedure can be conducted as either an experimental method by a test coupon or as an analytical/numerical method following a micromechanics-based framework. Most classical approaches to this problem investigate the mechanical response of an RVE when a far-field stress or strain is applied. In other words, it assumes that the composite material's size is much larger than the particles. However, with the miniaturization of experiments, the size of testing samples may fall in the range that is not significantly larger than particles. Therefore, how the test results represent the effective material behavior is an interesting question. Numerical methods have been commonly used in the stress analysis of an RVE or a unit cell containing a finite number of particles, which can represent the microstructure of the composite. However, due to the high computational cost, the material system is limited to a certain scale. A novel method has been explored by Ma and his colleagues [78,79] for two-dimensional inhomogeneity problems. By combining the boundary element method (BEM) and the equivalent inclusion method (EIM), an RVE with a large number of inhomogeneities can be considered. This section will introduce an EIM-based boundary integral formulation to investigate multiple inhomogeneities embedded in a matrix with finite domain under test loading. Then, a global equation system can be assembled to solve the unknown boundary displacements and the eigenstrain in each inhomogeneity. With that informations, the local displacement and stress fields can be derived by Green's function technique.

Consider a domain D is subjected to body force $\boldsymbol{b}$ and surface traction $\boldsymbol{t}$. In elasticity, the virtual work principle reads

$$\int_{\Omega} \sigma_{ij} u'_{i,j} \, \mathbf{dx} = \int_{\partial\Omega} t_i u'_i \, \mathbf{dx} + \int_{\Omega} b_i u'_i \, \mathbf{dx} \tag{10.50}$$

where u'_i is an admissible virtual displacement.

Use the true displacement u_i as a virtual displacement. Let σ'_{ij} be a new stress state associated with another equilibrium state caused by $\boldsymbol{t}'$ and $\boldsymbol{b}'$. Applying the virtual work principle in the above equation, one can obtain

$$\int_{\partial D} t_i u'_i \, \mathbf{dx} + \int_{D} b_i u'_i \, \mathbf{dx} = \int_{\partial D} t'_i u_i \, \mathbf{dx} + \int_{D} b'_i u_i \, \mathbf{dx} \tag{10.51}$$

which is the well-known Maxwell–Betti's reciprocal theorem. The two loading cases are shown in Figure 10.12.

Using the equilibrium $b'_i = -\sigma'_{ji,j}$, one can obtain the following equation:

$$\int_{D} \sigma'_{ji,j} u_i \, \mathbf{dx} + \int_{D} b_i u'_i \, \mathbf{dx} = \int_{\partial D} t'_i u_i \, \mathbf{dx} - \int_{\partial D} t_i u'_i \, \mathbf{dx} \tag{10.52}$$

Use the fundamental solution to the following equilibrium equation for the new stress state under a unit point force at $\mathbf{x}'$,

$$\sigma'_{ji,j}(\mathbf{x}) + \delta_i(\mathbf{x} - \mathbf{x}') = 0 \tag{10.53}$$

where $\delta_i(\mathbf{x} - \mathbf{x}')$ is the Dirac delta function indicating a unit point force in the i direction acting at position $\mathbf{x}'$ with the component d_i. Therefore, d_i exhibits the following properties:

$$\begin{cases} \int_{D} \delta_i(\mathbf{x} - \mathbf{x}') \, \mathbf{dx} = d_i \\ |\mathbf{d}| = \sqrt{d_i d_i} = 1 \end{cases}$$

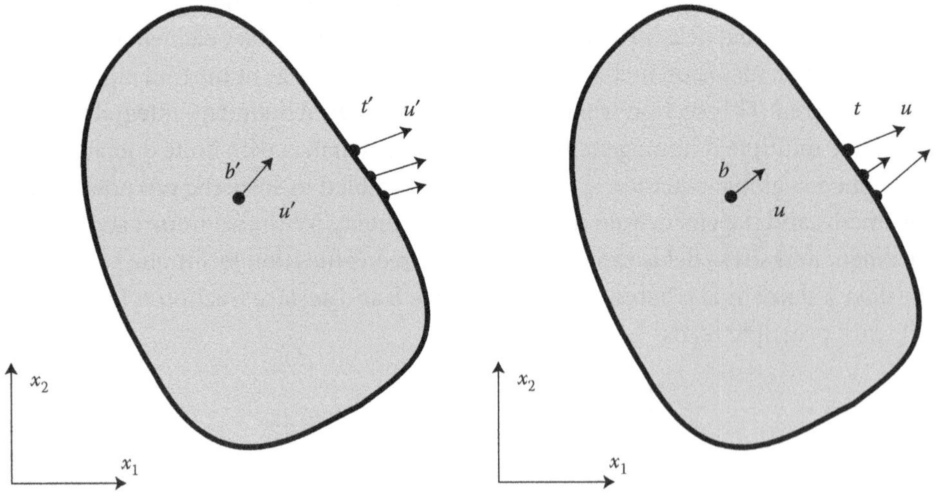

FIGURE 10.12 Application of the Betti reciprocal theorem for two loading cases.

This fictitious stress state can be obtained by the classic solution for an infinite domain containing a point force associated with the corresponding stress vector on the boundary. Using the property of the Dirac delta function, one can write

$$\int_D \sigma'_{ji,j}(\mathbf{x})u_i(\mathbf{x})\,\mathrm{d}\mathbf{x} = -\int_D \delta_i(\mathbf{x}-\mathbf{x}')u_i(\mathbf{x})\,\mathrm{d}\mathbf{x} = -d_iu_i(\mathbf{x}') \tag{10.54}$$

Now, to solve the displacement at any point $\mathbf{x}'$, one can use the fundamental solution for a fictitious unit point force at that point. The fictitious elastic field will depend on both variables of the fictitious loading point $\mathbf{x}'$ and the field point $\mathbf{x}$.

Combining Equations 10.52 and 10.54, one can obtain the displacement at any point $\mathbf{x}'$ as follows:

$$u_i(\mathbf{x}')d_i = \int_{\partial D} u'_i(\mathbf{x},\mathbf{x}')t_i(\mathbf{x})\,\mathrm{d}\mathbf{x} - \int_{\partial D} t'_i(\mathbf{x},\mathbf{x}')u_i(\mathbf{x})\,\mathrm{d}\mathbf{x} + \int_D u'_i(\mathbf{x},\mathbf{x}')b_i(\mathbf{x})\,\mathrm{d}\mathbf{x} \tag{10.55}$$

Using Green's function technique, the fictitious elastic field can be obtained as follows:

$$u'_j(\mathbf{x},\mathbf{x}') = G_{ij}(\mathbf{x},\mathbf{x}')d_i \tag{10.56}$$

$$t'_j(\mathbf{x},\mathbf{x}') = t_{ij}(\mathbf{x},\mathbf{x}')d_i \tag{10.57}$$

where $G_{ij}(\mathbf{x},\mathbf{x}')$ and $t_{ij}(\mathbf{x},\mathbf{x}')$ represent the displacement and traction, respectively, in the j direction at $\mathbf{x}$ caused by a unit point force f_i acting at $\mathbf{x}'$ in the i direction.

$$G_{ij}(\mathbf{x},\mathbf{x}') = \frac{1}{16\pi(1-\nu)\mu}\left[\frac{(3-4\nu)}{r}\delta_{ij} + \frac{r_{,i}r_{,j}}{r}\right] \tag{10.58}$$

$$t_{ij}(\mathbf{x},\mathbf{x}') = \frac{-1}{8\pi(1-\nu)r^2}\left[((1-2\nu)\delta_{ij} + 3r_{,i}r_{,j})r_{,m}n_m - (1-2\nu)(r_{,i}n_j - r_{,j}n_i)\right] \tag{10.59}$$

where $r = \left|\mathbf{x}-\mathbf{x}'\right|$, $r_{,i} = \frac{\partial(x_mx_m)^{\frac{1}{2}}}{\partial x_i}$, $r_{,ij} = \frac{\partial^2(x_mx_m)^{\frac{1}{2}}}{\partial x_i\partial x_j}$ and $\boldsymbol{n}$ is the unit normal vector. Notice that one can easily see the reciprocal property of the above tensors, that is,

$$G_{ij}(\mathbf{x},\mathbf{x}') = G_{ij}(\mathbf{x}',\mathbf{x}) \quad \text{and} \quad t_{ij}(\mathbf{x},\mathbf{x}') = t_{ij}(\mathbf{x}',\mathbf{x}) \tag{10.60}$$

Finally, the following boundary integral equation can be obtained:

$$d_iu_i(\mathbf{x}') = d_i\int_{\partial D} G_{ij}(\mathbf{x},\mathbf{x}')t_j(\mathbf{x})\,\mathrm{d}\mathbf{x} - d_i\int_{\partial D} t_{ij}(\mathbf{x},\mathbf{x}')u_j(\mathbf{x})\,\mathrm{d}\mathbf{x} + d_i\int_D G_{ij}(\mathbf{x},\mathbf{x}')b_j(\mathbf{x})\,\mathrm{d}\mathbf{x} \tag{10.61}$$

Therefore, to obtain the displacement at any point $\mathbf{x}'$ in a specific direction, one can apply a unit point force along that direction represented by d_i and derive the displacement through the above integrals of $\mathbf{x}$. For the simplicity of notations, we can switch $\mathbf{x}$ and $\mathbf{x}'$ in the above equation, so that $\mathbf{x}'$ stands for the integral variable and $\mathbf{x}$ the point to solve for displacement. In addition, the above equation is applicable to an arbitrary fictitious unit force, so that d_i at the both sides of the above equation can be eliminated. Therefore, one can obtain

$$u_i(\mathbf{x}) = \int_{\partial D} G_{ij}(\mathbf{x},\mathbf{x}')t_j(\mathbf{x}')\,\mathrm{d}\mathbf{x}' - \int_{\partial D} t_{ij}(\mathbf{x},\mathbf{x}')u_j(\mathbf{x}')\,\mathrm{d}\mathbf{x}' + \int_{D} G_{ij}(\mathbf{x},\mathbf{x}')b_j(\mathbf{x}')\,\mathrm{d}\mathbf{x}' \tag{10.62}$$

where the reciprocal property in Equation 10.60 is used. For an equilibrium state in the absence of body force, the last term does not exist, so that the displacement can be obtained by the two boundary integral terms. In this case, the expression of the stress tensor can be obtained:

$$\sigma_{ij} = -\int_{\partial D} R_{ijm}t_m\,\mathrm{d}\mathbf{x} + \int_{\partial D} S_{ijm}u_m\,\mathrm{d}\mathbf{x} \tag{10.63}$$

where

$$R_{ijm} = \frac{-1}{8\pi(1-\nu)}\left[\frac{3r_{,i}r_{,j}r_m}{r^2} + \frac{(1-2\nu)(\delta_{mi}r_{,j} + \delta_{mj}r_{,i} - \delta_{ij}r_{,m})}{r^2}\right] \tag{10.64}$$

$$\begin{aligned} S_{ijm} &= \frac{-\mu}{4\pi(1-\nu)}\left\{\frac{3r_{,p}n_p}{r^3}\left[(1-2\nu)r_{,m}\delta_{ij} + \nu(r_{,i}\delta_{mj} + r_{,j}\delta_{mi}) - 5r_{,m}r_{,i}r_{,j}\right]\right\} \\ &\quad - \frac{\mu}{4\pi(1-\nu)}\left[\begin{array}{l} \frac{3\nu}{r^3}\left(r_{,j}r_m n_i + r_{,i}r_{,m}n_j\right) - \frac{(1-4\nu)}{r^3}\delta_{ij}n_m \\ +\frac{3\nu}{r^3}\left(r_{,i}r_{,j}n_m + \delta_{mi}n_j + \delta_{mj}n_i\right) \end{array}\right] \end{aligned} \tag{10.65}$$

Using Green's function technique or the fundamental solution, one can write the displacement fields in the domain D in terms of the integrals of the eigenstrain ε^*_{kl} with the tensorial Green's function $\mathbf{G}$ as follows:

$$u_m(\mathbf{x}) = -\int_{\Omega} C_{ijkl}\varepsilon^*_{kl}(\mathbf{x}')G_{jm,i}(\mathbf{x},\mathbf{x}')\,\mathrm{d}\mathbf{x}' \tag{10.66}$$

Consider a bounded domain D containing an inclusion Ω with a continuously distributed eigenstrain, which is subjected to a certain boundary conditions. For simplicity, up to linear terms in the polynomial form are considered for the eigenstrain as

$$\varepsilon^*_{kl} = \varepsilon^0_{kl} + \varepsilon^1_{klp}x_p + \cdots \quad \mathbf{x} \in \Omega \tag{10.67}$$

The displacement caused by loads on the boundary and eigenstrain in the inclusions can be obtained by the combination of Equations 10.62 and 10.66 as follows:

$$u_m(\mathbf{x}) = \int_{\partial D} G_{mj}(\mathbf{x},\mathbf{x}')t_j(\mathbf{x}')\,d\mathbf{x}' - \int_{\partial D} t_{mj}(\mathbf{x},\mathbf{x}')u_j(\mathbf{x}')\,d\mathbf{x}' + C_{ijkl}D_{jmi}(\mathbf{x})\varepsilon^0_{kl} + C_{ijkl}D_{jmip}(\mathbf{x})\varepsilon^1_{klp} \tag{10.68}$$

The strain can also be obtained by applying the relationship between displacement and strain

$$\varepsilon_{mn}(\mathbf{x}) = -C^{-1}_{mnkl}\int_{\partial D} R_{klp}(\mathbf{x},\mathbf{x}')t_p(\mathbf{x}')\,d\mathbf{x}' + C^{-1}_{mnkl}\int_{\partial D} S_{klp}(\mathbf{x},\mathbf{x}')u_p(\mathbf{x}')d\mathbf{x}' + C_{ijkl}D_{jmin}(\mathbf{x})\varepsilon^0_{kl} + C_{ijkl}D_{jminp}(\mathbf{x})\varepsilon^1_{klp} \tag{10.69}$$

where

$$D_{jmin} = -\frac{1}{4}\left[\frac{\delta_{mj}}{4\pi\mu}\Phi_{,in} + \frac{\delta_{mi}}{4\pi\mu}\Phi_{,jn} + \frac{\delta_{nj}}{4\pi\mu}\Phi_{,im} + \frac{\delta_{ni}}{4\pi\mu}\Phi_{,jm} - \frac{1}{4\pi\mu(1-\nu)}\Psi_{,ijmn}\right]$$

$$D_{jminp} = -\frac{1}{4}\left[\frac{\delta_{mj}}{4\pi\mu}\Phi_{p,in} + \frac{\delta_{mi}}{4\pi\mu}\Phi_{p,jn} + \frac{\delta_{nj}}{4\pi\mu}\Phi_{p,im} + \frac{\delta_{ni}}{4\pi\mu}\Phi_{p,jm} - \frac{1}{4\pi\mu(1-\nu)}\Psi_{p,ijmn}\right]$$

and the tensors of R_{klp} and S_{klp} have been shown in Equations 10.64 and 10.65.

For the convenience of derivation later, the first derivative of strain with respect to **x** can also be obtained as follows:

$$\varepsilon_{mn,q} = C^{-1}_{mnkl}\int_{\partial\Omega} R_{klpq}t_p - C^{-1}_{mnkl}\int_{\partial\Omega} S_{klpq}u_p + C_{ijkl}D_{jminq}\varepsilon^0_{kl} + C_{ijkl}D_{jminpq}\varepsilon^1_{klp} \tag{10.70}$$

where

$$R_{klpq} = -\frac{1}{8\pi(1-\nu)}\left(\frac{3r_{,kq}r_{,l}r_{,p}}{r^2} + \frac{3r_{,k}r_{,lq}r_{,p}}{r^2} + \frac{3r_{,k}r_{,l}r_{,pq}}{r^2} - \frac{6r_{,k}r_{,l}r_{,p}r_{,q}}{r^3}\right) - \frac{(1-2\nu)}{8\pi(1-\nu)}\left(\frac{\delta_{pk}r_{,lq}}{r^2} + \frac{\delta_{pl}r_{kq}}{r^2} - \frac{\delta_{kl}r_{,pq}}{r^2}\right) + \frac{(1-2\nu)}{4\pi(1-\nu)}\left(\frac{\delta_{pk}r_{,l}r_{,q}}{r^3} + \frac{\delta_{pl}r_{,k}r_{,q}}{r^3} - \frac{\delta_{kl}r_{,p}r_{,q}}{r^3}\right) \tag{10.71}$$

$$S_{klpq} = \frac{9Gr_{,n}n_n r_{,q}}{4\pi(1-\nu)r^4}\left[(1-2\nu)r_{,p}\delta_{kl} + \nu(r_{,k}\delta_{lp} + r_{,l}\delta_{kp} - 5r_{,p}r_{,k}r_{,l})\right] + \frac{-3Gr_{,nq}n_n}{4\pi(1-\nu)r^3}\left[(1-2\nu)r_{,p}\delta_{kl} + \nu(r_{,k}\delta_{pl} + r_{,l}\delta_{pk}) - 5r_{,p}r_{,k}r_{,l}\right] + \frac{-3Gr_{,n}n_n}{4\pi(1-\nu)r^3}\left[(1-2\nu)r_{,pq}\delta_{kl} + \nu(r_{,kq}\delta_{pl} + r_{,lq}\delta_{pk})\right.$$

$$-5(r_{,pq}r_{,k}r_{,l}+r_{,p}r_{,kq}r_{,l}+r_{,p}r_{,k}r_{,lq})\big]+\frac{9G\nu r_{,q}}{4\pi(1-\nu)r^4}\left[r_{,l}r_{,p}n_k+r_{,k}r_{,p}n_l\right]$$

$$+\frac{-3G\nu}{4\pi(1-\nu)r^3}\left[r_{,lq}r_{,p}n_k+r_{,l}r_{,pq}n_k+r_{,kq}r_{,p}n_l+r_{,k}r_{,pq}n_l\right]$$

$$+\frac{3(1-2\nu)Gr_{,q}}{4\pi(1-\nu)r^4}\left[3r_{,k}r_{,l}n_p+\delta_{pk}n_l+\delta_{pl}n_k\right]$$

$$+\frac{-G(1-2\nu)}{4\pi(1-\nu)r^3}\left[3r_{,kq}r_{,l}n_p+3r_{,k}r_{,lq}n_p\right]+\frac{-3G(1-4\nu)r_{,q}}{4\pi(1-\nu)r^4}\delta_{kl}n_p \tag{10.72}$$

$$D_{jminq}=\frac{1}{4}\left[\frac{\delta_{mj}}{4\pi\mu}\Phi_{,inq}+\frac{\delta_{mi}}{4\pi\mu}\Phi_{,jnq}+\frac{\delta_{nj}}{4\pi\mu}\Phi_{,imq}+\frac{\delta_{ni}}{4\pi\mu}\Phi_{,jmq}-\frac{1}{4\pi\mu(1-\nu)}\Psi_{,ijmnq}\right]$$

$$D_{jminpq}=-\frac{1}{4}\left[\frac{\delta_{mj}}{4\pi\mu}\Phi_{p,inq}+\frac{\delta_{mi}}{4\pi\mu}\Phi_{p,jnq}+\frac{\delta_{nj}}{4\pi\mu}\Phi_{p,imq}+\frac{\delta_{ni}}{4\pi\mu}\Phi_{p,jmq}-\frac{1}{4\pi\mu(1-\nu)}\Psi_{p,ijmnq}\right]$$

Now consider the inhomogeneity problem for a bound domain D with stiffness $\mathbf{C}^0$ containing an inhomogeneity Ω with stiffness $\mathbf{C}^1$, which is subjected to a certain load on the boundary ∂D. Because of the material mismatch, the fundamental solution in Section 3.1 cannot be directly applied in the above material domain. Owing to Eshelby's original idea of EIM, the inhomogeneity problem can be simulated by an inclusion problem that the inhomogeneity can be replaced by an inclusion with a certain continuously distributed eigenstrain, which satisfies the stress equivalent condition that will be shown in Equation 10.74 later. Therefore, the displacement field can be obtained by Equation 10.68 with the effects of loading on the boundary.

To simulate the material mismatch, the stress equivalent condition is applied here. For a given inhomogeneity, a local coordinate is set up with the origin located at the inhomogeneity center. Due to continuity of the elastic field, one can expand the total strain by using the Taylor expansion of the local coordinate with the origin at the center of the inhomogeneity,

$$\varepsilon_{kl}(\mathbf{x})=\varepsilon_{kl}(\mathbf{0})+\varepsilon_{kl,q}(\mathbf{0})x_q+\cdots,\quad \mathbf{x}\in\Omega \tag{10.73}$$

In order to make the stresses in the inclusion problem equivalent to those in the inhomogeneity, the following equation should be satisfied.

$$C^0_{ijkl}(\varepsilon_{kl}-\varepsilon^*_{kl})=C^1_{ijkl}\varepsilon_{kl} \tag{10.74}$$

By substituting Equation 10.67 and Equation 10.73 into Equation 10.74, one can compare the coefficient for each rank of the $\mathbf{x}$ terms. For example, the following equations for the zeroth and first-order terms of $\mathbf{x}$ can be, respectively, written as

$$C^0_{ijkl}\left[\varepsilon_{kl}(\mathbf{0})-\varepsilon^0_{kl}\right]=C^1_{ijkl}\varepsilon_{kl}(\mathbf{0}) \tag{10.75}$$

$$C^0_{ijkl}\left[\varepsilon_{kl,q}(\mathbf{0})-\varepsilon^1_{klq}\right]=C^1_{ijkl}\varepsilon_{kl,q}(\mathbf{0}) \tag{10.76}$$

Although only up to linear terms are considered here for simplicity, the method can be extended to higher-order terms for higher accuracy. Because introducing one higher-order term will lead to one set of equations in the above stress equivalent condition, the problem is mathematically closed. It can be straightforwardly implemented in the next subsection for numerical simulation, in which quadratic terms have already been considered following the above procedure.

The above equation can be extended to multiple inhomogeneities with different sizes, stiffness, or orientations as follows: a local coordinate is set up in the same fashion as above, the eigenstrain on each inhomogeneity can be expanded in the polynomial form at a certain order, which can be different for different inhomogeneities depending on the accuracy requirement. Comparing the coefficient of the stress field on each inhomogeneity in the polynomial form, one can set up a linear equation system similarly to Equations 10.75 and 10.76. By solving the linear system, the eigenstrain for each particle can be obtained, and then the strain and displacement fields in the domain D can be solved similarly to Equations 10.68 and 10.69.

Since the ellipsoid is a versatile shape which can represent many kinds of inhomogeneity shapes by changing its aspect ratios, such as fibers, penny shape of cracks, slit-like cracks, spheroids, and spheres [3], the proposed inclusion based BEM formulation can be used to study the interaction among these types of inhomogeneities or voids or cracks. In this paper, because the particles are spheres with similar size, as the first step to demonstrate this method, the paper focuses on a solid containing spherical particles, so that the orientation of particles are not considered, that is, all local coordinates can share the same orientation as the global coordinate. Extension of this method to general multiphysical problems is underway.

The above formulation has been implemented in a C++ package—iBEM, which stands for a BEM code addressing inhomogeneity or inclusion problems. The programming structure is presented in Figure 10.13, in which each step is elaborated as follows:

1. Given a solid containing a certain number of inhomogeneities in the matrix with perfectly bonded interfaces, the external surface is discretized into boundary elements. This step can be treated as a preprocessing. Since only surface is meshed, the number of elements will be reduced significantly compared with the FEM or the general BEM.

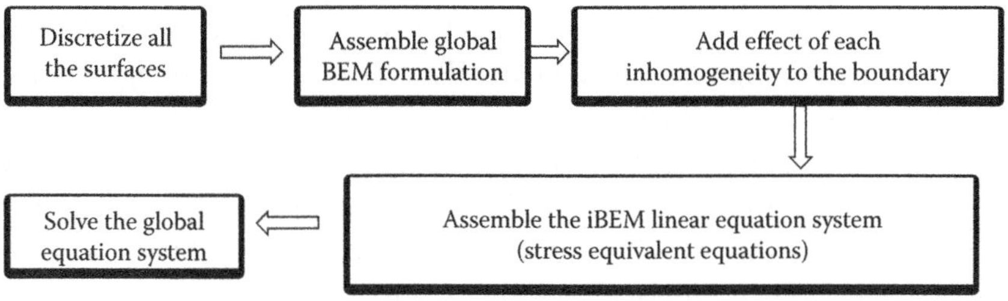

FIGURE 10.13 Implementation flow of the iBEM—a software package introducing inhomogeneity/inclusion problems into the BEM code.

2. Go through all the collection points on the surface to calculate individual element matrix for each collection point based on basic boundary element integration. This step is to complete the boundary element formulation. Therefore, the effect of boundary will be considered in this step.

3. Go through each inhomogeneity to build up stress equivalent into global matrix. This is the key step of this proposed iBEM algorithm. Not only the interaction between particles but also the effect from boundary will be taken into account through Equations 10.75 and 10.76. Notice that if higher-order terms of eigenstrain are considered, more equations for each inhomogeneity are required. Completeness of the equations can be achieved. The number of unknown for each inhomogeneity with linear terms of eigenstrain is 24 and the unknown for each boundary node is 3. Therefore, the total unknowns are 24 × *number of inhomogeneity*, 3 × *number of boundary nodes*. The traditional BEM will provide three equations for each node, which makes 3 × *number of boundary nodes* equations. In addition, the equivalent of stress will also provide 24 × *number of inhomogeneity* equations for the whole system. In the present method, the number of equations and unknowns are the same. Therefore, a closed linear equation system can be formed for numerical solution. If higher-order terms of eigenstrain is used, more unknowns in each particle can be used but the same number of equations will also be obtained.

4. Solve the global linear equation system to obtain unknown boundary response and unknown eigenstrain. The full LU decomposition with the pivot method has been implemented to calculate the numerical results from the linear equation system.

5. Use the eigenstrain and boundary traction/displacement to calculate the elastic fields, including the stress, strain, and displacement fields for analysis. This step can be treated as a post-processing. As the unknown boundary response and eigenstrains are all obtained, from Equation 10.68, one can get the displacement at any field point; therefore, from Equation 10.69 the inner strain can also be calculated.

In order to verify the proposed method, a case study is conducted to investigate two voids located in a simulation box. A cylinder model is built up with a diameter of 1 and a height of 2. Two voids are located in the cylinder. The diameter of both voids are 0.4. One is centered at the position (0,0,0.65). The other is centered at the position (0,0,1.35). Young's modulus is 16.9 GPa and Poisson's Ratio is 0.25. The cylinder is fixed in all directions at the bottom and subjected to a uniform compressive load of 10 MPa at the top. The configuration is shown in Figure 10.14. A model of this cylinder is also generated in ABAQUS for finite element simulation of an axisymmetric problem with 10,570 elements. The mesh is highly refined to get high accuracy. In the iBEM, only a mesh for the surface of cylinder is used, so that 1423 elements are used. Notice that ABAQUS uses a 2D mesh whereas iBEM uses a 3D mesh with a much lower number. Since a linear shape function is used here to interpolate the curved surface, errors in the boundary geometry can be incurred but they will be reduced with more elements used.

The elastic field distributed along three lines are selected to make comparison with the FEM results. For the ease of description, both Cartesian coordinate and cylindrical coordinate notations will be used with *x*, *y*, *z* and *r*, θ, *z* in consistent with the literature. Here, *x*, *y*, and *z* are the same as the index form of x_1, x_2, and x_3 in the derivation, respectively. The first line is starting from (0,0,0) along with *z* direction to (0,0,2). All points share the same *x* and *y* coordinates but with a varying *z* coordinates. The second and third lines are looped around both voids. The angle in the comparison figure is from the local polar coordinate of each void. The configuration of the simulation model is shown as Figure 10.14. Some results using iBEM with linear eigenstrain distributions are shown in Figures 10.15 through 10.20.

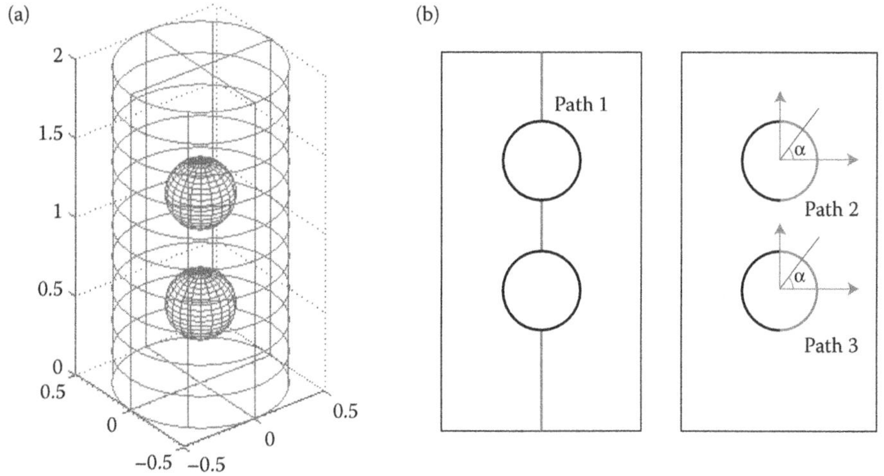

FIGURE 10.14 Configuration of the simulation model: (a) 3D geometry and (b) the path to show the points of interest.

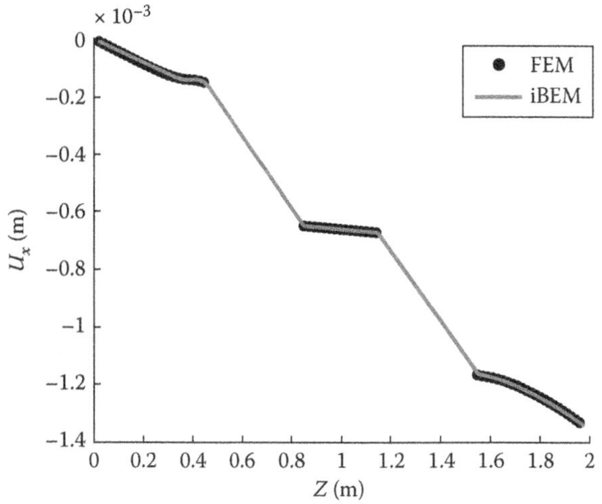

FIGURE 10.15 Comparison of displacement between the FEM and iBEM results along Path One.

Based on comparison of the displacements, very good agreement between FEM and iBEM is observed. In Figure 10.15, as the coordinate z increases, the displacement in the z direction also increases. The gap between 0.45 and 0.85, 1.15 and 1.55 are because of the presence of voids. Figures 10.16a and 10.17a show that displacements u_x are symmetric to the z axis, due to this being an axisymmetric problem about the z axis. The displacement comparison in Figures 10.16b and 10.17b show that due to an increase of the z coordinate, the displacement in the z direction also increases. The displacement in the x direction for each hole is relatively small compared with the displacement in the z direction, this is because our loading condition is purely compression. Also the maximum displacement in the x direction for each hole appears at the point where the x coordinate is at the maximum value, this is due to the effect of Poisson's ratio.

Figure 10.18 illustrates that the stress σ_{zz} will approach 10 MPa as the z coordinate increases, which satisfies the boundary condition at the loading end. However, in

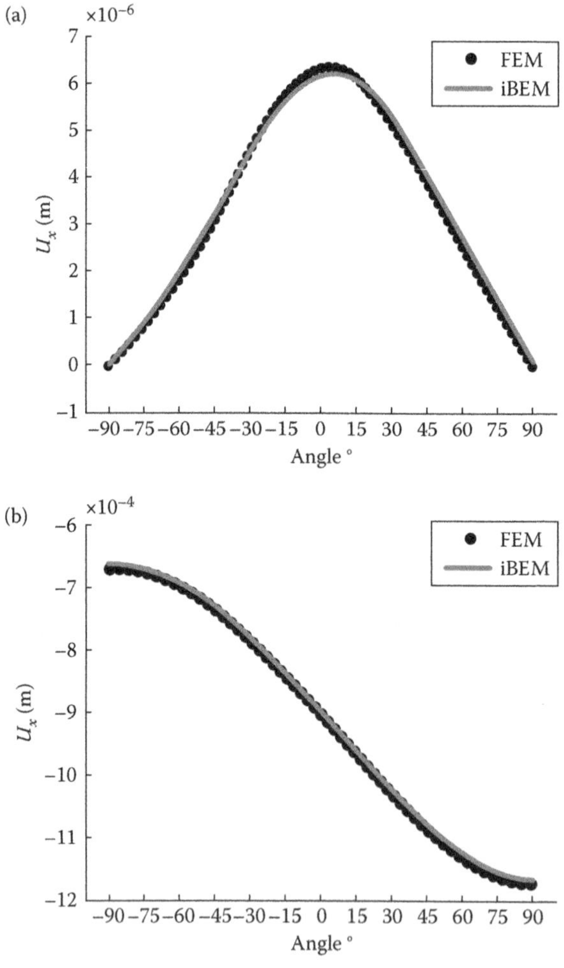

FIGURE 10.16 Comparison of displacement in (a) the x direction and (b) the z direction between FEM and iBEM results along Path Two.

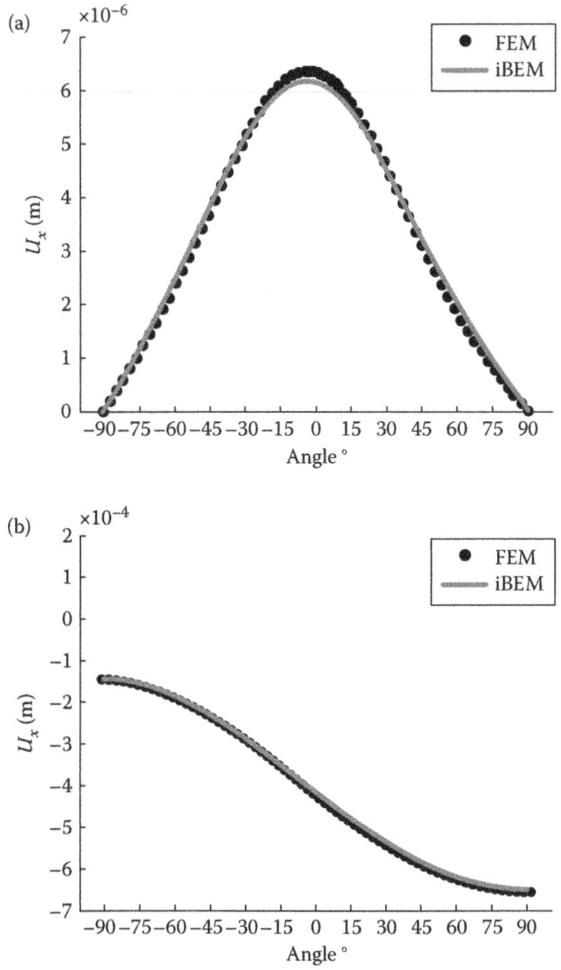

FIGURE 10.17 Comparison of displacement in (a) the x direction and (b) the z direction between the FEM and iBEM results along Path Three.

Figure 10.18, considerable differences in the stress exists in the neighborhood of the void. When higher-order terms are considered, more accurate results are expected. Even though only up to linear terms are considered, the present model can still provide fairly good prediction of the stress distribution. Figures 10.19 and 10.20 demonstrate the stress around the top and bottom voids. Since it is a traction free surface around the void, only stress $\sigma_{\theta\theta}$ exists, which maximizes at $\theta = 0$ due to stress concentration. When θ is changed far away from 0, the stress $\sigma_{\theta\theta}$ decreases to zero first and then reaches positive values, which indicates that the top and bottom of each void is under tension, which may lead to a different failure mode.

In general, EIM addresses the material inhomogeneity problem by replacing the inhomogeneity with matrix material and introducing a nonmechanical strain (eigenstrain) to make the constitutive law of Equation 10.74 equivalently satisfied. Therefore, the BVP is transferred to a homogeneous solid with an eigenstrain in the particle domains, which can be

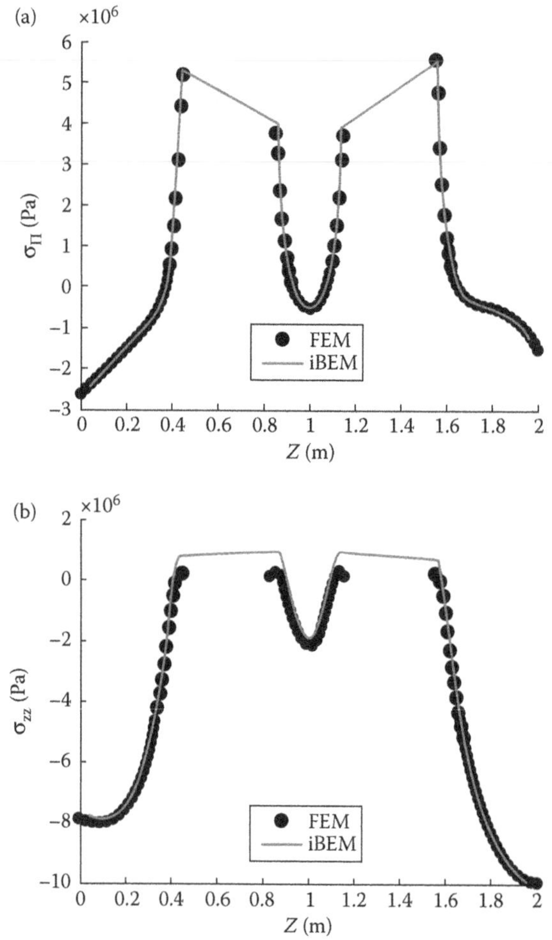

FIGURE 10.18 Stress comparison (a) σ_{xx} and (b) σ_{zz} between FEM and iBEM results along Path One.

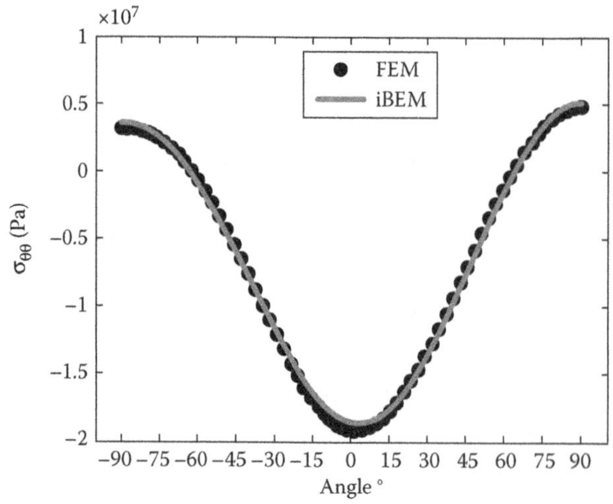

FIGURE 10.19 Stress comparison $\sigma_{\theta\theta}$ between the FEM and iBEM results along Path Two.

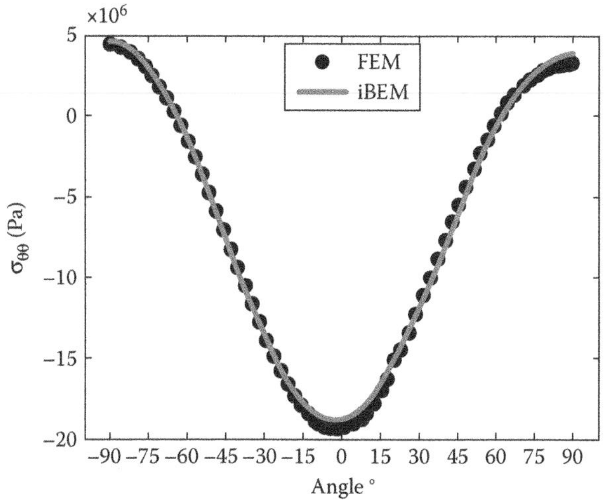

FIGURE 10.20 Stress comparison $\sigma_{\theta\theta}$ between the FEM and iBEM results along Path Three.

analytically solved with elliptic integrals for ellipsoidal particles. Compared with the boundary integral method (BIM), the present method uses eigenstrain to represent the material mismatch so that it does not need to mesh the particle interface into elements. For the single inhomogeneity case, the BEM may handle with this problem very well if the smoothly curved surface is carefully addressed. However, when the number of particles approaches 10, 50, or 100, and if the particle size changes in a large range, it is formidable to use the BEM, but the present method works very well. Therefore, the iBEM program can immediately be used in virtual experiments for mechanical behavior of particulate composites.

10.6 EXERCISES

1. From the fundamental solution for bi-semi-infinite domain, derive Rongved's solution for a semi-infinite domain with a fixed surface containing a point force at $\mathbf{x}'$.

$$
\begin{aligned}
G_{ij}\left(\mathbf{x},\mathbf{x}'\right) &= \frac{1}{4\pi\mu^0}\delta_{ij}\left(\phi-\bar{\phi}\right)-\frac{1}{16\pi\mu^0(1-v^0)}\left(\psi_{,ij}-Q_{im}Q_{jn}\bar{\psi}_{,mn}\right)\\
&+\frac{x_3}{8\pi\mu^0(1-v^0)(3-4v^0)}\\
&\left[Q_{jm}\bar{\psi}_{,im3}+4(1-v^0)\delta_{j3}\bar{\phi}_{,i}+2(1-2v^0)\delta_{i3}Q_{jm}\bar{\phi}_{,m}-Q_{jm}x_3\bar{\phi}_{,im}\right]
\end{aligned}
$$

2. Derive the displacement field a semi-infinite domain with a fixed surface containing an inclusion with eigenstrain $\boldsymbol{\epsilon}^*$. The radius of the inclusion is r_0 and its distance to the surface is $b > r_0$.

References

1. R. Hill, Elastic properties of reinforced solids: Some theoretical principles, *Journal of the Mechanics and Physics of Solids* 11, 5, 357–372, 1963.
2. J. R. Willis, Variational and related methods for the overall properties of composites, *Advances in Applied Mechanics* 21, 1–78, 1981.
3. T. Mura, Micromechanics of defects in solids, M. Nijhoff; Distributors for the U.S. and Canada, Kluwer Academic Publishers, 1987.
4. S. Nemat-Nasser and M. Hori, Micromechanics: Overall properties of heterogeneous materials, Elsevier, 1999.
5. J. D. Eshelby, The determination of the elastic field of an ellipsoidal inclusion and related problems, *Proceedings of the Royal Society of London Series A-Mathematical and Physical Sciences* 241, 1226, 376–396, 1957.
6. J. D. Eshelby, The elastic field outside an ellipsoidal inclusion, *Proceedings of the Royal Society of London Series A-Mathematical and Physical Sciences* 252, 1271, 561–569, 1959.
7. R. Hill, The elastic behaviour of a crystalline aggregate, *Proceedings of the Royal Society of London Series A-Mathematical and Physical Sciences* 65, 5, 349–355, 1952.
8. Z. Hashin and S. Shtrikman, On some variational principles in anisotropic and nonhomogeneous elasticity, *Journal of the Mechanics and Physics of Solids* 10, 4, 335–342, 1962.
9. Z. Hashin and S. Shtrikman, A variational approach to the theory of the elastic behaviour of multiphase materials, *Journal of the Mechanics and Physics of Solids* 11, 2, 127–140, 1963.
10. L. J. Walpole, On the overall elastic moduli of composite materials, *Journal of the Mechanics and Physics of Solids* 17, 14, 235–251, 1969.
11. J. J. McCoy, On the displacement field in an elastic medium with random variations of material properties, *Recent Advances in Engineering Sciences* 5, 235–254, 1970.
12. J. R. Willis, Bounds and self-consistent estimates for the overall properties of anisotropic composites, *Journal of the Mechanics and Physics of Solids* 25, 3, 185–202, 1977.
13. S. Torquato, Random heterogeneous media: Microstructure and improved bounds on effective properties, *Applied Mechanics Reviews* 44, 2, 37–76, 1991.
14. P. P. Castaeda and J. R. Willis, The effect of spatial distribution on the effective behavior of composite materials and cracked media, *Journal of the Mechanics and Physics of Solids* 43, 12, 1919–1951, 1995.
15. T. Mori and K. Tanaka, Average stress in matrix and average elastic energy of materials with misfitting inclusions, *Acta Metallurgica* 21, 5, 571–574, 1973.
16. Y. H. Zhao, G. P. Tandon and G. J. Weng, Elastic moduli for a class of porous materials, *Acta Mechanica* 76, 1–2, 105–131, 1989.
17. Y. P. Qiu and G. J. Weng, On the application of Mori-Tanaka's theory involving transversely isotropic spheroidal inclusions, *International Journal of Engineering Science* 28, 11, 1121–1137, 1990.

18. R. Hill, A self-consistent mechanics of composite materials, *Journal of the Mechanics and Physics of Solids* 13, 4, 213–222, 1965.
19. B. Budiansky, On the elastic moduli of some heterogeneous materials, *Journal of the Mechanics and Physics of Solids* 13, 4, 223–227, 1965.
20. R. McLaughlin, A study of the differential scheme for composite materials, *International Journal of Engineering Science* 15, 4, 237–244, 1977.
21. Z. Hashin, The differential scheme and its application to cracked materials, *Journal of the Mechanics and Physics of Solids* 36, 6, 719–734, 1988.
22. R. M. Christensen and K. H. Lo, Solutions for effective shear properties in three phase sphere and cylinder models, *Journal of the Mechanics and Physics of Solids* 27, 4, 315–330, 1979.
23. H. S. Chen and A. Andreas, The effective elastic moduli of composite materials containing spherical inclusions at non-dilute concentrations, *International Journal of Solids and Structures* 14, 5, 349–364, 1978.
24. H. S. Chen and A. Andreas, The solution of the equations of linear elasticity for an infinite region containing two spherical inclusions, *International Journal of Solids and Structures* 14, 5, 331–348, 1978.
25. J. W. Ju and T. M. Chen, Micromechanics and effective moduli of elastic composites containing randomly dispersed ellipsoidal inhomogeneities, *Acta Mechanica* 103, 1–4, 103–121, 1994.
26. J. W. Ju and T. M. Chen, Effective elastic moduli of two-phase composites containing randomly dispersed spherical inhomogeneities, *Acta Mechanica* 103, 1–4, 123–144, 1994.
27. S. Torquato, *Random Heterogeneous Materials: Microstructure and Macroscopic Properties*, Springer, 2001.
28. T. Iwakuma and S. Nemat-Nasser, Composites with periodic microstructure, *Computers and Structures* 16, 1–4, 13–19, 1983.
29. K. P. Walker, E. H. Jordan and A. D. Freed, equivalence of green's function and the fourier series representation of composites with periodic microstructure, *Micromechanics and Inhomogeneity*, Springer, New York, 1990.
30. A. Levy and J. M. Papazian, Tensile properties of short fiber-reinforced SiC/Al composites: Part II. finite-element analysis, *Metallurgical Transactions A* 21, 1, 411–420, 1990.
31. C. C. Swan and I. Kosaka, Homogenization-based analysis and design of composites, *Computers and Structures* 64, 1–4, 603–621, 1997.
32. J. Aboudi, S. M. Arnold and B. A. Bednarcyk, Micromechanics of composite materials: A generalized multiscale analysis approach, Butterworth-Heinemann, 2012.
33. J. Fish, Discrete-to-continuum Scale Bridging, Multiscaling in Molecular and Continuum Mechanics: Interaction of Time and Size from Macro to Nano, 85–102, Springer, Netherlands, 2007.
34. W. J. Drugan and J. R. Willis, A micromechanics-based nonlocal constitutive equation and estimates of representative volume element size for elastic composites, *Journal of the Mechanics and Physics of Solids* 44, 4, 497–524, 1996.
35. N. I. Muskhelishvili, Some fundamental problems of the mathematical theory of elasticity, Nauka, Moscow, 1966.
36. J. M. Qu and M. Cherkaoui, Frontmatter, Wiley Online Library. 2006.
37. A. F. Stevenson, Note on the existence and determination of a vector potential, *Quarterly of Applied Mathematics*, 12, 2, 194–198, 1954.
38. J. Hadamard, Le probleme de Cauchy et les quations aux drives partielles linaires hyperboliques, Paris, 1932.
39. M. Z. Wang, Application of the finite part of a divergent integral in the theory of elasticity. *Applied Mathematics and Mechanics* 6, 12, 1071–1078, 1985.
40. L. Z. Sun, Micromechanics and overall elastoplasticity of discontinuously reinforced metal-matrix composites, Ph.D. dissertation, UCLA, 1998.

41. Y. J. Liu and H. M. Yin, Stress concentration of a microvoid embedded in an adhesive layer during stress transfer, *Journal of Engineering Mechanics* 140, 10, 0401-4075, 2014.
42. Y. J. Liu, G. Song and H. M. Yin, Boundary effect on the elastic field and effective elasticity of a semi-infinite solid containing particles, *Proceedings of the Royal Society of London Series A-Mathematical and Physical Sciences* v10.1098/rspa.2015.0174, 20150174-20150174. Online publication date: Jun. 2015.
43. H. M. Yin and L. Z. Sun, Magnetic properties of randomly dispersed magnetic particulate composites: A theoretical study, *Physical Review B* 72, 5, 054-409, 2005.
44. H. M. Yin and L. Z. Sun, Effective magnetic permeability of composites containing chain-structured particles, *Acta Materialia* 54, 9, 2317–2323, 2006.
45. H. M. Yin and L. Z. Sun, Magnetoelasticity of chain-structured ferromagnetic composites, *Applied Physics Letters* 86, 26, 261-901, 2005.
46. H. M. Yin, L. Z. Sun and J. S. Chen, Magneto-elastic modeling of composites containing chain-structured magnetostrictive particles, *Journal of the Mechanics and Physics of Solids* 54, 5, 975–1003, 2006.
47. H. M. Yin and L. Z. Sun, Magnetoelastic modelling of composites containing randomly dispersed ferromagnetic particles, *Philosophical Magazine* 86, 28, 4367–4395, 2006.
48. W. Voigt, Ueber die Beziehung zwischen den beiden Elasticitätsconstanten isotroper Körper, *Annalen der Physik* 274, 12, 573–587, 1889.
49. A. Reuss, Berechnung der Fliegrenze von Mischkristallen auf Grund der Plastizittsbedingung fr Einkristalle, *ZAMM-Journal of Applied Mathematics and Mechanics / Zeitschrift für Angewandte Mathematik und Mechanik* 9, 1, 49–58, 1929.
50. E. Krner, Berechnung der elastischen Konstanten des Vielkristalls aus den Konstanten des Einkristalls, *Zeitschrift fr Physik* 151, 4, 504–518, 1958.
51. S. Li and G. Wang, *Introduction to Micromechanics and Nanomechanics*, World Scientific, 2008.
52. J. Fish, *Practical Multiscaling*, John Wiley & Sons, 2013.
53. A. A. Griffith, The phenomena of rupture and flow in solids, *Proceedings of the Royal Society of London. Series A, Containing Papers of a Mathematical and Physical Character* 221, 163–198, 1921.
54. A. A. Wells, Application of fracture mechanics at and beyond general yielding, *British Welding Journal* 10, 11, 563–570, 1963.
55. G. R. Irwin, plastic zone near a crack and fracture toughness, *Proceedings of Seventh Material Research Conference* Syracuse University, Syracuse NY, 63–87, 1960.
56. V. Volterra, Sur l'quilibre des corps lastiques multiplement connexes, *Annales scientifiques de l'Ecole Normale superieure*, Socit mathmatique de France 24, 401–517, 1907.
57. J. R. Rice, A path independent integral and the approximate analysis of strain concentration by notches and cracks, *Journal of Applied Mechanics* 35, 2, 379–386, 1968.
58. E. Orowan, Plasticity of crystals, *Z. Physics* 89, 8–10, 605–659, 1934.
59. M. Polanyi, Lattice distortion which originates plastic flow, *Z. Physics* 89, 8–10, 660–662, 1934.
60. G. I. Taylor, The mechanism of plastic deformation of crystals. Part I. theoretical, *Proceedings of the Royal Society of London. Series A, Containing Papers of a Mathematical and Physical Character* 145, 855, 362–387, 1934.
61. G. I. Taylor, The mechanism of plastic deformation of crystals. Part II. comparison with observations, *Proceedings of the Royal Society of London. Series A, Containing Papers of a Mathematical and Physical Character* 145, 855, 388–404, 1934.
62. J. Lemaitre, A continuous damage mechanics model for ductile fracture, *Journal of Engineering Materials and Technology* 107, 1, 83–89, 1985.
63. S. Murakami, Notion of continuum damage mechanics and its application to anisotropic creep damage theory, *Journal of Engineering Materials and Technology* 105, 2, 99–105, 1983.

64. Y. H. Zhao and G. J. Weng, The effect of debonding angle on the reduction of effective moduli of particle and fiber-reinforced composites, *Journal of Applied Mechanics* 69, 3, 292–302, 2002.
65. H. T. Liu, L. Z. Sun and J. W. Ju, An interfacial debonding model for particle-reinforced composites, *International Journal of Damage Mechanics* 13, 2, 163–185, 2004.
66. G. H. Paulino, H. M. Yin and L. Z. Sun, Micromechanics-based interfacial debonding model for damage of functionally graded materials with particle interactions, *International Journal of Damage Mechanics* 15, 3, 267–288, 2004.
67. J. Boussinesq, Applications of potentials for the study of equilibrium and movement of elastic solids, Paris: Gautier-Villars, 1885.
68. R. D. Mindlin, Force at a point in the interior of a semi-infinite solid, *Journal of Applied Physics* 7, 5, 195–202, 1936.
69. L. Rongved, Force at point in the interior of a semi-infinite solid with fixed boundary, *Journal of Applied Mechanics* 22, 545–546, 1955.
70. L. Rongved, Force interior to one of two joined semi-infinite solids, *Proceedings of the Second Midwestern Conference on Solid Mechanics*, 1, 13, Purdue University-Indiana Research Series, 1955.
71. H. Y. Yu and S. C. Sanday, Elastic fields in joined half-spaces due to nuclei of strain, *Proceedings of the Royal Society of London Series A-Mathematical and Physical Sciences* 434, 1892, 503–519, 1991.
72. L. J. Walpole, An elastic singularity in joined half-spaces, *International Journal of Engineering Science* 34, 6, 629–638, 1996.
73. L. J. Walpole, An inclusion in one of two joined isotropic elastic half-spaces, *IMA Journal of Applied Mathematics* 59, 2, 193–209, 1997.
74. Y. J. Liu and H. M. Yin, Equivalent inclusion method-based simulation of particle sedimentation toward functionally graded material manufacturing, *Acta Mechanica* 4–5, 1429–1445, 2014.
75. T. Mura, H. M. Shodja and Y. Hirose, Inclusion problems, *Applied Mechanics Reviews* 49, 10s, S118–S127, 1996.
76. R. Avazmohammadi, R. Hashemi, H. M. Shodja and M. H. Kargarnovin, Ellipsoidal domain with piecewise nonuniform eigenstrain field in one of joined isotropic half-spaces, *Journal of Elasticity* 98, 2, 117–140, 2010.
77. H. M. Yin, D. J. Yang, G. Kelly and J. Garant, Design and performance of a novel building integrated PV/Thermal system for energy efficiency of buildings, *Solar Energy* 87, 0, 184–195, 2013.
78. H. Ma, C. Yan and Q. H. Qin, Eigenstrain formulation of boundary integral equations for modeling particle-reinforced composites, *Engineering Analysis with Boundary Elements* 33, 3, 410–419, 2009.
79. H. Ma, Z. Guo and Q. H. Qin, Two-dimensional polynomial eigenstrain formulation of boundary integral equation with numerical verification, *Applied Mathematics and Mechanics* 32, 5, 551–562, 2011.

Index

O

P

R

S

Printed and bound by CPI Group (UK) Ltd, Croydon, CR0 4YY
22/10/2024
01777611-0006